UMTS NETWORK PLANNING, OPTIMIZATION, AND INTER-OPERATION WITH GSM

UMTS NETWORK PLANNING, OPTIMIZATION, AND INTER-OPERATION WITH GSM

Moe Rahnema

IEEE PRESS
IEEE Communications Society, Sponsor

John Wiley & Sons (Asia) Pte Ltd

Copyright © 2008   John Wiley & Sons (Asia) Pte Ltd, 2 Clementi Loop, # 02-01,
Singapore 129809

Visit our Home Page on www.wiley.com

All Rights Reserved. No part of this publication may be reproduced, stored in a retrieval system or transmitted in any form or by any means, electronic, mechanical, photocopying, recording, scanning, or otherwise, except as expressly permitted by law, without either the prior written permission of the Publisher, or authorization through payment of the appropriate photocopy fee to the Copyright Clearance Center. Requests for permission should be addressed to the Publisher, John Wiley & Sons (Asia) Pte Ltd, 2 Clementi Loop, #02-01, Singapore 129809, tel: 65-64632400, fax: 65-64646912, email: enquiry@wiley.com.sg.

Designations used by companies to distinguish their products are often claimed as trademarks. All brand names and product names used in this book are trade names, service marks, trademarks or registered trademarks of their respective owners. The Publisher is not associated with any product or vendor mentioned in this book. All trademarks referred to in the next of this publication are the property of their respective owners.

This publication is designed to provide accurate and authoritative information in regard to the subject matter covered. It is sold on the understanding that the Publisher is not engaged in rendering professional services. If professional advice or other expert assistance is required, the services of a competent professional should be sought.

Other Wiley Editorial Offices

John Wiley & Sons, Ltd, The Atrium, Southern Gate, Chichester, West Sussex, PO19 8SQ, UK

John Wiley & Sons Inc., 111 River Street, Hoboken, NJ 07030, USA

Jossey-Bass, 989 Market Street, San Francisco, CA 94103-1741, USA

Wiley-VCH Verlag GmbH, Boschstr. 12, D-69469 Weinheim, Germany

John Wiley & Sons Australia Ltd, 42 McDougall Street, Milton, Queensland 4064, Australia

John Wiley & Sons Canada Ltd, 6045 Freemont Blvd, Mississauga, ONT, L5R 4J3, Canada

Wiley also publishes its books in a variety of electronic formats. Some content that appears in print may not be available in electronic books.

IEEE Communications Society, Sponsor
COMMS-S Liaison to IEEE Press, Mostafa Hashem Sherif

Library of Congress Cataloging-in-Publication Data

Rahnema, Moe.
  Umts network planning, optimization, and inter-operation with GSM / Moe Rahnema.
    p. cm.
  Includes bibliographical references and index.
  ISBN 978-0-470-82301-9 (cloth)
  1. Global system for mobile communications. I. Title.
  TK5103.483.R35 2008
  621.3845–dc22   2007039290

ISBN 978-0-470-82301-9 (HB)

Typeset in 10/12pt Times by Thomson Digital Noida, India.

Contents

Preface	xv
Acknowledgments	xix

1 Introduction — 1
1.1 Overview of 3G Standards and WCDMA Releases — 1
1.2 3G Challenges — 3
1.3 Future Trends — 5

2 UMTS System and Air Interface Architecture — 7
2.1 Network Architecture — 8
   2.1.1 The Access Stratum — 8
   2.1.2 The Non-Access Stratum and Core Network — 9
   2.1.3 UTRAN Architecture — 9
   2.1.4 Synchronization in the UTRAN — 10
   2.1.5 UE Power Classes — 11
2.2 The Air Interface Modes of Operation — 11
2.3 Spectrum Allocations — 12
2.4 WCDMA and the Spreading Concept — 12
   2.4.1 Processing Gain and Impact on C/I Requirement — 13
   2.4.2 Resistivity to Narrowband Interference — 14
   2.4.3 Rake Reception of Multipath Signals and the Efficiency — 15
   2.4.4 Variable Spreading and Multi-Code Operation — 16
2.5 Cell Isolation Mechanism and Scrambling Codes — 17
2.6 Power Control Necessity — 17
2.7 Soft/Softer Handovers and the Benefits — 18
2.8 Framing and Modulation — 19
2.9 Channel Definitions — 19
   2.9.1 Physical Channels — 20
      2.9.1.1 Uplink Physical Channels — 20
      2.9.1.2 Downlink Physical Channels — 22
   2.9.2 Frame Timing Relationships — 28
      2.9.2.1 DPCCH and DPDCH on Uplink and Downlink — 28
      2.9.2.2 Uplink-Downlink Timing at UE — 28
      2.9.2.3 HS-SCCH/HS-PDSCH Timing Relationship — 28

		2.9.3	Transport Channels	29
		2.9.4	Channel Mappings	30
		2.9.5	Logical Channels	30
	2.10	The Radio Interface Protocol Architecture		31
		2.10.1 The RLC Sub-layer		33
		2.10.2 The MAC Protocol Functions		34
		2.10.3 RRC and Channel State Transitions		34
		2.10.4 Packet Data Convergence Sub-layer (PDCP)		36
		2.10.5 The Broadcast Multicast Control (BMC) Protocol		37
	2.11	The Important Physical Layer Measurements		37
		2.11.1 UE Link Performance Related Measurements		37
			2.11.1.1 CPICH RSCP	37
			2.11.1.2 UTRA Carrier RSSI	38
			2.11.1.3 CPICH Ec/No	38
			2.11.1.4 BLER	38
			2.11.1.5 UE Transmitted Power on One Carrier	38
			2.11.1.6 UE Transmission Power Headroom	38
		2.11.2 UTRAN Link Performance Related Measurements		38
			2.11.2.1 Received Total Wide Band Power	38
			2.11.2.2 SIR	39
			2.11.2.3 Transmitted Carrier Power	39
			2.11.2.4 Transmitted Code Power	39
			2.11.2.5 Transport Channel BER	39
			2.11.2.6 Physical Channel BER	39
		References		40
3	**Multipath and Path Loss Modeling**			**41**
	3.1	Multipath Reception		42
		3.1.1 Delay Spread		42
		3.1.2 Coherence Bandwidth		43
		3.1.3 Doppler Effect		45
		3.1.4 Small-scale Multipath Effects		45
		3.1.5 Channel Coherence Time		46
	3.2	3GPP Multipath Channel Models		48
	3.3	ITU Multipath Channel Models		49
	3.4	Large-Scale Distance Effects		51
		3.4.1 Lognormal Fading		51
		3.4.2 Path Loss Models		52
			3.4.2.1 The Free-space Path Loss Model	53
			3.4.2.2 The Two-ray Ground Reflection Path Loss Model	53
			3.4.2.3 Okumura-Hata Path Loss Models	54
			3.4.2.4 COST 231 Hata Model	55
			3.4.2.5 Two-Slope Extension to Hata Path Loss Models	55
			3.4.2.6 COST 231 Walfisch-Ikegami Path Loss Model	56

			3.4.2.7 Ray Tracing Models	57

            3.4.2.7   Ray Tracing Models   57
            3.4.2.8   Indoor Path Loss Modeling   58
    3.4.3   Model Tuning and Generalized Propagation
            Models   59
            3.4.3.1   The Model Tuning Process   60
            3.4.3.2   Map Data Requirement   61
            3.4.3.3   Model Resolution Requirement   62
3.5 Far-Reach Propagation Through Ducting   62
    References   63

4 Formulation and Analysis of the Coverage-capacity and Multi-user Interference Parameters in UMTS — 65

4.1 The Multi-user Interference   65
4.2 Interference Representation   67
    4.2.1 Noise Rise   67
    4.2.2 Load Factor   67
    4.2.3 Geometric Factor   68
    4.2.4 The f Factor   68
4.3 Dynamics of the Uplink Capacity   68
4.4 Downlink Power-capacity Interaction   71
    4.4.1 The General Power-capacity Formula on Downlink   71
    4.4.2 Downlink Effective Load Factor and Pole Capacity   73
    4.4.3 Single Service Case and Generalization
          to Multi-service Classes   74
    4.4.4 Implications of Downlink Power-capacity
          Analysis   75
4.5 Capacity Improvement Techniques   76
4.6 Remarks in Conclusion   77
    References   78

5 Radio Site Planning, Dimensioning, and Optimization — 81

5.1 Radio Site Locating   82
5.2 Site Engineering   83
    5.2.1 Pilot and Common Channel Power Settings   83
    5.2.2 Pilot Coverage Verification   85
    5.2.3 RACH Coverage Planning   86
    5.2.4 Site Sectorisation   87
    5.2.5 Controlling Site Overlap and Interference   87
5.3 Link Budgeting for Dimensioning   89
    5.3.1 Uplink Link Budgeting and Static Analysis   90
          5.3.1.1 Uplink Load Factor Formulation   91
          5.3.1.2 Base Station Sensitivity Estimation   93
          5.3.1.3 Soft Handover Gain Estimation   94
          5.3.1.4 The Uplink Link Budgeting Formulation   96

	5.3.2 Downlink Load and Transmit Power Checking	99
	5.3.3 Downlink Link Budgeting for the Pilot Channel (P-CPICH)	100
	5.3.4 HS-PDSCH Link Budget Analysis	101
	5.3.5 Setting Interference Parameters	102
5.4	Simulation-based Detailed Planning	104
	5.4.1 Uplink Simulation Iterations	105
	5.4.2 Downlink Simulation Iterations	106
	5.4.3 Area Coverage Probabilities	110
5.5	Primary CPICH Coverage Analysis	111
5.6	Primary and Secondary CCPCH Coverage Analysis	111
5.7	Uplink DCH Coverage Analysis	112
5.8	Pre-launch Optimization	113
5.9	Defining the Service Strategy	113
5.10	Defining Service Requirements and Traffic Modeling	113
5.11	Scrambling Codes and Planning Requirements	115
5.12	Inter-operator Interference Protection Measures	116
	5.12.1 The Characterizing Parameters	116
	5.12.2 Effects on Downlink and Uplink	118
	5.12.3 The Avoidance Measures	118
	References	119

6 The Layered and Multi-carrier Radio Access Design — 121
6.1	Introduction	121
6.2	Service Interaction Analysis	122
6.3	Layered Cell Architectures	126
	6.3.1 Carrier Sharing	126
	6.3.2 Multi-carrier Design	127
	References	128

7 Utilization of GSM Measurements for UMTS Site Overlay — 129
7.1	Introductory Considerations	129
7.2	Using GSM Measurements to Characterize Path Losses in UMTS	130
	7.2.1 Local Cumulative Path Loss Distribution	132
	7.2.2 Model Tuning	132
7.3	Neighbor-Cell Overlap and Soft Handover Overhead Measurement	132
7.4	Interference and Pilot Pollution Detection	134
	References	135

8 Power Control and Handover Procedures and Optimization — 137
8.1	Power Control	137
	8.1.1 Open Loop Power Control	138
	8.1.1.1 Uplink Open Loop Power Control	138
	8.1.1.2 Downlink Open Loop Power Control	139

		8.1.2	Fast Closed Loop Power Control (Inner-loop PC)	139

        8.1.2 Fast Closed Loop Power Control (Inner-loop PC)   139
            8.1.2.1 Closed Loop Fast Power Control Specifics on Uplink   140
            8.1.2.2 Closed Loop Fast Power Control Specifics on Downlink   141
        8.1.3 Outer-Loop Power Control   142
            8.1.3.1 Estimating the Received Quality   143
            8.1.3.2 Settings of the Maximum and Average Target E_b/N_0 144
            8.1.3.3 Power Control in Compressed Mode   144
        8.1.4 Power Control Optimization   145
    8.2 Handover Procedures and Control   145
        8.2.1 Neighbor Cell Search and Measurement Reporting   146
            8.2.1.1 Intra-frequency HO Measurements   146
            8.2.1.2 Inter-frequency and Inter-system HO Measurements   146
            8.2.1.3 UE Internal Measurements   147
            8.2.1.4 BTS Measurements   147
        8.2.2 Hard Handover   148
        8.2.3 Soft (and Softer) Handovers   149
            8.2.3.1 WCDMA SHO Algorithm and Procedures   149
            8.2.3.2 Measurement Reporting in Support of SHO   150
            8.2.3.3 SHO Gains   153
            8.2.3.4 SHO Performance Optimization   154
    References   157

9 Radio Resource and Performance Management   **159**
    9.1 Admission Control   160
        9.1.1 Processing Admission Control   160
        9.1.2 Radio Admission Control   160
            9.1.2.1 Uplink Radio Admission   161
            9.1.2.2 Downlink Radio Admission   163
    9.2 Congestion/Load Control   164
        9.2.1 Congestion Detection Mechanisms   165
        9.2.2 Congestion Resolving Actions   165
    9.3 Channel Switching and Bearer Reconfiguration   166
    9.4 Code Resource Allocation   168
        9.4.1 Code Allocation on the Uplink   169
        9.4.2 Code Allocation on the Downlink   169
    9.5 Packet Scheduling   170
        9.5.1 Time Scheduling   170
        9.5.2 Code Division Scheduling   171
        9.5.3 Scheduling on the HS-DSCH Channel   171
        9.5.4 Integration with Load Control   173
    References   173

10 Means to Enhance Radio Coverage and Capacity   **175**
    10.1 Coverage Improvement and the Impact   176
    10.2 Capacity Improvement and the Impact   176
    10.3 HSDPA Deployment   177

10.4	Transmitter Diversity		177
	10.4.1	Transmit Diversity Benefits and Gains	178
	10.4.2	Mobile Terminal Requirements	178
10.5	Mast Head Amplifiers		179
	10.5.1	MHA Benefit on System Coverage	180
	10.5.2	MHA Impact on System Capacity	181
10.6	Remote Radio Heads (RRH)		181
	10.6.1	RRH Benefits	181
10.7	Higher Order Receiver Diversity		182
	10.7.1	Operation and Observed Benefits	182
	10.7.2	Impact to Downlink Capacity	183
	10.7.3	Diversity Reception at Mobile Terminal	184
10.8	Fixed Beam and Adaptive Beam Forming		184
	10.8.1	Implementation Considerations and Issues	184
	10.8.2	Gains of Beam Forming	185
10.9	Repeaters		185
	10.9.1	Operating Characteristics	186
	10.9.2	Repeater Isolation Requirements	187
	10.9.3	Repeater Coverage and Capacity Evaluation	187
	10.9.4	Impact on System Capacity	187
10.10	Additional Scrambling Codes		188
10.11	Self-Organizing Networks		188
	References		189

11 Co-planning and Inter-operation with GSM — 191

11.1	GSM Co-location Guidelines		191
	11.1.1	The Isolation Requirements	191
	11.1.2	Isolation Mechanisms	192
	11.1.3	Inter-modulation Problems and Counter-measures	193
	11.1.4	Antenna Configuration Scenarios	195
11.2	Ambient Noise Considerations		201
11.3	Inter-operation with GSM		201
	11.3.1	Handover between the Operator's GSM and UMTS Networks	202
	11.3.2	Handover with other UMTS Operators	203
	References		203

12 AMR Speech Codecs: Operation and Performance — 205

12.1	AMR Speech Codec Characteristics and Modes		205
12.2	AMR Implementation Strategies		207
	12.2.1	AMR Network Based Adaptation	207
	12.2.2	AMR Source Controlled Rate Adaptation	208
12.3	Tradeoffs between AMR Source Rate and System Capacity in WCDMA		209
12.4	AMR Performance under Clean Speech Conditions		210
12.5	AMR Performance under Background Noise and Error Conditions		210

12.6	Codec Mode Parameters		211
	12.6.1 Compression Handover Threshold		211
	12.6.2 AMR Adaptation Parameters		211
12.7	The AMR-Wideband (WB)		212
12.8	AMR Bearer QoS Requirements		212
	References		213

13 The Terrestrial Radio Access Network Design — 215

13.1	RNC Planning and Dimensioning		215
13.2	Node Interconnect Transmission		216
	13.2.1 Node B to RNC		216
		13.2.1.1 Using ATM Virtual Paths	218
		13.2.1.2 Using Microwave Links	218
		13.2.1.3 Using Leased Lines	218
		13.2.1.4 Sharing GSM Transmission Facilities	221
	13.2.2 RNC to Core Network Nodes		221
13.3	Link Dimensioning		223
	13.3.1 Protocol Overhead		223
	13.3.2 Dimensioning of Node B–RNC Link (Iub)		224
		13.3.2.1 Sizing the Voice Links	225
		13.3.2.2 Sizing Data Links	226
	13.3.3 RNC–MSC Link Dimensioning		226
	13.3.4 RNC to SGSN Link Dimensioning		227
	13.3.5 SGSN to RNC Link Dimensioning		227
		13.3.5.1 No Service Priorities Implemented	227
		13.3.5.2 Service Priorities Implemented	229
	References		230

14 The Core Network Technologies, Design, and Dimensioning — 231

14.1	The Core Network Function		231
14.2	The IP Core Network Architecture		232
	14.2.1 The Serving GPRS Support Node (SGSN)		233
		14.2.1.1 SGSN Node Architectures	234
	14.2.2 Gateway GPRS Support Node (GGSN)		234
	14.2.3 The HLR		235
		14.2.3.1 HLR Implementation Architecture	235
	14.2.4 The Core Network Protocol Architecture in GPRS		235
	14.2.5 SS7 Over IP Transport Option (SS7oIP)		237
14.3	Mobility Management in GPRS		237
	14.3.1 Location and Routing Area Concepts		238
	14.3.2 User States in Mobility Management		238
	14.3.3 MS Modes of Operation		239
14.4	IP Address Allocation		239
14.5	Core Network in WCDMA		240
14.6	IP Multimedia Subsystem (IMS)		240

14.7	Roaming in Mobile Networks		241
	14.7.1 Mobility Handling Mechanisms in Roaming		242
14.8	Soft Switching		242
	14.8.1 Benefits of Soft Switching		243
	14.8.2 Transition to Soft Switching		244
14.9	Core Network Design and Dimensioning		245
	14.9.1 Traffic Model		245
	14.9.2 The No Traffic Information Scenario		246
	14.9.3 Dimensioning of SGSN, GGSN, and the Interfaces		247
	14.9.4 Active PDP Contexts and Impact of Call Mix on Dimensioning		247
	14.9.5 Signaling Traffic and Link Dimensioning Guidelines		248
		14.9.5.1 Signaling between SGSNs and GGSNs	248
		14.9.5.2 Signaling between SGSN and HLR	248
		14.9.5.3 Signaling between SGSN and MSC/VLR	249
		14.9.5.4 Signaling between GGSN and HLR	250
	14.9.6 Protocol Overheads		250
14.10	Core Network Transport Technologies		250
	14.10.1 Dedicated Private Lines		251
		14.10.1.1 Advantages and Disadvantages of Private Lines	251
		14.10.1.2 Sizing Criteria for Private Lines	251
	14.10.2 ATM Virtual Circuits		252
		14.10.2.1 ATM Advantages and Disadvantages Compared to Private Lines	252
		14.10.2.2 Sizing Parameters and Issues	252
	14.10.3 Frame Relay		253
		14.10.3.1 Frame Relay Advantages and Disadvantages Compared to ATM [26]	253
		14.10.3.2 Sizing Parameters and Issues	254
	14.10.4 IP Transport		254
	14.10.5 Transport Technology Selection for Core Network		255
	References		256

15 UMTS QoS Classes, Parameters, and Inter-workings — 257

15.1	The QoS Concept and its Importance		257
15.2	QoS Fundamental Concepts		258
15.3	QoS Monitoring Process		259
15.4	QoS Categories in UMTS		260
	15.4.1 Conversational Traffic		261
	15.4.2 Streaming Traffic		261
		15.4.2.1 Streaming Packet Switched QoS	261
	15.4.3 Interactive Traffic		262
	15.4.4 Background Traffic		262
15.5	Instant Messaging		262
15.6	UMTS Bearer Service Attributes		262
	15.6.1 Ranges of UMTS Bearer Service Attributes		263
	15.6.2 Ranges of Radio Access Bearer Service Attributes		264

15.7	UMTS QoS Mechanisms		264
15.8	UMTS QoS Signaling		265
15.9	UMTS–Internet QoS Inter-working/Mapping		267
15.10	End-to-End QoS Delay Analysis		267
15.11	ATM QoS Classes		268
15.12	More on QoS Mechanisms in IP Networks		269
15.13	IP Precedence to ATM Class of Service Mapping		270
15.14	Web Traffic Classification for QoS		271
15.15	QoS Levels of Agreement		271
	References		271

16 The TCP Protocols, Issues, and Performance Tuning over Wireless Links — 273

- 16.1 The TCP Fundamentals — 274
    - 16.1.1 TCP Connection Set Up and Termination — 275
    - 16.1.2 Congestion and Flow Control — 275
        - 16.1.2.1 Slow Start Congestion Control Phase — 276
        - 16.1.2.2 Congestion Avoidance Phase — 276
        - 16.1.2.3 TCP Congestion Algorithm Bottlenecks in Wireless Networks — 277
    - 16.1.3 TCP RTO Estimation — 277
    - 16.1.4 Bandwidth-Delay Product — 278
- 16.2 TCP Enhanced Lost Recovery Options — 279
    - 16.2.1 Fast Retransmit — 279
    - 16.2.2 Fast Recovery — 279
    - 16.2.3 Selective Acknowledgement (SACK) — 280
    - 16.2.4 The Timestamp Option — 280
- 16.3 TCP Variations as used on Fixed Networks — 280
    - 16.3.1 TCP Tahoe — 280
    - 16.3.2 TCP Reno — 280
    - 16.3.3 TCP New Reno — 281
    - 16.3.4 TCP SACK — 281
- 16.4 Characteristics of Wireless Networks and Particularly 3G — 281
    - 16.4.1 BLER, Delays, and Delay Variations — 281
    - 16.4.2 Delay Spikes — 282
    - 16.4.3 Dynamic Variable Bit Rate — 282
    - 16.4.4 Asymmetry — 283
- 16.5 TCP Solutions Proposed for Wireless Networks — 283
    - 16.5.1 Link Layer Solutions — 283
        - 16.5.1.1 The RLC Solution for TCP Connections — 284
    - 16.5.2 TCP Parameter Tuning — 288
        - 16.5.2.1 TCP Rxwnd tuning — 288
        - 16.5.2.2 TCP Maximum Segment Sizing (MSS) — 289
        - 16.5.2.3 Initial Transmission Window — 290
    - 16.5.3 Selecting the Proper TCP Options — 290
        - 16.5.3.1 TCP SACK Option for Wireless — 290

		16.5.3.2 TCP Timestamp Option	291
		16.5.3.3 Fast Retransmit and Recovery	291
	16.5.4	Conventional TCP Implementation Options	292
	16.5.5	Split TCP Solutions	292
	16.5.6	Indirect TCP (I-TCP)	293
	16.5.7	Mobile TCP Protocol	293
	16.5.8	Mobile-End Transport Protocol	293
	16.5.9	The Proxy Solutions	293
	16.5.10	TCP End-to-End Solutions	294
		16.5.10.1 Probing TCP	294
		16.5.10.2 TCP Santa Cruz	295
		16.5.10.3 Wireless TCP (WTCP99)	295
16.6	Application Level Optimization		295
	References		296

17 RAN Performance Root Cause Analysis and Trending Techniques for Effective Troubleshooting and Optimization — 299

17.1	RAN KPIs		299
17.2	Measurement Guidelines		300
	17.2.1	Live Network Traffic	300
	17.2.2	Drive Testing	301
17.3	Correlation Based Root Cause Analysis		303
	17.3.1	Correlative Analysis Based on a priori Knowledge	303
	17.3.2	Correlation Analysis Based on Data Clustering	306
		17.3.2.1 Data Reduction and Clusterization Based on Self-organized Maps (SOM)	306
		17.3.2.2 Clustering of SOM Data Based on the K-means Algorithm	308
17.4	Applications to Network Troubleshooting and Performance Optimization		309
	17.4.1	Formation of Vector PIs	309
	17.4.2	Data Scaling	310
	17.4.3	Clustering of Performance Data (Building Performance Spectrum)	310
	17.4.4	Clustering Cells into Behavioral Classes	311
	Appendix		312
	References		313

Abbreviations — 315

Index — 323

Preface

The continuing explosive growth in mobile communication is demanding more spectrally efficient radio access technologies than the prevalent second generation (2G) systems such as GSM to handle just the voice traffic. We are already witnessing high levels of mobile penetration exceeding 70% in some countries. It is anticipated that by 2010 more than half of all communications will be carried out by mobile cellular networks. On the other hand, the information revolution and changing life habits are bringing the requirement of communicating on a multimedia level to the mobile environment. But the data handling capabilities and flexibility of the 2G cellular systems are limited. The third generation (3G) systems based on the more spectrally efficient wideband CDMA and a more flexible radio channel structure are needed to provide the high bit rate services such as image, video, and access to the web with the necessary quality and bandwidth. This has promoted the inception of a global 3G standard that will bring higher capacities and spectral efficiencies for supporting high data rate services, and the flexibility for mixed media communication. The 3G mobile communication network referred to here as UMTS (Universal Mobile Telecommunication System) is based on the Wideband Code Division Multiple Access (WCDMA) and is the main 3G radio access standard in the world. UMTS has been deployed in Europe, and is being deployed in the USA, Japan, Korea, and in many other parts of Asia around the same frequency band of 2 GHz. The present book provides a detailed description of the WCDMA air interface, the detailed radio planning, and the optimization and capacity improvement mechanisms for the FDD-mode, the QoS classes, and the end-to-end parameter interworking mechanisms, as well as an adequate coverage of the terrestrial and the core network design, dimensioning, and end-to-end data transfer optimization mechanisms based on the TCP protocol.

Chapter 1 provides a snapshot description of the evolution of the UMTS releases, highlights the main features introduced in each release, and then briefly discusses the challenges facing the network operators in the planning and optimization of 3G networks, their inter-operation with existing GSM networks, and the trends of future network evolutions.

Chapter 2 provides a detailed and comprehensive overview of the UMTS architecture, network elements, interfaces, and code division multiple access spread spectrum concepts and issues. The chapter also covers the UMTS air interface channel organization and protocols, contains an overview of specific mechanisms that impact 3G radio performance such as power control and handovers, and ends with a description of the key WCDMA link performance indicators used in radio network planning and optimization.

Chapter 3 is a detailed and comprehensive overview of multipath radio channel statistical parameters that impact communication system and network design and a description of 3GPP and ITU multipath channel models. It also presents the numerous path loss channel models and parameters for various environments, and discusses in fair detail guidelines for path loss model tuning based on RF measurements and obtaining adequate path loss prediction model resolutions, which is of particular importance in 3G network planning. This chapter is also a general useful reference for RF path loss prediction and RF channel model development for RF professionals concerned with mobile communication.

Chapter 4 presents the key 3G radio network parameters modeling the multi-user load and the interference geometries. It also derives the theoretical formulation for base station power, the uplink and downlink load factors, and the pole capacities, as well as presenting sample numerical results to illustrate the concepts and deriving conclusions and implications to guide optimal radio network planning in WCDMA. This chapter provides the necessary theoretical background and concepts for the next chapter, which focuses on the detailed practical radio network planning.

Chapter 5 presents the detailed processes and formulations for radio network planning and dimensioning. This chapter presents the guidelines for selecting radio base station sites based on the results of the latest research activities, derives the link budget formulas for the traffic, the pilot, and the HSDPA channels, presents a detailed iterative link budgeting static analysis approach, and provides sample link budgeting templates and examples. Then follows a presentation of flowcharts for the iterative Monte Carlo simulation processes for detailed radio capacity and coverage verification. The chapter also presents engineering design guidelines for site sectorisation and engineering, antenna selections, pilot and control channel power settings, traffic requirements analysis, and radio dimensioning and site placement coordination with other operators to mitigate inter-operator interferences.

Chapter 6 presents further guidelines for optimal radio network planning based on layered radio architectures. The layered radio architectures implemented on single and/or multiple frequency carriers are a necessity mechanism to provide optimum capacity and service coverage in the multi-service scenarios of 3G networks. This chapter discusses how this is achieved and provides practical guidelines for designing layered multi-carrier radio architectures.

Chapter 7 presents the cost-effective and realistic 3G planning models and strategies for incumbent GSM operators. It discusses how the existing GSM operators can utilize RF path loss measurements collected by their GSM networks to obtain site re-engineering guidelines, and realistic path loss models for 3G site co-location scenarios to minimize interference geometries.

Chapter 8 discusses and presents the various power control and handover mechanisms and related measurements and parameters for WCDMA. Power control and handover (soft handover) are two very important and basic mechanisms in 3G networks, and understanding them and the impact of related parameters, and their optimization on network performance, are critical to proper radio network planning and optimization. This chapter provides the detailed guidelines for tuning these mechanisms where possible.

Chapter 9 focuses on the typical strategies and algorithms that are implemented by vendors for the management and control of traffic load and the allocation of radio resources to achieve coverage and quality for each service category. These strategies are based on measurements defined in the 3GPP Standards and include admission and load/congestion control functions, allocation of radio resources to different services, and the related measurements used in the process. The chapter also discusses guidelines for setting the decision thresholds for measurements used in the control and admission of each traffic category into the network, so that the overall desired coverage and quality can be achieved for the multi-service environment of 3G.

Chapter 10 introduces and discusses various additional coverage and capacity improvement techniques beyond what is discussed in Chapter 5 on radio site planning and optimization. The mechanisms introduced here include antenna receive and transmit diversities, use of mast head amplifiers, repeaters, optimal site configurations, etc. The chapter includes practical examples and case studies.

Chapter 11 introduces the reader to issues involved in co-planning WCDMA with existing GSM networks and their optimal inter-operation. The issues addressed include inter-system interference and avoidance guidelines, antenna sharing configuration examples, and inter-system handover parameter tuning for resource pooling and overall network capacity and coverage optimization.

In Chapter 12, the AMR speech codecs for GSM and 3G networks are introduced. The various implementation options and performance under varying background noise conditions are discussed, and the tradeoffs in the AMR source coding rate and capacity in WCDMA are quantitatively evaluated and presented along with the associated control parameters for guiding the radio optimization process. The chapter also discusses the wideband AMR, which uses higher sampling rates to achieve superior voice quality.

Chapter 13 covers the guidelines for the design and dimensioning of the terrestrial access network in 3G. Strategies for dimensioning the Iub and Iu links, and sharing access links with existing GSM networks using alternative transport technologies, are also discussed.

Chapter 14 introduces the reader to the core networks in WCDMA, with a detailed discussion of the protocols and transport technologies involved. The chapter also presents dimensioning guidelines for various core network elements, links based on practical traffic models, and protocol overhead accounts. Furthermore, the chapter discusses some of the recent trends for distributed core network elements based on the separation of call and mobility control from the actual user information transport as paving the way to an all-IP core. It discusses soft switching and presents practical migration strategies for migration to soft switching core architectures. The chapter also discusses the IMS service platform and the flexibility for multimedia traffic handling and service support.

Chapter 15 presents the WCDMA end-to-end Quality of Service (QoS) architecture, signaling flows, QoS service classification, and parameters/attributes. The chapter also discusses key QoS implementation mechanisms in the core and the mapping of QoS related attributes and parameters across the radio access, the Iu, and the core network to achieve end-to-end performance.

Chapter 16 provides the reader with a thorough detailed discussion of the important TCP (transmission control protocol) and its adaptation to the wireless links, particularly for UMTS and GPRS networks. The chapter presents and discusses the issues involved in using the conventional TCP in the mobile communication environment and presents the appropriate variations of the protocol and complementary measures such as tuning relevant parameters within TCP and the underlying radio link control protocol to adapt the performance for achieving optimal data throughputs and reduced delays.

Finally in Chapter 17, the reader is introduced to efficient time saving and practical methodologies for measuring and monitoring the network performance and finding the root cause problems for quick troubleshooting in the perplexing multi-service and highly interactive 3G radio environment. The performance trending and troubleshooting techniques discussed in this chapter are equally applicable to GSM and other network technologies.

This book is an outgrowth of the author's years of experience and consulting in the wireless telecommunication field starting from low earth orbit satellites at Motorola to GSM, to GPRS planning in the USA and Asia, and to extensive investigation, studies, and development of radio and core network planning for 3G in the USA and Europe. The book has a heavy focus on the radio/RF planning aspects of 3G networks, but is also intended to benefit significantly professionals involved in core network planning, dimensioning, and end-to-end optimization aspects, RF propagation channel modeling professionals, university students, and new researchers to the field, as well as provide insight for advanced developments in equipment manufacturing.

Acknowledgments

The author would like to thank the many pioneering researchers in the industry, and in academia, whose efforts have helped to create the groundwork for this book. Without this work, it would never have been possible to put together such a publication. I would also like to thank the many clients who have provided consulting opportunities to me in this industry. This has helped me gather the knowledge, experience, and insight to be able to put together a book of this scope.

Special thanks are also due to the staff in the Singapore and UK offices of John Wiley & Sons who have provided excellent support in the production of this book and helped me to meet a reasonable schedule. In particular, I would like to thank James Murphy and Ann-Marie Halligan for arranging the initial review of the proposal and for securing the approval for its publication, as well as their subsequent coordination of the Wiley production team. This team includes Diane Tan whose tireless efforts to arrange and obtain the copyright permissions for the referenced material in this book were instrumental in making its timely publication possible. The author is especially thankful to Roger Bullen, Andrew Finch, and Sarah Hinton for their guidance and assistance in the editorial review of the manuscript and the proposing of useful refinements.

I would also like to thank my friend from academia, Dr. Behnam Kamali, of Mercer University, Georgia, USA, who encouraged and supported me in this effort by providing copies of some of the references as requested.

The author welcomes any comments and suggestions for improvement or changes that could be implemented in possible future editions of this book. The email address for gathering such information is mroi_us@yahoo.com.

1

Introduction

The information revolution has created a new 'post-industrial paradigm', which has transformed the way we live and work, the way we create arts, and the way we make new products and provide services. Clearly information technology has completely changed from a network of oral and print mechanisms to one that is electronic, visual, and multimedia. Along with this development has come the natural evolution of the speed with which information is transferred, from months and days to nanoseconds. This was made possible with the merging of the computer and the telephone, which prompted the emergence of a communications revolution. And it is needless to say that the wireless technologies have played and will continue to play the crucial role in this revolution, as they are the most convenient, efficient, and personal means of communicating information.

The tremendous growth in wireless communication technology over the past decade, along with reduced costs, has created major changes in people's communication habits, social and business networking, and lifestyles. This has in turn led to an explosive growth in the number of subscribers and the traffic placed on wireless networks. It is notable that the growth in voice traffic alone, which is currently the main source of revenues for most operators, is in many cases placing a huge burden on the existing capacity limited second generation (2G) systems such as GSM and other TDMA networks. On the other hand new bandwidth consuming applications, such as access to information on the move, video messaging, music downloading, mobile location based content retrieval, and voice calls with simultaneous access to data or images, are or will soon place new demands on capacity. The best answer to this explosive demand in capacity on the move is the provision of new spectrums and the deployment of advanced spectrally efficient multiple access techniques that can efficiently offer multiple type services from a wide range of bit rate characteristics and quality requirements on demand over the radio link. The Wideband CDMA (W-CDMA) technology is currently one such efficient and flexible radio access technology adopted for the implementation of the third generation (3G) wireless networks.

1.1 Overview of 3G Standards and WCDMA Releases

UMTS (Universal Mobile Telecommunication System) is the European 3G Standard based on W-CDMA technology, and is normally the solution generally preferred by countries that used GSM for the 2G network. UMTS is managed by the 3GPP organization, which also became responsible for the GSM continued standardization from July 2000. CDMA2000 is another significant 3G standard that is an outgrowth of the earlier 2G CDMA standard IS-95. CDMA2000's primary proponents are mainly in the Americas, Japan, and Korea, though UMTS is being tested and deployed at this time in the Americas by T-Mobile and Cingular. CDMA2000 is managed by 3GPP2, which is separate and independent from UMTS's 3GPP. The various types of transmission technology used in CDMA2000 include 1xRTT, CDMA2000-1xEV-DO, and 1xEV-DV. China has also come up with a Standard of its own, referred to as TD-SCDMA, which has been developed by the companies Datang and Siemens, for which field trails have been taking place in Beijing and Shanghai.

The first commercial UMTS network was deployed by Japan's NTT DoCoMo in 2001. Since then UMTS networks have been deployed in more than 20 countries including Germany, France, UK, Malaysia, Netherlands, Norway, Singapore, Spain, and Bahrain. The 3G networks based on WCDMA continue to be deployed in more and more countries. This situation is demanding that more and more radio planning professionals become more familiar with the WCDMA technology to design and launch high quality 3G networks. This book has been written with a heavy emphasis on radio planning and optimization principles for RF engineering professionals. The book also contains four extensive chapters (Chapters 13 to 16), which discuss the end-to-end QoS (Quality of Service) inter-working, and the design, dimensioning, and optimization of the access network, the core network, and the Transmission Control Protocol (TCP) protocol for wireless networks. Therefore this book is expected to benefit protocol and core network engineering professionals as well, and provide a good reference for the end-to-end network planning and optimization. 3G network planning involves a number of new challenges over the 2G networks, which relate to the underlying WCDMA radio access, the multi-service requirements, and opportunities to make use of new technologies in the core network such as the split connection and call control architectures (soft switching) for the design of efficient scaleable and flexible network architectures. These challenges are briefly outlined in this introductory chapter, and then discussed in greater detail and depth in the remaining chapters. This book is also expected to be highly valuable for graduate level students and new researchers in the field with an interest in the WCDMA technologies for network planning and optimization.

The development of the UMTS specifications based on W-CDMA in 3GPP has taken several phases. The first release of the UMTS specifications is known as 3GPP R99, which was functionally frozen in December 1999. The 3GPP R99 implementation offers the same services with those of GSM Phase 2+ (GPRS/EDGE). That is, all the same supplementary services are available; teleservices and bearer services have different implementation but this is not visible to the subscriber. The 3G network in this phase may offer some other services not available in GSM, for example, a video call. The second phase known as 3GPP Release 4 introduces all-IP in the core network allowing separation of call control and signaling from the actual connection or media used on the core network (CN) side to transport circuit switched (CS) services such as voice. In the CN CS domain actual user data flow passes through Media Gateways (MGW), which are elements that maintain the connection and perform switching functions when required. The whole process is controlled by a separate element evolved from MSC/VLR called MSC server. One MSC server can handle numerous MGWs thus making the CN CS domain scalable. This approach is also referred to as soft switching. Release 4 Specifications were frozen in March 2001. The 3GPP Release 5 then introduces a new element called the IP Multimedia Subsystem (IMS) for unifying the methods to perform IP based multimedia services. Multimedia service is a scenario in which more than one service type component is combined on one physical connection to a user such as voice along with image or video. In Release 5 of the 3GPP specifications, the notion of all-IP is introduced, extending IP transport to the access network as well. This extends the IP mode communication all the way to the radio access network including the circuit switched domain. So, a voice call from UE to PSTN is transported through UTRAN as packets and from the GGSN the VoIP is routed to the PSTN via IMS, which provides the required conversion functions. Release 5 also introduces Wideband AMR, as well as HSDPA. HSDPA service is a new evolution in the air interface for providing high-speed data rates on the downlink. HSDPA provides integrated voice on a dedicated channel and high-speed data on a downlink shared channel on the same carrier, which allows data rates of up to 14.4 Mbps. HSDPA is primarily deployed for dense urban and indoor coverage. Release 5 Specifications were frozen in June 2002. A similar enhancement is introduced on the uplink side in Release 6 for offering high-speed data rates on the uplink, HSUPA. Release 6 also includes wireless LAN/UMTS inter-working, Multimedia Broadcast/Multicast Service (MBMS), network sharing, and the Push services.

From the user terminal point of view, the network is basically the same in the various developmental phases, except for some new service capabilities such as HSDPA that will require new capabilities in the terminal for using the service. The major changes introduced by the various releases of the UMTS specifications occur within the network and are in the transport technologies, and the new flexibilities and efficiencies provided in operating the network. For instance, release 1999 uses ATM as the transport technology, whereas in 3GPP R4, and R5, ATM is swapped withv IP.

1.2 3G Challenges

The current deployment of UMTS networks is not in many cases ubiquitous and is only concentrated in the congested urban business areas. They are used to provide either the special higher rate data services or increased capacity for handling the voice traffic in specific locations and are therefore complementary and supplemental to the GSM networks. The GSM networks are anticipated to stay around and even continue to grow and expand for at least the next five years given the huge investments already made by the operators in the GSM infrastructure networks and their fine capability to handle voice, though not with the same spectral efficiency as the WCDMA. This means that the island deployment of UMTS networks will be the trend for some time to come, and hence the requirement for the seamless roaming, handover, and inter-operation with the existing GSM networks to provide service coverage continuity and load sharing. Therefore, the elaborate inter-operability and coordination mechanisms and features provided by the equipment need to be exploited by the network planners to effectively result in the pooling of the resources, and hence result in the most efficient utilization of the limited expensive radio spectrums. Moreover, for uniform service quality provisioning, prior optimization of existing GSM/GPRS networks may be necessary to provide the same service quality as in WCDMA in inter-system roaming.

On the other hand, the incumbent GSM operators can exploit their existing GSM network infrastructures in multiple ways to facilitate cost effective optimal planning of UMTS in their networks. These include substantial radio base station co-location to save on site costs, sharing of access transmission facilities to achieve higher trunking efficiencies, and use of network provided radio propagation measurements. Site co-location brings new challenges for efficient antenna sharing solutions, and antenna placement configurations that can provide the proper isolation between the GSM and UMTS systems. The W-CDMA system particularly can be impacted by interference caused by GSM systems, if proper RF isolation measures are not taken in co-siting scenarios. Meanwhile before any co-site deployments, the existing GSM radio access facilities can be used to obtain radio propagation related measurements to characterize path loss and interference geometries to guide link budgeting and site engineering for a UMTS overlay scenario. Interference is a major factor that impacts both the coverage and capacity in CDMA based networks due to the tight frequency re-use of 1. Therefore, having an accurate realistic picture of the RF interference geometry resulting from candidate sites that are selected in sets is highly critical before actual deployment, to make sure that adequate cell isolation is obtained.

In addition to the concerns over interference caused from other cells, intra-cell multi-user interference inherent to CDMA systems results in a dependency between the cells coverage and its capacity (load). This situation makes the radio network planning more complex than in 2G systems. This additional complexity means that site location and dimensioning can no longer be performed based on coverage consideration alone, or capacity adjustment left to later stages. The traffic profile and distribution will have to enter the planning phase from the very beginning to make sure that sites are positioned and dimensioned to achieve a proper balance between the expected coverage and the capacity (load) to be handled. Coverage is also not uniform for all services due to the bit-rate and quality dependent power requirements. Therefore layered architectures based on the use of micro-cells, indoors, and macro-cells need to be implemented to provide efficiently the necessary coverage and capacity for multi-rate services. The coverage limitations imposed by the cell capacity also result in the consideration of using multi-carriers to split the traffic and alleviate the multi-user interference within the same cell, and/or between the cell layers, in order to provide high capacity and throughput. The multi-layer and multi-band architectures will require efficient inter-layer and inter-band handover and traffic distribution mechanisms, and the RF planning and dimensioning of each layer.

Meanwhile, recent developments in the core network technologies have provided new design options to separate the call control and signaling hardware from the media switching fabrics. This allows the design of core network architectures that are scaleable for easy network expansion and flexible to handle the multimedia services efficiently. The separation of call control and signaling from the media switching functions, based for instance on soft switching, also paves the way for migration of the core networks to all-IP transport. This provides the framework for the cost efficient long term growth of 3G services.

Yet another challenge in the optimization of 3G networks is the proper tuning and selection of specific versions of transport protocols such as TCP. TCP is used for the reliable end-to-end delivery of IP based data thatare expected to be a significant service in 3G networks. There are certain versions of TCP that are more suitable to handling packet loss and large delay variations, which can occur over the lossy radio links in the mobile environment. Furthermore the parameters within the radio link protocols and selected TCP version should be tuned to utilize efficiently the allocated limited radio spectrum, and achieve high throughput and QoS for IP based application with their unlimited demand for capacity.

The high throughput and capacity demand of the services anticipated for the 3G networks and the interference-limiting environment of the CDMA based systems require highly skilled radio planning practices and the use of spectral efficiency measures. These include receiver and transmitter antenna diversity mechanisms, efficient site sectorisation, selecting antennas with optimum beamwidths and positioning for proper tilt and orientation to provide maximum cell isolation, iterative link budgeting for reasonable cell load plans, providing for adequate cell overlap for soft handover gains while also minimizing the overheads, and use of vendor equipment capacity improvement mechanisms such as interference cancellations, rate adaptations, and smart efficient packet scheduling algorithms that can operate in near real time. Moreover, the higher complexity of the 3G multi-access dynamics, and the multitude of diverse services with varying QoS requirements and performance metrics, bring new challenges in developing effective methods for performance measurement, problem classification, and root cause analysis.

1.3 Future Trends

The future trends in the development of 3G networks and beyond can be viewed from two perspectives: evolutionary developments and introduction of disruptive technologies. The evolutionary developments include enhancement of the current 3G networks with the features in the frozen and evolving releases, and making use of the capabilities supported by the current specifications for development of advanced services. The enhanced services may include the implementation of diverse location based services (providing information that is highly localized) for various applications, push services, and multimedia services based on IMS (IP Multi-media Sub-System) capabilities supported in Releases 5 and up. IMS entities placed in the core network elements allow the creation and deployment of IP-based multimedia services in 3G mobile networks. IMS facilitates the integration of real-time and non-real-time services within a single session and provides the capability for services to interact with each other. Efficient handling of resources is a key requirement, because the network must satisfy the QoS requirements of data flows from the different media. The IMS architecture separates the service layer from the network layer, facilitating interoperability between 3G mobile networks and fixed networks such as the PSTN and the Internet.

The implementation of MIMO (Multi-Input Multi-Output) antenna technologies on the radio access side for at least data terminals is viewed as another evolutionary trend for enhancing the radio access capacity of 3G networks. MIMO is expected to result in huge capacity improvement of the current 3G technologies.

The evolutionary developments include also integration of 3G networks with wireless local area networks (WLAN), WiMAX, etc. to provide seamless roaming between alternative network access technologies while using the same IP core infrastructure. In a highly integrated network access scenario, alternative access networks can be used based on user mobility and throughput requirements. Such developments are already in progress based on UMA technologies that provide the wireless networks with access to the unlicensed spectrum bands used for WiMAX, Bluetooth, and 802.11 WLANs. UMA enables GSM/GPRS/3G handsets equipped with Wi-Fi or Bluetooth to access the wireless core networks using the unlicensed air interface when available. As such, UMA represents an extension of the cellular network for mobile operators, which can support voice and data services in homes, offices, and hotspots such as coffee houses, convention centers, hotels, school campuses, and libraries, which have a high demand for wireless data services. However, the criticality of the UMA access integration or the competition of the WiMAX camp over WCDMA will depend on rapid agreements on the technology roadmap beyond HSDPA/HSUPA.

The introduction of disruptive technologies such as the use of OFDM (Orthogonal Frequency Division Multiple access) for multi-access on the radio side has been a proposal under consideration for the next generation of wireless networks beyond 3G's HSDPA and HSUPA. OFDM was developed initially by the Bell Labs in the 1960s, and is now a de facto choice for most of the next-generation wireless technologies and used by Flarion's technology, WiMAX, and WiBro. It is also currently the basis for digital audio broadcast and some fixed wired transmission techniques such as xDSL. Since OFDM provides for the transmission of a number of sub-channels, each at a lower symbol rate, it helps to eliminate the inter-symbol interference that would be present in the wideband channel of WCDMA, and greatly simplifies channel equalization. With much higher bandwidths of 10–20 MHz, OFDM can be combined with advanced space division multiple access techniques of Multiple Input Multiple Output (MIMO) antennas, to achieve much higher peak data rates with less complex implementations. These considerations are driving 3GPP to consider OFDM for its long-term evolution project. The fact that newer wireless LAN standards that employ 20 MHz radio channels are also based on OFDM is making this a favored access technology for radio systems that have extremely high peak data rates.

2

UMTS System and Air Interface Architecture

A major differentiation of 3G systems over the 2G cellular networks is the provision of multiple data-rate services with varying quality requirements and a higher spectrally efficient and flexible radio interface. The benefit of a higher spectrally efficient and flexible air interface particularly pertains to UMTS, which has adopted the WCDMA for its air interface. The WCDMA, as one air interface specification for the 3G systems under the ITU umbrella of the IMT-2000 (International Mobile Telephony 2000) systems, has been or is being used widely in Europe, many parts of Asia including Japan and Korea, and even in North America recently. In North America, because the original spectrum specified for the IMT-2000 at around 2 GHz had originally been assigned to 2G systems (PCS band at around 1900 MHz), 3G systems using the WCDMA air interface are implemented through the re-farming of portions of the spectrum allocated to 2G systems. A second alternative for countries where the global IMT-2000 spectrum has not been available is the use of the CDMA-2000 air interface. This is a multi-carrier evolution of the CDMA based 2G systems, IS-95, which is implemented over 1.25 MHz bands. However, the WCDMA air interface allows the use of 5 MHz wide channels compared to the 1.25 MHz-wide channels used in the CDMA 2000 version of the 3G air interface specifications. This results in many benefits, which include higher trunking efficiency over the radio band, higher multipath diversity gains that improve the coverage, and more flexible and efficient higher data rate implementations. This chapter introduces the architecture and the specific mechanisms of the WCDMA air interface used to implement efficiently multiple data-rate services and utilize the radio capacity. A brief discussion is given of the overall UMTS network architecture and the core network elements definitions and functions. This introduction is expected to provide the reader with the necessary background for what follows in the rest of the book: the in-depth coverage and study of the important WCDMA mechanisms, functions, and procedures for the planning and optimization of UMTS networks. We begin with a brief discussion of the overall UMTS network architecture, elements, and terminologies as related to the radio access and the core network.

UMTS Network Planning, Optimization, and Inter-Operation with GSM   Moe Rahnema
© 2008 John Wiley & Sons, (Asia) Pte Ltd

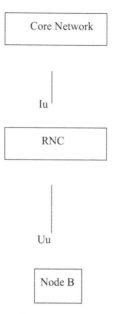

Figure 2.1 High level UMTS architecture.

2.1 Network Architecture

The three major elements in the UMTS architecture on a high level are the user equipment (UE), which houses the UMTS subscriber identity module (USIM), the UMTS Terrestrial Radio Access Network (UTRAN), and the Core network (CN). These three elements and their interfaces are shown in Figure 2.1 [1]. The Uu and Iu are used to denote the interfaces between the UE and ITRAN, and UTRAN and core network, respectively. The UTRAN and core network are in turn composed of several elements. The UTRAN elements are discussed in this section. The core network elements are discussed in Chapter 14.

The protocols over Uu and Iu interfaces are divided into two categories: the User plane and the Control plane protocols. The User plane protocols implement the actual Radio Access Bearer (RAB) services carrying user data through the access stratum. A bearer is an information transmission path of defined capacity, delay, bit error rate, etc. A bearer service is a type of telecommunication service that provides the capability of transmission between access points. A bearer service includes all the necessary aspects, such as User plane data transport and QoS management, needed to provide a certain QoS.

Node B also has some limited resource management functionality.

The Control plane protocols are used for controlling the radio access bearers and the connections between the UE and the network. These functions include service request, control of transmission resources, handovers, and mobility management.

2.1.1 The Access Stratum

The Access Stratum (AS) [2] consists of a functional grouping, which includes all the layers embedded in the URAN, part of the layers in the User Equipment (UE), and the

UMTS System and Air Interface Architecture

infrastructure (IF) edge nodes. Its boundaries border on the layers that are independent of the access technique and the ones that are dependent on it. The AS contains all access specific functions such as handling and coordinating the use of radio resources between the UE and the UTRAN, providing flexible radio access bearers, radio bandwidth on demand, and handover and macro-diversity functions. The AS offers services through the Service Access Points (SAPs) to the Non-Access Stratum (NAS). The RAB is a service provided by the AS to the NAS for the transfer of data between the UE and the core network.

2.1.2 The Non-Access Stratum and Core Network

The Non-Access Stratum refers to the set of protocols and functions that are used to handle user services, Call Control for data and voice such as provided by Q.931 and ISUP, UE identification through IMSI, core network related signaling such as session management, supplementary services management, and establishment and release of PDP contexts. In the packet services domain, there is a one-to-one relation between a PDP context and a Radio Access bearer. The RAB is a service that the AS provides to the NAS for the transfer of user data between the UE and the core network. A bearer is described by a set of parameters called attributes, which define the quality of service or traffic aspects of the service as described in Chapter 15. In a way, the NAS deals with functions and parameters, which have end-to-end significance.

2.1.3 UTRAN Architecture

The UTRAN consists of a set of Radio Network Subsystems (RNS) connected to the core network through the Iu interface. The UTRAN architecture with element connections is shown in Figure 2.2 [3]. An RNS consists of a Radio Network Controller (RNC) and one or more Node Bs. The RNS can be either a full UTRAN or only a part of a UTRAN.

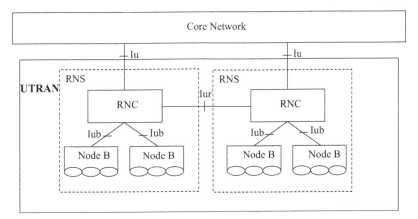

Figure 2.2 UTRAN Architecture.
[3GPP™ TSs and TRs are the property of ARIB, STIS, ETSI, CCSA, TTA and TTC who jointly own the copyright in them. They are subject to further modifications and are therefore provided to you 'as is' for information purposes only. Further use is strictly prohibited]

An RNS manages the allocation and release of specific radio resources for establishing connections between a UE and the UTRAN. A Node B is connected to the RNC through the Iub interface. A Node B can support FDD mode, TDD mode, or dual-mode operation. Inside the UTRAN, the RNCs of the Radio Network Subsystems can be interconnected together through the Iur. Iu(s) and Iur are logical interfaces. Iur can be conveyed over direct physical connection between RNCs or virtual networks using any suitable transport network.

Node –B – also called the base station – houses the radio transceiver equipment (antennas, transceivers) and handles the radio transmission between the UE and one or more cells. The Node B interface to RNC is called Iub. The radio signaling over Iub is handled by the Node B application part. The RNC is generally a logical node in the RNS but is currently a separate hardware unit and controls the use and integrity of radio resources such as carriers, scrambling codes, spreading codes, and power for common and dedicated resources. An RNC is called a serving RNC (SRNC) with respect to a UE if it is in charge of radio connection of the UE. There is one serving RNS for each UE that has a connection to UTRAN. An RNC is referred to as a Drifting RNC (DRNC) with respect to a UE when it controls the cells (Node Bs) that the UE moves into in a handover process. An RNC is called a controlling RNC with respect to a specific set of Node Bs if it has the overall control of their logical resources. There is only one controlling RNC for any Node B.

The radio network signaling over the Iu and Iur are referred to as the 'Radio Access Network Application part', and 'Radio Network Subsystem Application Part', respectively.

2.1.4 Synchronization in the UTRAN

The control of synchronization in a cellular UMTS network is crucial, because it has a considerable effect on the QoS that can be offered to the user. Synchronization generally refers to two different aspects. These are timing or frame synchronization (also called frame alignment) between different network nodes, and frequency synchronization and stability. Frame synchronization or timing alignment in turn refers to the following two aspects:

- RNC-Node B synchronization
- Inter-Node B synchronization

The synchronization between RNC and Node B is concerned with finding out the timing difference between the two nodes for the purpose of setting good DL and UL offset values for transport channel synchronizing between RNC and their Node Bs. This helps to minimize the transmission delay and the buffering time for the DL transmission on the air interface. The knowledge of the frame timing relationship between two nodes is based on a measurement procedure called RNC-Node B node synchronization [4]. The fractional frequency error between the RNC and Node B clocks should not exceed a value of 1E-11 [5].

By Inter-Node B synchronization, is generally meant the achievement of a common reference timing phase between Node Bs. This ensures that the frame or slot boundaries are positioned at the same instant in adjacent cells. In UTRAN FDD mode, such a common reference timing phase among neighboring Node Bs is not required unlike the CDMA 2000 system, though it could help to reduce the inter-cell interference that results from a synchronous CDMA operation (the latter is not completely achievable due to multipath effects, see Chapters 3 and 4). This means that although the different Node Bs would be locked to the same reference frequency source, they would not be phase aligned. This asynchronous mode of Node B operation in UMTS eliminates the need for global time references such as a GPS signal. A benefit is that it makes the deployment of indoor and micro base stations in dense urban areas easier where there is usually hardly any line of sight to a GPS satellite. However, the Inter-Node B timing phase synchronization is required in UTRAN TDD mode, in order to transmit frames aligned in time between the different neighboring cells and thus prevent interference between the uplinks and downlinks of adjacent cells. The TDD systems place a limit of 2.5 µs phase variations for synchronization signals between any two Nodes B of the same area [4].

By frequency synchronization, is meant that the UTRAN nodes are all synchronized to a reference frequency source with certain accuracy and stability requirements. The frequency source is normally used to derive both the timing clock signals and the carrier frequency. Therefore an accurate clock reference is necessary to maintain precisely the carrier frequencies in order to not overlap spectrums assigned to other operators and to other carriers of a multi-carrier UMTS network. The 3GPP specifications [6, 7] require base station frequencies with an accuracy of ± 0.05 ppm observed over a period of one timeslot for the macro and micro cells. The 0.05 ppm limit will cause variation that equals 5×10^{-8} seconds for each second in the phase of the synchronization signal. In other words, the phase may vary at most 50 ns per second in respect to the nominal value.

This requirement is reduced to ± 0.1 ppm for pico-cells. The macro and micro cells are defined in 3GPP[C] as scenarios with BS to UE coupling losses equal to 70 dB and 53 dB. The pico-cell cell scenarios are characterized with a BS to UE coupling loss equal to 45 dB.

2.1.5 UE Power Classes

The UE power classes define the maximum power output capability according to the user equipment power class. The 3GPP Standard [1] has defined four UE power classes, the most common of which are power classes 4 and 3. The UE power classes are given in Table 2.1.

2.2 The Air Interface Modes of Operation

Two different modes of operation have been specified for the air WCDMA air interface used in UMTS. These are referred to as the FDD (frequency division duplex) and the TDD (time division duplex) modes. The FDD systems use different frequency bands for the uplink and downlink, which are separated by a duplex distance of 190 MHz. The TDD mode systems utilize the same frequency band for both uplink and downlink. The communication on the uplink and downlink in TDD are separated in the time domain, by assigning different time slots to uplink and downlink (time division multiplexing). The time slots assigned to the

Table 2.1 UE power class.

Power class	Maximum output power (dBm)	Tolerance (dB)
1	+33	+1/−3
2	+27	+1/−3
3	+24	+1/−3
4	+21	±2

uplink and downlink can be dynamically swapped depending on the traffic demand on each link. Therefore the TDD mode is best for asymmetric traffic (such as web downloading) where traffic in one direction is higher than in the other. In a way TDD can be used as a complement to FDD systems in hot spot traffic areas where there is a high asymmetry of the traffic demand in the uplink and downlink. One advantage of TDD is that it uses lower spreading codes, which allow easy multi-user detection at the base station to help cancel interference. However there are certain complications and disadvantages associated with TDD systems. The use of the same band on uplink and downlink in TDD systems means that new additional interference geometries develop, which include the interference between the downlink and uplinks of different base stations, and also mobile stations in different cells. Reducing this kind of interference would require slot synchronization between the base stations. Also, TDD systems do not support soft handovers, and are not able to track the fast fading as well as the FDD systems. TDD systems effectively use a 100 Hz power control rate, which is much smaller than the 1500 Hz power control used in FDD systems. However, the emphasis in this book is on the FDD systems. We will refer to TDD in the rest of the book wherever a specific function or issue would relate to the TDD mode.

2.3 Spectrum Allocations

In Europe and most of Asia, including Japan and Korea, the IMT-2000 bands of 2X60 MHz in 1920–1980 MHz for the uplink, and in 2110–2170 MHz for the downlink, are allocated for the WCDMA FDD mode. The uplink direction refers to the communication from the mobile to the base station, and the downlink in the direction from the base station to the mobile. The availability of the TDD spectrum varies from country to country. In Europe, 25 MHz are available for licensed TDD use in the 1900–1920 MHz and the 2020–2025 MHz bands. The allocation of the band between 470–600b MHz for UMTS/IMT-2000 is also under consideration in the ITU. The lower frequency bands help to more efficiently extend coverage to rural areas with low traffic densities (lower frequencies have lower path low attenuation rates). In Korea, where IS-95 is used for both cellular and PCS operation, the PCS spectrum allocation is different from the US PCS spectrum. This leaves the IMT-2000 spectrum for WCDMA fully available in Korea. In Japan, part of the IMT-2000 TDD spectrum is used by PHS, the cordless personal handy phone system. In USA, 3G systems using the WCDMA air interface (UMTS) are implemented by re-farming 3G systems within the existing PCS spectrum. This means replacing part of the existing 2G frequencies (used for GSM based PCS) with 3G systems. In China, the FDD spectrum allocations of 2X60 MHz are available.

The WCDMA carrier bandwidth is approximately 5 MHz, which is assigned to operators within the allocated bands in the area. The WCDMA carriers spacing is implemented in 200 kHz rasters and ranges from 4.2 to 5.4 MHz. An operator may obtain multiple 5 MHz carriers to increase capacity, for instance in the form of hierarchical layered radio access designs as discussed in Chapter 6.

2.4 WCDMA and the Spreading Concept

Generally CDMA systems are based on spread spectrum communication [8]. In CDMA systems, the communication of multiple users over the same radio channel band is separated through different spreading codes, which are assigned uniquely to each user at the time of connection set up [9]. The spreading codes are mutually orthogonal to each other, meaning that they have zero cross-correlation when they are perfectly synchronized. This means that in a perfectly ideal synchronous system, in which the transmissions from all users are received synchronously at the base station for instance, a particular user's signal can be detected by a cross-correlation of the received composite signal with the user's known spreading code. In a non-synchronous system, which is the case in practice, the cross correlation process results in a de-spreading of the wanted user's signal by providing a peak correlation value for that user's signal, and noise-like residue interference from the cross-correlation takes place with other user codes in the received composite signal. The noise like interference has been shown to be very similar to the white Gaussian noise with respect to the wanted user's signal when there are a sizeable number of users communicating [8, 9]. The multi-user interference results in the requirements for providing interference or load margins in the link budgets as discussed in Chapter 5.

WCDMA is a wideband Direct Sequence Code Division Multiple Access (DS-CDMA) system in which the user data bits are spread over a wide bandwidth by multiplying each user's data with quasi-random bit sequences called chips transmitted at the rate of 3.84 Mcps (Mega chips per second), which form the CDMA spreading codes. The spreading codes are also referred to as the channelization codes. One chip is defined as the period of the spreading code. The spreading operation multiplies the user data with the spreading code. This results in the replacement of each bit in the user's data with the spreading code (within a + or − multiplication factor). Because the chip rate is higher than the user's data rate, user data bandwidth is spread by the ratio of the chip rate to the data rate. The chip rate of 3.84 Mcps used in WCDMA results in a bandwidth of approximately 5 MHz, after the necessary filtering is done. This bandwidth is considerably larger than the coherence bandwidth (see Chapter 3) encountered in the typical mobile communication environment. Hence is coined the name 'wideband' for UMTS.

2.4.1 Processing Gain and Impact on C/I Requirement

The processing gain is the enhancement that takes place in the signal to noise ratio of the wanted user in the detection process of CDMA systems. The signal to noise ratio of the wanted signal is enhanced by the dispreading process in proportion to the ratio of the bandwidths. This bandwidth ratio is equal to the ratio of the chip rate to the user's data rate and is known as the processing gain [9, 10]. This processing gain is achieved simply because only a small portion of the noise in the wideband signal, equal to the ratio of the spread signal to the de-spread signal, is captured in the correlation and filtering operation. The noise is de-correlated with the wanted signal and is filtered out from the band that is de-spread from the signal. Since part of the noise is interference from other users' signals, the power spectral density of the interfering users within the wideband signal should be at most comparable to that of the wanted user to result in the processing gain. This requires that the signals of all users be received at or near the same power at the receiver. This is achieved by the power control of the mobile transmitters in almost real time.

The processing gain achieved in CDMA systems allows for the detection of signals that are buried under the noise floor, and hence significantly lowers the carrier-to-noise ratio requirements. The processing gain in fact lowers the carrier to interference ratio, C/I, required for a service by the spreading factor, W/R, where W is the chip rate of the spreading code (3.84 Mcps for WCDMA), and R is the service bit rate. This is seen simply from the consideration that

$$\frac{C}{I} = \left(\frac{E_b}{N_0}\right) \cdot \frac{R}{W} \qquad (1)$$

where E_b/N_0 is the required signal to noise ratio for the service.
From this

$$\text{processing gain} = \frac{W}{R} \qquad (2)$$

For AMR speech at 12.2 kbps bit rate, the required E_b/N_0 is around 5 dB. This means that the C/I required for acceptable speech quality in WCDMA will be

$$5\,\text{dB} - 10.\log_{10}(W/R) = 5 - 25 = -20\,\text{dB}$$

Compare this to the C/I required for acceptable speech quality in GSM (for same codec), which is normally in the range of 9 to 12 dB depending on whether frequency hoping is used or not. Therefore the processing gain achieved in WCDMA lowers the C/I requirements by a large amount.

2.4.2 Resistivity to Narrowband Interference

WCDMA is resistive to interference from a narrowband signal whose bandwidth is much smaller in comparison and is uncorrelated to the wanted signal. In the detection process, the composite received signal is multiplied by the spreading code of a wanted user. This causes a de-spreading of the narrowband interference power over the band of the WCDMA signal determined by the code chip rate. At this point the power spectral density of the narrowband interfering signal has been reduced by the ratio of the chip rate (3.84 Mcps in WCDMA) and the bandwidth of the narrowband signal. Subsequent filtering to pull out the wanted user's signal results in capturing only the portion of the reduced interfering power that lies within the band of the wanted signal. This amount will be insignificant depending on the ratio of the bandwidths,and the power of the interfering signal compared to the power of the wanted signal. To quantify the resistivity to the narrowband interferer, assume that the power of the received wanted signal and the narrowband interferer are P_{sig}, and P_{int}, respectively. For simplifying the analysis, assume the only noise or interference present is due to the narrowband signal. Then, the signal-to-interference power ratios before and after the de-spreading operations are

$$\left(\frac{C}{I}\right)_{Bef} = \frac{P_{sig}}{P_{int}} \qquad (3)$$

$$\left(\frac{C}{I}\right)_{Aft} = \frac{P_{sig}}{\frac{P_{int}}{W}BW} \qquad (4)$$

where *BW* is the bandwidth of the wanted signal. Substituting Equation (3) into Equation (4) gives

$$\left(\frac{C}{I}\right)_{Aft} = \left(\frac{C}{I}\right)_{Bef} \cdot \frac{W}{BW} \tag{5}$$

Thus the improvement achieved against the interferer is seen to be *W/BW*, which is just the processing gain. If this gain is enough to result in the required value for the service $\left(\frac{C}{I}\right)_{Aft}$, the resistivity to the narrowband signal is achieved. This will of course depend on the initial $\left(\frac{C}{I}\right)_{Bef}$ as seen by Equation (5). Based on this, although WCDMA can be resistive to certain narrowband interferes, it is not always immune to interference from spill over of adjacent GSM bands, depending on how much out-of-band emissions are received from a nearby GSM system. If the narrowband signal significantly exceeds the UMTS signal power input to the low noise amplifier, it can overshadow the desired signal and prevent it from being properly decoded and detected. Interference between GSM and UMTS is further discussed in Chapter 11.

2.4.3 Rake Reception of Multipath Signals and the Efficiency

In the mobile communication environment, signal propagation takes place through reflections, diffraction, and scattering of signals from multiple obstacles, such as buildings, hills, vehicles, etc. This results in the reception of multiple delayed versions of the signal known as the multipath effect [11]. Each version undergoes a different delay depending on the path through which the signal travels from the transmitter to the receiver. The multipath signals whose relative delays are at least the size of a chip period, that is 0.26 µs (the inverse of 3.84 Mcps), can be resolved and combined coherently to obtain diversity gain against multipath. Each resolvable multipath signal component may in turn be a combined version of un-resolvable multiple delayed versions of the signal, which can be weakened (by about 20 to 30 dB) or strengthened depending on the relative phases between the constituent signal components. The resolvable multipath signal components with significant power are identified by a Rake receiver [12] and combined coherently to boost up the received signal power and obtain diversity gain against multipath fading. The Rake receiver is based on the autocorrelation properties of the spreading codes. It uses multiple parallel correlators, matched filters, and code generators to detect the resolvable multipath signal components.

The efficiency of a Rake receiver is determined by how well and how many of the resolvable multipath components it is able to find and lock on to. The number of resolvable multipath components that can be detected and combined is determined by the number of fingers in the Rake receiver. A good Rake receiver is able to find and lock on to the dominant resolvable multipath components (with highest powers) at sufficient speeds to meet the channel changing rates. The multipath channel characterization and parameters are discussed in detail in Chapter 3.

WCDM provides improved diversity against multipath due to the higher chip rate used compared to the 2G CDMA systems, IS-95. IS-95 systems use a chip rate of 1.25 Mcps that results in a chip period of 0.8 µs. This is much larger than the 0.26 µs chip period of WCDMA systems, and means that the multipath components must be relatively delayed by at least 0.8 µs to be resolvable by a Rake receiver. Considering that the typical multipath delay spreads in the mobile communication environment are from about 0.5 to 2 μs, the WCDMA's smaller chip period of 0.26 μs makes it capable of more accurately resolving more multipath components and hence achieving higher multipath diversity gain against fading.

2.4.4 Variable Spreading and Multi-Code Operation

The spreading codes used in WCDMA are a special set of codes known as orthogonal variable spreading factors (OVSF) codes. These codes include member codes with different lengths resulting in spreading factors ranging from 4 to 512. Therefore, the user data rates obtainable with one code are in the range of 1 to 936 kbps in the downlink. The downlink uses QPSK modulation with both the I and Q branches used for data transmissions. The QPSK modulation is also used in the uplink with the difference that the entire Q branch is assigned to transmitting signaling information. Thus the data rates achieved on the uplink are half the rates on the downlink. High rate connection can be obtained either through a shorter code (because the user data bits will fill more of the channel slot) or through assignment of multiple codes with the same spreading factor. The 3GPP Standards leave the option to the product vendor's discretion on how to map user data rates into various coding and channelization strategies.

The OVSF codes are organized in a tree structured manner as shown in Figure 2.3. Each level in the tree corresponds to a certain spreading factor. Once a code from a branch of the tree has been allocated, its sub-tree can no longer be allocated to other channels in order to keep mutual orthogonality between the assigned codes. Hence the numbers of available

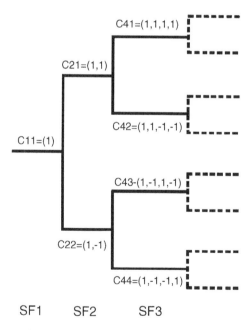

Figure 2.3 The Orthogonal Code Tree.

codes are limited in practice because it depends on the usage patterns. These codes lose their mutual orthogonality on the uplink due to the fact that they are transmitted asynchrously by mobiles in the cell. They lose some of their orthogonality on the downlink as a result of experiencing multipath depending on the severity of the channel. This results in a degree of intra-cell interference known as the downlink orthogonality factor. The assignment of these codes to users is done automatically by the network and does not involve the network planners. The network is normally capable of dynamically assigning codes with different spreading factors on each channel recurring slot to meet the varying user traffic requirements. This results in dynamic bandwidth assignment. More discussions on OSVF codes and their management by the network equipments are provided in Chapter 9.

2.5 Cell Isolation Mechanism and Scrambling Codes

Isolation between neighboring cells is achieved through the use of pseudo-noise (PN) scrambling codes. These codes are transmitted at the same rate as the WCDMA chip rate, and hence do not cause any additional spreading effects. The scrambling codes are generated by mostly linear type shift registers and have good cross-correlation properties, which make them suitable for cell and call separation in non-synchronous systems such as WCDMA. The scrambling codes are assigned normally one to each cell on the downlink (by the network planner), and one to each call on the uplink by the system. More details on these codes and their planning for the downlink side are given in Chapter 5.

2.6 Power Control Necessity

The uplink in CDMA based multiple access systems is prone to the near-far problem. This means that a single overpowered mobile can block all the other users in the cell. For example, if all users were to transmit with the same power, then a mobile that happens to be near the cell edge could have its signal received tens of dBs below the signal from a mobile near the base station. Then the stronger signal will bury the weaker signal in the interference, which it creates at the base station. This happens because the orthogonal spreading codes lose their orthogonality on the uplink due to asynchronous transmission from mobile stations in different locations in the cell (their signals are therefore received at the base station with different delays). Therefore tight power control is a critical necessity to WCDMA. The rate of this power control has to be fast to compensate for the large signal variations that occur due to the fast multipath Rayleigh fades (see Chapter 3). For that reason, the power control implemented in WCDMA is at the rate of 1500 times per second, which is faster than the speed of any fading due to the multipath Rayleigh phenomenon [13]. The power control tries to compensate for different path loss effects and the deep channel fades. The fast power control thus tries to equalize the channel effects for different users and ensure their signals are received at the base station with just the necessary power to meet the required E_b/N_0 for each. This is done in a closed loop fashion by having the mobile station adjust its power using the feedback received from the base station. The closed loop as opposed to the open loop strategy is used because of the frequency band separations between the uplink and downlink, which make the dynamic of channel behavior on one link uncorrelated with the other end (see Chapter 3). The closed loop fast power control is also called the inner loop power control.

There is also an outer loop power control, which is used to set and adjust the E_b/N_0 required for the connection according to service quality measurements that it performs (for example, by measuring the BLER or BER periodically). The E_b/N_0 required for a given service quality is generally dependent on the channel multipath profile, as well as the mobile speed. As these change, the outer loop power control prepares updated target E_b/N_0 based on real time quality measurements and sends the target value to the inner loop power control. The inner loop power control then uses the information to increase or reduce the transmit power in order to meet the indicated target E_b/N_0 at the receiving end. The outer loop power control helps to prevent excess transmit powers and hence interference in the network by setting the E_b/N_0 to just what is required for each channel condition rather than setting it to a fixed value for the worse case conditions. Since the E_b/N_0 required for the service should be determined after a possible soft handover, the outer loop power control is implemented in the RNC.

The fast closed loop power control is also implemented on the downlink. Here, the motivation is not the near-far problem of the uplink. The near-far problem does not exist on the downlink due to the fact that all transmissions are done by the base station, which is in one location. But the fast closed loop power control is also necessary on the downlink to help compensate for the deep multipath Rayleigh fades, which can cause errors that are not correctable by the interleaving and the error correcting codes. The fast power control helps to adjust the transmit powers to each mobile in real time according to just what is necessary to meet the required E_b/N_0 at the mobile station.

The detailed dynamics and the gains associated with fast power control are discussed in detail in Chapter 8.

2.7 Soft/Softer Handovers and the Benefits

In the cell overlapping areas normally in the cell borders, the mobile station may be connected to more than one cell simultaneously before it reverts to a single cell connection. This is referred to as soft or softer handover in CDMA systems, depending on whether the cells belong to different base stations (Node Bs) or are sectors of the same Node B. Soft handovers are called *make before break* because a connection to at least one cell is maintained at all times in the handover process. Soft handover (SHO) has close ties to power control. For the power control to work effectively, the system must ensure that each mobile station is connected to the base stations with the strongest signals at all times, otherwise a positive power feedback problem can destabilize the entire system. Handovers generally, and soft handovers in particular, are necessary for providing improved coverage and seamless roaming in the mobile communication environment.

In the downlink side, soft or softer handovers require the use of two separate channelization codes, thus resulting in more resource usage. This is necessary so that the mobile station can distinguish the signals from the different cells. The signals from the different cells are received via Rake receivers similar to the detection of multipath signals. In this case, the Rake receiver generates the codes used by the two cells taking part in the handover. In the uplink direction, no additional resources such as extra codes or powers are used, because the mobile signal is simply received and processed by Rake receivers in multiple sectors or cells. Softer handovers are processed within Node B using maximal ratio combining of received signal-to-noise ratios on each branch. Soft handovers are processed within the RNC using frame selection combining from the different Node Bs involved. In the frame selection combining performed in the RNC, the frame with the best reliability indicator as indicated by the respective Node B is selected. The frame reliability indicator is provided by the Node Bs for the outer loop power control.

Soft/softer handovers require additional Rake receivers in the base stations, additional transmission links between Node B and RNC on the Iub interface (for soft handovers), and additional Rake receiver fingers in the user equipment. Additional downlink codes are also used. In a good network design, soft and softer handovers occur on average in 20–30% and 5–15% of the connections, respectively [14, 15].

There is some interaction between power control loop and handover as two power control loops, one for each base station, are involved in the soft handover process. As well as details of soft handover parameters, gains and optimization are discussed in Chapter 8, which is dedicated to these two important subjects.

2.8 Framing and Modulation

WCDMA uses a frame structure of 10 ms in length, which is composed of 15 slots on both the uplink and downlink. Each slot has a duration of $T_{slot} = 2560$ chips, or 666 µs. The format and structure of each slot depends on the type of channel defined on the slot. A superframe of 72 frames is defined for derivation of paging channel groups. The modulation in WCDMA is QPSK on both the uplink and downlink. On the uplink, the in-phase branch of the QPSK modulation is used for transmission of user data, and forms the dedicated physical data channel, DPDCH. The Q branch is used for the transmission of control information for the I branch and forms the dedicated physical control channel, DPCCH. On the downlink, QPSK is also the modulation scheme used. However, the control channels for signaling are time multiplexed within both the I and the Q branches, which are both used to carry the user data. This results in more efficient use of channelization codes on the downlink, which are a limited resource.

2.9 Channel Definitions

There are basically three different types of channels defined in WCDMA, referred to as the physical channels, transport channels, and logical channels. The physical channels offer information transfer services to medium access control (MAC) and higher layers. The physical channels have been designed to support variable bit rate transport channels and offer bandwidth-on-demand capability. The transport channels are mapped into the physical channels and carry the data that is generated at higher layers. The transport channels define *how* and with what format data are transferred over the physical channels. The transport channels define the bit rates, the multiplexing structure used, etc. Coded composite transport channels (CCTrCH) can also be obtained by multiplexing several different services within the same radio connection on a physical channel. One physical control channel is assigned to a CCTrCH channel for carrying related signaling and control information.

The logical channels are simply defined by the type of information that is transferred over a transport channel. For instance, they sort and prioritize signaling information by functional use.

2.9.1 Physical Channels

The physical channels are defined by a specific carrier frequency, scrambling code, channelization (spreading) code, time start & stop (giving a duration), and on the uplink, the relative phase (0 or $\pi/2$) which determines the I or Q branch of the QPSK modulation.

2.9.1.1 Uplink Physical Channels

There are two types of uplink physical channels called dedicated and common physical channels.

Uplink Dedicated Physical Channels
The uplink dedicated physical channels, include the uplink Dedicated Physical Data Channel (uplink DPDCH), the uplink Dedicated Physical Control Channel (uplink DPCCH), and the uplink Dedicated Control Channel associated with HS-DSCH transmission (uplink HS-DPCCH). The DPDCH, DPCCH, and HS-DPCCH are I/Q code multiplexed [16].

DPDCH and DPCCH
The uplink DPDCH is used to carry the user data. There may be zero, one, or several uplink DPDCHs on each radio link as determined by the TFCI in the associated DPCCH channel. The slot structures for the DPDCH and DPCCH are shown in Figure 2.4.

In the control slot structure on the DPCCH, the pilot bits are a set of bit patterns known to the Node B that are used for channel estimation in the coherent detection process. The TPC field, transmission power control, is a 2-bit pattern coded as 11 or 00 and informs the downlink transmitter to increase or decrease the power (by the power control step decided in the RNC) on the downlink (the closed loop power control operation). The FBI, feedback

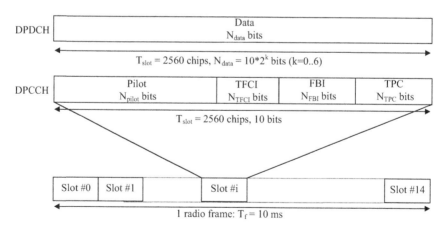

Figure 2.4 Uplink slot structures for DPDCH and DPCCH.
[3GPP™ TSs and TRs are the property of ARIB, STIS, ETSI, CCSA, TTA and TTC who jointly own the copyright in them. They are subject to further modifications and are therefore provided to you 'as is' for information purposes only. Further use is strictly prohibited]

information filed, is used to indicate the downlink channel status to Node B, and is utilized when transmit antenna diversity is implemented (see Chapter 10). The TFCI field (transport format control identifier) informs the receiver about the transport channels that are multiplexed on the associated DPDCH channel. In other words, the TFCI informs the receiver about the number of code channels multiplexed on DPDCH and their rates (spreading factors used) as decided by the RNC. The spreading factors used are selected from the range 256, 128, 64, 32, 16, 8, and 4, which corresponds to data rates of 15, 30, 60, 120, 240, 480, and 960, respectively. There are two types of uplink dedicated physical channels: those that include TFCI (e.g., for several simultaneous services) and those that do not include TFCI (e.g., for fixed-rate services).

The TFCI is decided by the MAC layer from the different transport format combinations sets (TFCS) that are given by the RAN. A transport format combination set (TFCS) defines the valid set of transport format combinations (TFC). A transport format set (TFS) is a group of transport formats that are allocated for a given transport channel. The assignment of the TFCS is done by radio resource management (RRM) at layer 3. When mapping data onto layer 1, MAC chooses between the different transport format combinations given in the TFCS.

HS-DPCCH

The frame (slot) structure of the HS-DPCCH is given in Figure 2.5. The HS-DPCCH is used to carry uplink feedback signaling related to downlink HS-DSCH transmission. The HS-DSCH-related feedback signaling consists of Hybrid-ARQ Acknowledgement (HARQ-ACK) and Channel-Quality Indication (CQI) [17]. Each sub-frame of length 2 ms (3×2560 chips) consists of 3 slots, each 2560 chips long. The HARQ-ACK is carried in the first slot of the HS-DPCCH sub-frame. The CQI is carried in the second and third slots of a HS-DPCCH sub-frame. There is at most one HS-DPCCH on each radio link. The HS-DPCCH can only exist together with an uplink DPCCH.

The spreading factor of the HS-DPCCH is 256, that is, there are 10 bits per uplink HS-DPCCH slot.

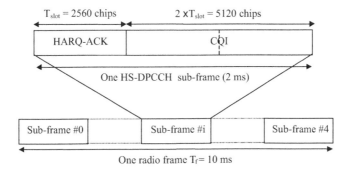

Figure 2.5 Frame structure for the uplink HS-DPCCH.
[3GPP^{TM} TSs and TRs are the property of ARIB, STIS, ETSI, CCSA, TTA and TTC who jointly own the copyright in them. They are subject to further modifications and are therefore provided to you 'as is' for information purposes only. Further use is strictly prohibited]

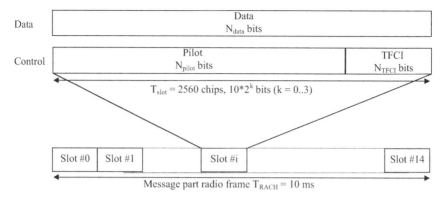

Figure 2.6 Structure of the RACH radio frame message part.
[3GPP™ TSs and TRs are the property of ARIB, STIS, ETSI, CCSA, TTA and TTC who jointly own the copyright in them. They are subject to further modifications and are therefore provided to you 'as is' for information purposes only. Further use is strictly prohibited]

Uplink Common Physical Channels
The uplink common physical channel is the Physical Random Access Channel (PRACH), which is used for initial network access in the slotted Aloha mode by the mobile station. It is also used for sending small short packets of user data on the uplink. The random access transmission consists of one or several *preambles* 4096 chips long and a *message* 10 ms or 20 ms long. Each preamble is 4096 chips long and consists of 256 repetitions of a signature that is 16 chips long. There are a maximum of 16 available signatures [17].

The structure of the random access message part (RACH) is shown in Figure 2.6. The 10 ms message part radio frame is split into 15 slots, each $T_{slot} = 2560$ chips long. Each slot consists of two parts, a data part to which the RACH transport channel is mapped and a control part that carries the physical control information. The data and control parts are transmitted in parallel. A 10 ms message part consists of one message part radio frame, whereas a 20 ms message part consists of two consecutive 10 ms message part radio frames. The message part length is equal to the Transmission Time Interval (TTI) of the RACH transport channel in use. This TTI length is configured by higher layers. The data part consists of 10×2^k bits, where $k = 0,1,2,3$. This corresponds to a spreading factor of 256, 128, 64, and 32 respectively for the message data part. The control part consists of 8 known pilot bits to support channel estimation for coherent detection and 2 TFCI bits. This corresponds to a spreading factor of 256 for the message control part. The total number of TFCI bits in the random-access message is $15 \times 2 = 30$. The TFCI of a radio frame indicates the transport format of the RACH transport channel mapped to the simultaneously transmitted message part. In the case of a 20 ms PRACH message part, the TFCI is repeated in the second radio frame.

2.9.1.2 Downlink Physical Channels

The downlink physical channels also consist of downlink dedicated and downlink common physical channels. The dedicated ones are assigned to one UE at a time. These are discussed in the following sections.

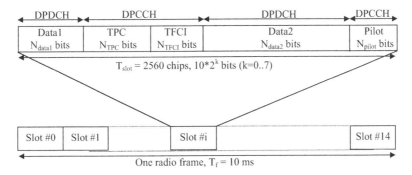

Figure 2.7 Frame structure for downlink DPCH.
[3GPP™ TSs and TRs are the property of ARIB, STIS, ETSI, CCSA, TTA and TTC who jointly own the copyright in them. They are subject to further modifications and are therefore provided to you 'as is' for information purposes only. Further use is strictly prohibited]

Downlink Dedicated Physical Channels (DPCH)

The frame structure of the downlink DPCH is shown in Figure 2.7. Each frame of length 10 ms is split into 15 slots, each $T_{slot} = 2560$ chips long, corresponding to one power-control period.

The parameter k determines the total number of bits per downlink DPCH slot. It is related to the spreading factor SF of the physical channel as $SF = 512/2^k$. The spreading factor may thus range from 512 down to 4.

There are basically two types of downlink Dedicated Physical Channels: those that include TFCI (e.g., for several simultaneous services) and those that do not include TFCI (e.g., for fixed-rate services). The UTRAN determines if a TFCI should be transmitted. It is mandatory for all UEs to support the use of TFCI in the downlink. The mapping of TFCI bits onto slots is described in Reference 17.

Downlink Physical Common Channels (PCCH)

The main common downlink physical channels consist of the common pilot channels, CPICH (primary and secondary), the primary and secondary common control physical channels (P-CCPCH, and S-CCPCH), the primary and secondary synchronization channels (P-SCH, and S-SCH), the acquisition indication channel (AICH), the paging indication channel (PICH), the shared control channel (HS-SCCH), and the physical downlink shared channel (HS-PDSCH) for high speed downlink data transfers. These are described in the following sections.

Common Pilot Channel (CPICH)

The CPICH is a fixed rate (30 kbps, $SF = 256$) downlink physical channel that carries a pre-defined bit sequence. Figure 2.8 shows the frame structure of the CPICH, which is the same for both primary and secondary CPICH.

When transmit diversity (open or closed loop) is used on any downlink channel in the cell, the CPICH is transmitted from both antennas using the same channelization and scrambling code. In this case, the pre-defined bit sequence of the CPICH is different for Antenna 1 and Antenna 2.

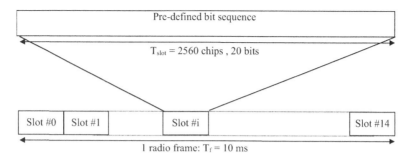

Figure 2.8 Downlink common pilot channels frame structure.
[3GPP™ TSs and TRs are the property of ARIB, STIS, ETSI, CCSA, TTA and TTC who jointly own the copyright in them. They are subject to further modifications and are therefore provided to you 'as is' for information purposes only. Further use is strictly prohibited]

The Primary and Secondary CPICH differ in their use and the limitations placed on their physical features.

Primary Common Pilot Channel (P-CPICH)
The Primary Common Pilot Channel (P-CPICH) has the following characteristics [16]:

- The same channelization code is always used for the P-CPICH.
- The P-CPICH is scrambled by the primary scrambling code.
- There is one and only one P-CPICH per cell.
- The P-CPICH is broadcast over the entire cell.

Secondary Common Pilot Channel (S-CPICH)
The Secondary Common Pilot Channel (S-CPICH) has the following characteristics [16]:

- An arbitrary channelization code of $SF = 256$ is used for the S-CPICH.
- A S-CPICH is scrambled by either the primary or a secondary scrambling code.
- There may be zero, one, or several S-CPICH per cell.
- A S-CPICH may be transmitted over the entire cell or only over a part of the cell.

The UE can use either of the pilot bits provided in the downlink dedicated physical channel for phase reference. The primary CPICH or a secondary CPICH may also be used as phase references as follows:

- By default (i.e., without any indication by higher layers) the UE may use the primary CPICH as a phase reference.
- The UE is informed by higher layers when it may use a secondary CPICH as a phase reference. In this case the UE does not use the primary CPICH as a phase reference. Indication that a secondary CPICH may be a phase reference is also applicable when open loop or closed loop TX diversity is enabled for a downlink physical channel.

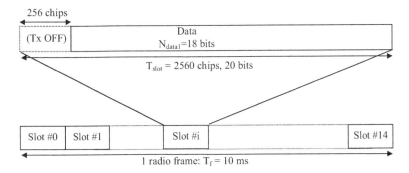

Figure 2.9 Frame structure for Primary CCPCH.
[3GPP^{TM} TSs and TRs are the property of ARIB, STIS, ETSI, CCSA, TTA and TTC who jointly own the copyright in them. They are subject to further modifications and are therefore provided to you 'as is' for information purposes only. Further use is strictly prohibited]

Primary Common Control Physical Channel (P-CCPCH)
The Primary CCPCH is a fixed rate (30 kbps, $SF = 256$) downlink physical channel used to carry the BCH transport channel. Figure 2.9 shows the frame structure of the Primary CCPCH. The frame structure differs from the downlink DPCH in that no TPC commands, no TFCI, and no pilot bits are transmitted. The Primary CCPCH is not transmitted during the first 256 chips of each slot. Instead, Primary SCH and Secondary SCH are transmitted during this period.

Secondary Common Control Physical Channel (S-CCPCH)
The Secondary CCPCH is used to carry the FACH and PCH. There are two types of Secondary CCPCH: those that include TFCI and those that do not include TFCI. It is the UTRAN that determines if a TFCI should be transmitted, hence making it mandatory for all UEs to support the use of TFCI. The set of possible rates for the Secondary CCPCH is the same as for the downlink DPCH. The frame structure of the Secondary CCPCH is shown in Figure 2.10.

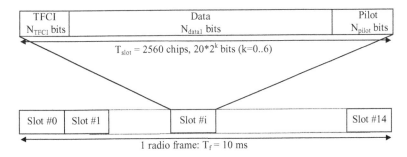

Figure 2.10 Frame structure for Secondary CCPCH.
[3GPP^{TM} TSs and TRs are the property of ARIB, STIS, ETSI, CCSA, TTA and TTC who jointly own the copyright in them. They are subject to further modifications and are therefore provided to you 'as is' for information purposes only. Further use is strictly prohibited]

The parameter k in Figure 2.10 determines the total number of bits per downlink Secondary CCPCH slot. It is related to the spreading factor SF of the physical channel as $SF = 256/2^k$. The spreading factor range is from 256 down to 4.

The main difference between a CCPCH and a downlink dedicated physical channel is that a CCPCH is not inner-loop power controlled. The main difference between the Primary and Secondary CCPCH is that the BCCH transport channel that is mapped to the Primary CCPCH (BCH) can only have a fixed predefined transport format combination, while the Secondary CCPCH supports multiple transport format combinations using TFCI.

Synchronization Channel (SCH)

The Synchronization Channel (SCH) is a downlink signal used for cell search. The SCH consists of two sub channels, the Primary and Secondary SCH. The 10 ms radio frames of the Primary and Secondary SCH are divided into 15 slots, each 2560 chips long. Figure 2.11 illustrates the structure of the SCH radio frame.

The Primary SCH consists of a modulated code of 256 chips in length, the Primary Synchronization Code (PSC) denoted c_p in Figure 2.11, transmitted once every slot. The PSC is the same for every cell in the system.

The Secondary SCH consists of repeatedly transmitting a sequence of 15 modulated codes of 256 chips in length, the Secondary Synchronization Codes (SSC), transmitted in parallel with the Primary SCH. The SSC is denoted $c_s^{i,k}$ in Figure 2.11, where $i = 0, 1, \ldots, 63$ is the number of the scrambling code group, and $k = 0, 1, \ldots, 14$ is the slot number. Each SSC is chosen from a set of 16 different codes of length 256. This sequence on the Secondary SCH indicates to which of the code groups the cell's downlink scrambling code belongs.

Acquisition Indicator Channel (AICH)

The Acquisition Indicator channel (AICH) is a fixed rate (SF = 256) physical channel used to carry Acquisition Indicators (AI). Acquisition Indicator AI_s corresponds to signatures on the PRACH for identifying the intended UEs. The spreading factor (SF) used for channelization of the AICH is 256. The phase reference for the AICH is the Primary CPICH. If an Acquisition Indicator is set to $+1$, it represents a positive acknowledgement. If an Acquisition Indicator is set to -1, it represents a negative acknowledgement.

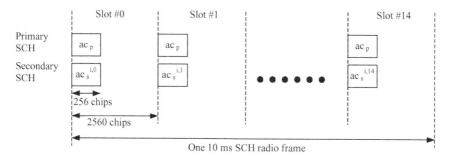

Figure 2.11 Structure of the SCH channels.
[3GPP™ TSs and TRs are the property of ARIB, STIS, ETSI, CCSA, TTA and TTC who jointly own the copyright in them. They are subject to further modifications and are therefore provided to you 'as is' for information purposes only. Further use is strictly prohibited]

Paging Indicator Channel (PICH)
The Paging Indicator Channel (PICH) is a fixed rate (SF = 256) physical channel used to carry the paging indicators. The PICH is always associated with an S-CCPCH to which a PCH transport channel is mapped.

One PICH radio frame of length 10 ms consists of 300 bits ($b_0, b_1, \ldots, b_{299}$). Of these, 288 bits ($b_0, b_1, \ldots, b_{287}$) are used to carry paging indicators. The remaining 12 bits are not formally part of the PICH and are not transmitted (DTX). The part of the frame with no transmission is reserved for possible future use.

In each PICH frame, Np paging indicators $\{P_0, \ldots, P_{Np-1}\}$ are transmitted, where Np = 18, 36, 72, or 144.

The P_I calculated by higher layers for use for a certain UE is associated to the paging indicator P_q, where q is computed as a function of the PI computed by higher layers.

Shared Control Channel (HS-SCCH)
The HS-SCCH is a fixed rate (60 kbps, SF = 128) downlink physical channel used to carry downlink signaling related to HS-DSCH transmission. Figure 2.12 illustrates the sub-frame structure of the HS-SCCH.

High Speed Physical Downlink Shared Channel (HS-PDSCH)
The High Speed Physical Downlink Shared Channel (HS-PDSCH) is used to carry the High Speed Downlink Shared Channel (HS-DSCH).

An HS-PDSCH corresponds to one channelization code of fixed spreading factor SF = 16 from the set of channelization codes reserved for HS-DSCH transmission by the RAN. Multi-code transmission is allowed, which means the UE can be assigned multiple channelization codes in the same HS-PDSCH sub-frame, depending on its capability. The sub-frame and slot structure of HS-PDSCH are shown in Figure 2.13.

An HS-PDSCH may use QPSK or 16QAM modulation symbols. In Figure 2.13, M is the number of bits per modulation symbols, that is M = 2 for QPSK and M = 4 for 16QAM modulation. The number of modulation symbols that can be carried by each of the three slots (that make up 2 ms) is 2560/16 = 160 symbols, where 16 is the spreading factor. Then the number of data bits carried is 160 × M.

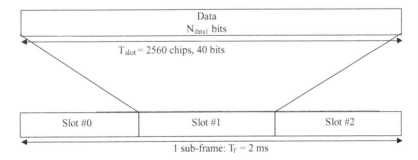

Figure 2.12 Sub-frame structure of the HS-SCCH.
[3GPP™ TSs and TRs are the property of ARIB, STIS, ETSI, CCSA, TTA and TTC who jointly own the copyright in them. They are subject to further modifications and are therefore provided to you 'as is' for information purposes only. Further use is strictly prohibited]

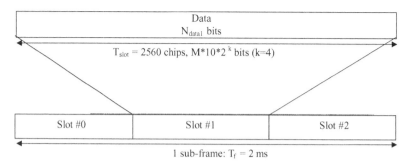

Figure 2.13 Structure of the HS-PDSCH.
[3GPP™ TSs and TRs are the property of ARIB, STIS, ETSI, CCSA, TTA and TTC who jointly own the copyright in them. They are subject to further modifications and are therefore provided to you 'as is' for information purposes only. Further use is strictly prohibited]

2.9.2 Frame Timing Relationships

2.9.2.1 DPCCH and DPDCH on Uplink and Downlink

In the uplink, the DPCCH and all the DPDCHs transmitted from one UE have the same frame timing. In the downlink, the DPCCH and all the DPDCHs carrying CCTrCHs of dedicated type to one UE also have the same frame timing.

2.9.2.2 Uplink-Downlink Timing at UE

At the UE, the uplink DPCCH/DPDCH frame transmission takes place approximately T_0 chips after the reception of the first detected path (in time) of the corresponding downlink DPCCH/DPDCH. T_0 is a constant defined to be 1024 chips. The first detected path (in time) is defined implicitly by the relevant tests in Reference 18. More information about the uplink/downlink timing relation and meaning of T_0 can be found in Reference 19.

2.9.2.3 HS-SCCH/HS-PDSCH Timing Relationship

Figure 2.14 shows the relative timing between the HS-SCCH and the associated HS-PDSCH for one HS-DSCH sub-frame. The HS-PDSCH starts $\tau_{HS\text{-}PDSCH} = 2 \times T_{slot} = 5120$ chips after the start of the HS-SCCH.

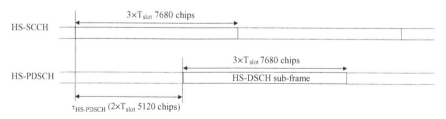

Figure 2.14 Timing relationship between HS-SCCH and HS-PDSCH.
[3GPP™ TSs and TRs are the property of ARIB, STIS, ETSI, CCSA, TTA and TTC who jointly own the copyright in them. They are subject to further modifications and are therefore provided to you 'as is' for information purposes only. Further use is strictly prohibited]

2.9.3 Transport Channels

The transport channels in WCDMA are classified into the dedicated and common categories. The dedicated transport channels are dedicated to specific UE in which the UE is identified by the physical channel (that is, code and frequency for FDD and code, time slot, and frequency for TDD). In 3GPP, all transport channels are defined as unidirectional, that is, uplink, downlink, or a relay-link. Depending on the services, a UE can be assigned simultaneously several transport channels in the downlink or in the uplink. In the common transport channels, particular UEs are addressed when there is a need for in-band identification of a UE. The transport channels may carry User plane information or Control plane information. To each transport channel, there is an associated Transport Format (for transport channels with a fixed or slow changing rate) or an associated Transport Format Set (for transport channels with a fast changing rate). A Transport Format is defined as a combination of encodings, interleaving, bit rate, and mapping onto physical channels [20]. A Transport Format Set is a set of Transport Formats informing the receiver about what transport channels are simultaneously active in the current physical channel. As an example, a variable rate DCH has a Transport Format Set (one Transport Format for each rate), whereas a fixed rate DCH has a single Transport Format.

The dedicated transport channel types are:

- **Dedicated Channel (DCH):** A channel dedicated to one UE used in uplink or downlink. The DCH channel can be used for the transmission of packet or circuit data. Since the set up time for this channel is of the order of about 250 ms, it is suitable for transmission of data on long sessions.
- **Enhanced Dedicated Channel (E-DCH):** A channel dedicated to one UE used in uplink only. The E-DCH is subject to Node B controlled scheduling and HARQ.

The common transport channel types for the FDD mode are [20]:

- **Random Access Channel (RACH):** A contention based uplink channel used for transmission of relatively small amounts of data, for example, for initial network access or non-real-time traffic data. Since this channel is used for initial network access, it should have coverage throughout the cell.
- **Common Packet Channel (uplink):** This is an extension to the RACH channel and is intended for uplink transmission of packet-based user data. The CPCH transmission may last several frames in contrast to one or two frames for the RACH. The main differences of this channel with RACH are the use of fast power control and a physical layer-based collision detection mechanism.
- **Forward Access Channel (FACH):** This is a common downlink channel without closed loop power control used for transmission of relatively small amount of data. In addition, FACH is used to carry broadcast and multicast data.
- **Broadcast Channel (BCH):** A downlink channel used for broadcast of system and cell-specific information such as the random access time slots, random access codes, etc. into an entire cell.
- **Paging Channel (PCH):** A downlink channel used for broadcast of control information into an entire cell allowing efficient UE sleep mode procedures. Currently identified

information types are paging and notification. Another use could be UTRAN notification of change of BCCH information.
- **High Speed Downlink Shared Channel (HS-DSCH):** A downlink channel shared between UEs by allocation of individual codes, from a common pool of codes assigned for the channel.

The common transport channels needed for the basic network operation are RACH, FACH, and PCH transport channels. The use of HS-DSCH and CPCH are optional and can be decided by the network. However, for high-rate data services in the multiple- Mbps on the downlink direction (file downloading, high-speed web searching, etc.), the use of the HS-DSCH is necessary.

2.9.4 Channel Mappings

The mapping of the transport channels to physical channels is given in Figure 2.15 [21].

2.9.5 Logical Channels

A logical channel is simply defined by the type of information that is communicated over a transport channel [22]. For instance, the system information broadcast on the BCH transport channel is referred to simply as the broadcast control channel (BCCH). Then BCCH is defined as a logical channel. Thus the 'control and system information' nature of the data transmitted over the BCH channel defines it as the logical BCCH. The specific data rate formats, spreading codes, etc., which are used on the primary common control physical channel (P-CCPCH) to build the BCH, make the 'BCH' a transport channel out of the physical channel P-CCPCH. In another example, specific paging information that is

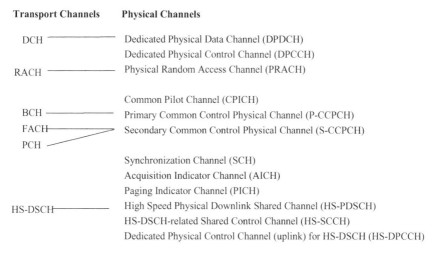

Figure 2.15 Transport to physical channel mappings.
[3GPP™ TSs and TRs are the property of ARIB, STIS, ETSI, CCSA, TTA and TTC who jointly own the copyright in them. They are subject to further modifications and are therefore provided to you 'as is' for information purposes only. Further use is strictly prohibited]

transmitted over the PCH transport channel (which was derived from the secondary common control physical channel, S-CCPCH) is then called the paging control channel (PCCH) and the PCCH is then defined as a logical channel. And so on.

Logical channels can be used either for the transfer of network control related information or user traffic. Examples of the logical control channels include BCCH mapped to BCH or FACH, common control channel (CCCH) mapped to RACH and FACH, dedicated control channel (DCCH) mapped to DCH, and Common Traffic Channel (CTCH) mapped to FACH. Examples of logical traffic channels include Dedicated Traffic Channel (DTCH) mapped to DCH, or RACH, and Common Traffic Channel (CTCH) mapped to FACH.

2.10 The Radio Interface Protocol Architecture

The radio interface has a 3-layered structure consisting of the physical (L1), the data link (L2), and the network layers (L3). The data link layer is split into a number of sub-layers that consist of a Medium Access Control (MAC), a Radio Link Control (RLC), a Packet Data Convergence Protocol (PDCP), and a Broadcast/Multicast Control (BMC) sub-layer. The network layer and the RLC are divided into Control (C–) and User (U–) planes. PDCP and BMC exist in the U-plane only [22].

The network layer is partitioned into a number of sub-layers in which the lowest sub-layer, called the Radio Resource Control (RRC), interfaces with layer 2 and terminates in the UTRAN. The next sub-layer provides 'Duplication avoidance' functionality as specified in Reference [23]. It terminates in the CN but is part of the Access Stratum and provides the Access Stratum Services to higher layers. The higher layer handles the signaling for Mobility Management (MM) and Call Control (CC) and is assumed to belong to the Non-Access Stratum (NAS) and not in the scope of 3GPP TSG RAN. On the general level, the protocol architecture is similar to the current ITU-R protocol architecture, ITU-R M.1035. The protocol termination points for the Control and User plane are shown in Figures 2.16 and 2.17, respectively.

The UTRA radio interface protocol architecture is shown in Figure 2.18. The service access point (SAP) between the MAC and the physical layer provides the transport channels, whereas the SAPs between RLC and the MAC sub-layer provide the logical channels. The RLC layer provides three types of SAPs, one for each RLC operation mode (UM, AM, and

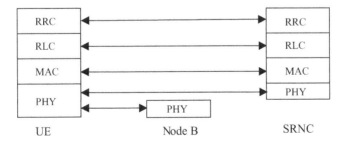

Figure 2.16 Protocol termination for DCH, Control plane.
[3GPP™ TSs and TRs are the property of ARIB, STIS, ETSI, CCSA, TTA and TTC who jointly own the copyright in them. They are subject to further modifications and are therefore provided to you 'as is' for information purposes only. Further use is strictly prohibited]

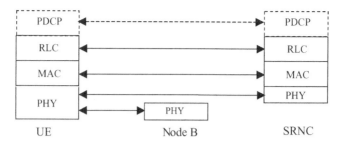

Figure 2.17 Protocol termination for DCH, User plane.
[3GPP^{TM} TSs and TRs are the property of ARIB, STIS, ETSI, CCSA, TTA and TTC who jointly own the copyright in them. They are subject to further modifications and are therefore provided to you 'as is' for information purposes only. Further use is strictly prohibited]

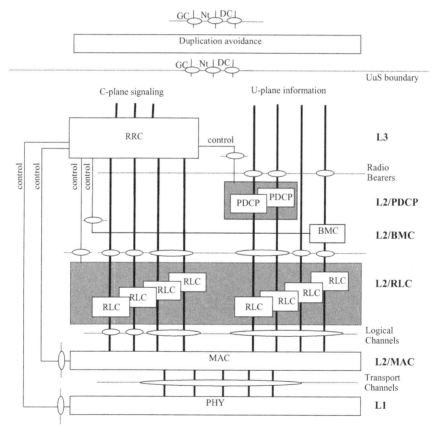

Figure 2.18 Radio interface protocol architecture (SAPs are marked by circles).
[3GPP^{TM} TSs and TRs are the property of ARIB, STIS, ETSI, CCSA, TTA and TTC who jointly own the copyright in them. They are subject to further modifications and are therefore provided to you 'as is' for information purposes only. Further use is strictly prohibited]

TM). The service provided by layer 2 is referred to as the radio bearer. The C-plane radio bearers, which are provided by RLC to RRC, are denoted as signaling radio bearers.

There are primarily two kinds of signaling messages transported over the radio interface: the RRC generated signaling messages and NAS messages generated in the higher layers. On establishment of the signaling connection between the peer RRC entities, three or four UM/AM signaling radio bearers may be set up. Two of these bearers are set up for transport of RRC generated signaling messages – one for transferring messages through an unacknowledged mode RLC entity and the other for transferring messages through an acknowledged mode RLC entity. One signaling radio bearer is set up for transferring NAS messages set to 'high priority' by the higher layers. An optional signaling radio bearer may be set up for transferring NAS messages that are set to 'low priority' by the higher layers. Subsequent to the establishment of the signaling connection, zero to several TM signaling radio bearers may be set up for transferring RRC signaling messages using transparent mode RLC.

2.10.1 The RLC Sub-layer

The RLC (radio link control) sub-layer provides several services to the upper layer (usually a network layer), including segmentation and reassembly of the PDU (protocol data units) into and from SDUs (service data units) for transmission on the User or Control plane, re-ordering and duplication avoidance in the delivery of the received SDUs (service data units), and optional acknowledgement of correctly received PDUs. The RLC has three different modes of operation. These are referred to as the transparent mode (TM), acknowledged mode (AM), and unacknowledged mode (UM). The transfer mode of the RLC is determined by the transfer mode configured for the RAB (radio access bearer). The transfer mode of the RAB is determined by the admission control within the serving RNC using the RAB attributes and possibly information from the core network domain.

The transparent mode transmits upper layer PDUs without adding any protocol information. The acknowledge mode (AM) uses the ARQ protocol (automatic request for retransmission) for requesting to the peer entity the re-transmission of erroneously received data to guarantee data delivery.

In acknowledged mode, the RLC basically provides an enhancement of the reliability of the radio link over what is provided by the error correction codes and interleaving. Due to fading and resulting burst errors, not all errors are correctable by interleaving and error correcting codes. When uncorrected errors are detected through the CRC (cyclic redundancy check bits) in the data frame, a re-transmission of the frame is requested by the RLC when it is set to run in the acknowledge mode. In this way the RLC can hide the errors on the radio link from the TCP (transmission control protocol), and improves its efficiency when run on the wireless communication links.

In unacknowledged mode, the re-transmission protocol is not used and data delivery is not guaranteed. The results of CRC error detection are provided to the RLC in all modes. The number of re-transmissions performed in the AM RLC provides information on the link quality, which can affect the quality target in the outer-loop power control. The information is also useful for network planning and optimization purposes. The protocol header overheads added by the RLC sub-layer are 16 bits for the AM, 8 bits for UM, and 0 bits for the TM.

2.10.2 The MAC Protocol Functions

The MAC (Medium Access Control) protocol represents the control entity for the mapping of logical to transport channels, and provides the interface between the two. The MAC maps the logical channels, used for communication with the RLC layer, to the transport channels that exist between MAC and the physical layer. This also means that the MAC layer is responsible for multiplexing/de-multiplexing data from/to upper layers. A MAC entity exists for each different channel type, for example, a MAC_d entity handles the mapping of dedicated logical channels to the dedicated transport channels. MAC_c performs a similar function for handling the common channels, and likewise, MAC_h handles the transport channels on the high-speed shared downlink channel. The different MAC entities can also interact with each other. For instance, if a dedicated logical channel is to use a common transport channel (such as a specific user trying to send a short message through the RACH channel), the MAC_d entity transfers data to the MAC_c entity, which then maps the data to the relevant common transport channel. The MAC entities are responsible for dynamically sharing the transport resources between different logical connections sharing a transport channel. The MAC selects the appropriate TFs (transport formats) from a TFCS (transport format combination set) assigned by the RRC for each transmission time interval.

The MAC entities also perform the multiplexing and de-multiplexing of RLC PDUs into and from transport channels using common physical channels, and handle the dynamic scheduling of transmissions to different users based on priorities such as on the high-peed shared downlink channel. Traffic volume measurements and their reporting to the RRC is another function of the MACs. They track and measure the buffer sizes in the RLC sub-layers, and convey the information to the RRC. The RRC processes the information and decides to command the MAC to perform a switching of a designated logical channel from a common transport channel to a dedicated one or vice versa based on the indicated traffic level. This function performed by the MAC but under the control of the RRC is referred to as channel type switching [24].

2.10.3 RRC and Channel State Transitions

The RRC protocol forms an important aspect of radio resource management and control between the UE and the UTRAN. Its understanding is therefore highly recommended for those interested in radio network performance analysis, tuning, and troublshooting. The RRC defines the procedures and the signaling used to control radio channel states, mobility functions, and the communication of related information in the UE connected state. These procedures include the broadcast of information provided by the Non-Access Stratum (Core Network) and the Access Stratum; establishment, maintenance, and release of RRC connections between the UE and the UTRAN; establishment, reconfiguration, and release of Radio Bearers; assignment, reconfiguration, and release of radio resources for the RRC connection; handling of related mobility functions; routing of higher layer PDU's; controlling of requested QoS; UE measurement reporting and control of the reporting; setting up of the outer loop power control on downlink; and control of ciphering. The RRC also handles paging notifications, the initial cell selections, and re-selections in the idle mode, and the arbitration of radio resources allocations on the uplink dedicated channels. These procedures

UMTS System and Air Interface Architecture

are further discussed in Reference [22]. The RRC is also responsible for congestion control and admission functions, as discussed in Chapter 9 on radio resource management.

The transitions between the UE channel states are controlled by the RRC. At power on, the UE stay in the Idle Mode until a request to establish an RRC connection is transmitted to the network. The transition to the UTRAN Connected Mode from the Idle Mode can only be initiated by the UE by transmitting a request for an RRC Connection. The event is triggered either by a paging request from the network or by a request from upper layers in the UE. In the Idle Mode, the RRC has no information about individual UEs and can only address all UEs in a cell or all UEs monitoring a paging occasion as further explained in Reference [25]. When the UE receives a message from the network that confirms the RRC connection establishment, the UE enters the CELL_FACH or CELL_DCH state of UTRAN Connected Mode. Upon connection establishment, the UE is assigned a Radio Network Temporary Identity (RNTI). This is then used as the UE identity when communicating on common transport channels. When the RRC connection is released, the signaling link and all the radio bearers between the UE and the UTRAN are released.

The RRC states and state transition diagram is shown in Figure 2.19 [25]. The RRC states are defined in the following list:

- **Cell_DCH State.** In this state, a dedicated physical channel is allocated to the UE in uplink and downlink, on either a permanent basis (needing a DCH release message) or based on time or amount-of-data. In this state, the UE is known on a cell level and

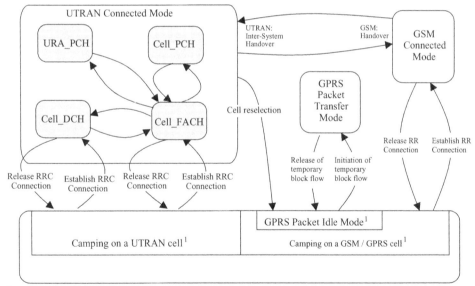

[1]The indicated division within the Idle Mode is only included for clarification and should not be interpreted as seperate states.

Figure 2.19 RRC states and state transitions, showing the Idle Mode and including the GSM and GPRS connected modes [25].
[3GPP[TM] TSs and TRs are the property of ARIB, STIS, ETSI, CCSA, TTA and TTC who jointly own the copyright in them. They are subject to further modifications and are therefore provided to you 'as is' for information purposes only. Further use is strictly prohibited]

performs measurements according to RRC measurement control messages. The UE can also use the shared transport channels such as a PDSCH. This state is entered from the Idle Mode through the setup of an RRC connection or by establishing a dedicated physical channel from the CELL_FACH state. An RRC connection release transitions this state to the idle mode. A transition from this state to a CELL_FACH state can occur either through expiration of an inactivity timer (T_{DCH}), at the end of the allocated time for the DCH, or via explicit signaling.

- **Cell_FACH State.** In this state, no dedicated physical channel is allocated to the UE, and the UE position is known at cell level. The UE can receive data through the FACH, which it continuously monitors, and is also assigned a default common or shared channel in the uplink (e.g., RACH or CPCH) that it can use to transmit small amounts of data. Upon expiration of an in-activity timer (T_{rf}), the UE transitions to CELL_PCH in order to decrease power consumption. The UE listens to the BCH transport channel of the serving cell as well as neighboring cells and from this identifies the need for any cell location updates. Measurement reports are also carried out. The UE can transition from this state to a Cell_DCH state via explicit signaling. In this state, the UE reports the observed traffic volume or the buffer status in the UE, which the network uses to evaluate the current allocation of resources for state transition decisions.
- **Cell_PCH State.** No dedicated physical channel is allocated to the UE, and the UE monitors the paging occasions according to the DRX cycle and receives paging information on the PCH. In this state, no uplink activity is possible, and the UE can only be paged by the network. The position of the UE is known by UTRAN on a cell level according to the cell where the UE last made a cell update in CELL_FACH state. The UE listens to the BCH transport channel of the serving cell for the decoding of system information messages and performs measurement reporting. The UE mobility is handled through cell reselection. The UE is transferred to CELL_FACH state either by a command (packet paging) from UTRAN or through any uplink access by the UE. The only overhead in this state is possible cell updating when the UE moves to a new cell. When the number of cell updates exceeds a configurable parameter, the UTRAN transitions the UE to the state URA_PCH to reduce the overhead involved. This transition is made via the CELL_FACH state.
- **URA_PCH State.** The URA_PCH state is characterized by no dedicated channels being allocated to the UE. Any activity can be initiated only by the network paging the UE or by an uplink access by the UE on the RACH, in which case it transitions to the Cell_FACH state. The UE uses DRX for monitoring a PCH for paging messages via an allocated PICH. The location of the UE is known on the UTRAN Registration area level according to the URA assigned to the UE during the last URA update in CELL_FACH state. Registration area updates are performed when necessary by temporarily transitioning to the Cell_FACH state. The UE listens to the BCH transport channel of the serving cell for the decoding of system information messages, as well as performing measurement reporting.

2.10.4 Packet Data Convergence Sub-layer (PDCP)

The PDCP sub-layer is used only for packet switched services and is implemented in the User plane within the UE and RNC. The purpose of this sub-layer is to perform compression functions on long protocol headers (such as in TCP/IP or other similar protocols) for efficient transfer over the radio interface. This is usually done by replacing redundant protocol headers with compressed coded versions. The PDCP performs the compression at the transmitting end and the decompression at the receiving end [26]. The PDCP protocol provides a more efficient transmission of user IP packets over the air interface. It results in a decrease of transmission power and reduced interference, as fewer bits need to be transmitted over the air interface. RFC 2507 supports compression of TCP and UDP headers (both IPv4 and IPv6). Typically, RFC 2507 compresses TCP/IPv4 header (40 bytes) or UDP/IPv4 header (28 bytes) to about 4–7 bytes. RFC 2507 is used by default to compress the TCP headers. RFC 2507 is used also to compress the RTP/UDP/IP headers if the Robust Header Compression (ROHC, RFC 3095) algorithm cannot be used.

The RTP/UDP/IP protocols are used with real time IP services. With this feature IP/UDP/RTP headers are compressed before transmission over the air interface and decompressed at the receiving end. Header compression and decompression are performed by the PDCP layer in the RNC and MS. The RTP/UDP/IP header compression algorithm has been specified in IETF and is called Robust Header Compression (ROHC/RFC3095). Similar to the IP Header Compression feature, this feature provides a more efficient transmission of user IP packets over the air interface.

2.10.5 The Broadcast Multicast Control (BMC) Protocol

This protocol also is implemented in the User plane. It is used to provide broadcast and multicast transmission services on the radio interface over common transport channels in transparent or unacknowledged modes. The transport channels used for this purpose have to be indicated to all the users involved via RRC generated system information, which is broadcast on the BCH [27].

2.11 Important Physical Layer Measurements

The 3GPP Standards [28] specify certain physical link layer related measurements, which should be carried out by both the UE and the UTRAN and reported to the higher layers. This section lists those measurements that are considered to be the most relevant to the radio link performance evaluation or its analysis. All measurements by the UE and the UTRAN are referenced to the antenna connectors of the UE, and the Node B, respectively.

2.11.1 UE Link Performance Related Measurements

2.11.1.1 CPICH RSCP

Received Signal Code Power (RSCP) is the received power on one code measured on the Primary CPICH. If Tx diversity is applied on the Primary CPICH the received code power from each antenna is separately measured and summed together in watts to a total received code power on the Primary CPICH. This measurement can be used to calculate the path loss on the downlink by subtracting the value from the transmitted power on the P-CPICH that is known to the operator. The power transmitted on the P-COICH can also be read from the system information broadcast on the BCCH transport channel.

2.11.1.2 UTRA Carrier RSSI

UTRA carrier RSSI is the received wide band power, including thermal noise and noise generated in the receiver, within the downlink bandwidth defined by the receiver pulse shaping filter.

2.11.1.3 CPICH Ec/No

This measurement is the received energy per chip of the primary CPICH divided by the power spectral density in the band. The CPICH Ec/No is identical to CPICH RSCP/UTRA Carrier RSSI. If Tx diversity is applied on the Primary CPICH, the received energy per chip (Ec) from each antenna is separately measured and summed together in watts to a total received chip energy per chip on the Primary CPICH, before calculating the Ec/No. This measurement is highly accurate as it is performed in baseband [14], and it is the most important UE measurement in WCDMA for network planning and optimization. This ratio is the nominal cell RF coverage indicator and is also used in handover decisions.

2.11.1.4 BLER

This measurement is the estimation of the transport channel block error rate (BLER). The BLER estimation is based on evaluating the CRC of each transport block associated with the measured transport channel after radio link combining.

2.11.1.5 UE Transmitted Power on One Carrier

This measurement can be used to help analyze the service coverage, and also in performance of the power control loop.

2.11.1.6 UE Transmission Power Headroom

UE transmission power headroom (UPH) is the ratio of the maximum UE transmission power and the corresponding DPCCH code power, and shall be calculated as follows:

$$UPH = P_{\max,tx}/P_{DPCCH}$$

where $P_{\max,tx} = \min\{Maximum\ allowed\ UL\ TX\ Power, P_{max}\}$ is the UE maximum transmission power; *Maximum allowed UL TX Power* is set by UTRAN; P_{max} is determined by the UE power class; and P_{DPCCH} is the transmitted code power on DPCCH.

2.11.2 UTRAN Link Performance Related Measurements

2.11.2.1 Received Total Wide Band Power

This measurement is the received wide band power, including noise generated in the receiver, within the bandwidth defined by the receiver pulse shaping filter. In the case of receiver diversity, the reported value is the linear average of the power in the diversity branches. This is an important indicator of the uplink load and interference floor.

2.11.2.2 SIR

Type 1
Signal to Interference Ratio (SIR) is defined as (RSCP/ISCP) × SF, where RSCP = Received Signal Code Power, the unbiased measurement of the received power on one code; ISCP = Interference Signal Code Power, the interference on the received signal; and SF = the spreading factor used on the DPCCH. The measurement is performed on the DPCCH of a Radio Link Set. In compressed mode the SIR is not measured in the transmission gap. If the radio link set contains more than one radio link, the reported value is the linear summation of the SIR from each radio link of the radio link set. If Rx diversity is used in the Node B for a cell, the SIR for a radio link is the linear summation of the SIR from each Rx antenna for that radio link.

Type 2
Again defined as (RSCP/ISCP) × SF, the measurement shall be performed on the PRACH control part. When cell portions are defined in the cell, the SIR measurement shall be possible in each cell portion.

$$SIR_{error}$$
$$R_{error} = SIR - SIR_{target\_ave}$$

where
SIR = the SIR measured by UTRAN, in dB; SIR_{target_ave} = the SIR_{target} averaged over the same time period as the SIR used in the SIR_{error} calculation and is the target SIR set by the closed loop power control.

2.11.2.3 Transmitted Carrier Power

The transmitted carrier power is the ratio between the total transmitted power on one DL carrier from one UTRAN access point, and the maximum transmission power possible to use on that DL carrier at the time of measurement.

2.11.2.4 Transmitted Code Power

Transmitted code power is the transmitted power on one channelization code on one given scrambling code on one given carrier.

2.11.2.5 Transport Channel BER

The transport channel BER is an estimation of the average bit error rate (BER) of the DPDCH data of a Radio Link Set during the latest TTI for the TrCH.

2.11.2.6 Physical Channel BER

The physical channel BER is an estimation of the average bit error rate (BER) on the DPCCH of a Radio Link Set over the latest TTI of the respective TrCH. This is also called the 'raw' or 'Uncoded' BER.

The combinations of the measurements of the 'UE transmitted power', the base station measured SIR, and BLER can be used in the troubleshooting and performance analysis of the uplink power control loop as pointed out in Reference 14.

References

[1] 3GPP TS 23.101, 'General UMTS Architecture'.
[2] 3GPP TS 23.110, 'Access Stratum Services and Functions'.
[3] 3GPP TS 25.401, 'RAN Overall Description' (Release 7).
[4] 3GPP TS 25.402, 'UTRAN Synchronization, Stage 2' (Release 5).
[5] 3GPP TS 25.431, 'UTRAN Iub Interface Layer 1'.
[6] 3GPP TS 25.104, 'UTRA (BS) FDD; Radio Transmission and Reception'.
[7] 3GPP TS 25.105, 'UTRA (BS) TDD; Radio Transmission and Reception'.
[8] Peterson, R.L., Ziemer, R.E. and Borth, D.E., *Introduction to Spread Spectrum Communications*, Prentice-Hall, Upper Saddle River, NJ, 1995.
[9] Viterbi, A.J., *CDMA Principles of Spread Spectrum Communications*. Addison-Wesley, Reading MA, 1995.
[10] Dehghan, S., Lister, D., Owen, R. and Jones, P., 'WCDMA capacity and planning issues'. *IEE Electronics & Communication Engineering Journal*, June 2000, pp. 101–118.
[11] Lee, W.C.Y., *Mobile Communication Engineering, Theory and Applications*, Second Edition, McGraw-Hill, 1998.
[12] Noneaker, D.L. and Parsley, M.B., 'Rake reception for a CDMA mobile communication system with multipath fading', in Glistic, S.G. and Leppanen, P.A., *Code Division Multiple Access Communications*. Kluwer Academia, Dordrecht, 1995, pp. 183–201.
[13] Holma, H. and Toskala, A., *WCDMA for UMTS*, John Wiley & Sons, Ltd, Chichester, 2000.
[14] Laiho, J., Wacker, A. and Novosad, T., *Radio Network Planning and Optimisation for UMTS*, John Wiley & Sons, Chichester, 2002.
[15] Lee, C.C. and Steele, R., 'Effects of soft and softer handoff on CDMA system capacity', *IEEE Trans. Vehic. Tech.*, **47**(3), 1998, pp. 830–841.
[16] 3GPP TS 25.213, 'Spreading and Modulation (FDD)'.
[17] 3GPP TS 25.212, 'Multiplexing and Channel Coding (FDD)'.
[18] 3GPP TS 25.133, 'Requirements for Support of Radio Resource Management'.
[19] 3GPP TS 25.214, 'Physical Layer Procedures'.
[20] 3GPP TS 25.302, 'Services Provided by the Physical Layer'.
[21] 3GPP TS 25.211-700, 'Physical Channels and Mapping of Transport Channels'.
[22] 3GPP TS 25.301-700, 'Radio Interface Protocol Architecture'.
[23] 3GPP TS 24.007, 'Mobile Radio Interface Signaling Layer 3; General Aspects'.
[24] 3GPP TS 25.321, 'Medium Access Control Protocol Specifications'.
[25] 3GPP TS 25.331, 'RAN RRC Protocol Specifications'.
[26] 3GPP TS 25.323, 'PDCP Protocol Specifications'.
[27] 3GPP TS 25.324, 'BMC Protocol Specifications'.
[28] 3GPP TS 25.215-700, 'Physical Layer Measurements (FDD)'.

3

Multipath and Path Loss Modeling

The radio channels in land mobile communication are burdened with particular propagation complications compared to radio systems with fixed and carefully positioned antennas. The small antenna heights of the mobile terminals, which can communicate from anywhere within the network coverage area, have little clearance to the base station sites. The obstacles in the vicinity of the mobile terminals cause multiple reflections and scattering of the signals transmitted from the radio sites. This results in special propagation characteristics, such as multipath signal receptions and fading, the geometry of which can change from place to place and with the movement of the mobile terminal as well as obstacles in the area. Furthermore the transmission path between the mobile terminal and the radio sites can vary from simple direct line-of-sight to one that is severely obstructed by buildings, foliage, and the terrain. This results in propagation characteristics that require modeling and parameterization at several different levels to analyze and quantify the impact to the mobile communication channels and the radio equipments. The results can then be used for the proper link budgeting of communication channels and the engineering of radio site configurations, as well as equipment design. Two important aspects of the mobile radio communication channel characteristics are the multipath signal profile and effects, and the channel path losses resulting from distance and shadowing through the terrain. These are sometimes referred to as the small-scale and large-scale environmental effects, respectively. The proper characterization of typical multipath channel profiles encountered in the mobile environment is essential to defining receiver performance requirements. One consequence of multipath reception is signal amplitude variations whose statistical characterization and parameterizations are also essential to engineering of radio links. Models appropriate to each distinctly definable category of terrain and environment for signal path loss estimations are another necessity for the engineering of mobile radio link budgets and coverage design. This chapter covers the basic fundamentals of signal propagation and characterization in the mobile environment and presents the key industry standard models and parameters for the characterization of multipath reception, and path losses due to distance and shadowing, as well as the practical guidelines for their use.

UMTS Network Planning, Optimization, and Inter-Operation with GSM   Moe Rahnema
© 2008 John Wiley & Sons, (Asia) Pte Ltd

3.1 Multipath Reception

In the mobile communication environment, there are usually a large number of signal reflecting obstacles such buildings and terrain irregularity between the base station and the mobile terminal. These obstacles cause signals transmitted by one end to arrive at the other of the communication link via multiple reflections. Moreover within the vicinity of the mobile terminal, small objects with dimensions of the order of the signal wavelength can cause the signal arrivals from multiple reflections to scatter in all directions. Thus the signal offered to a mobile terminal usually contains a large number of reflected radio waves that travel through different path lengths and hence arrive at the receiver with different time delays. For a typical wireless radio channel, the received signal usually consists of several discrete multipath components, sometimes referred to as *fingers* [1]. The different time delays or path lengths translate into different phases, which can then attenuate or reinforce each other depending on the relative phasing of the multipath signals. When the multiple rays are phased to attenuate each other, it results in signal fading.

A basic advantage of the CDMA based communication system is that it can resolve the multipath signals that are spaced in time by at least one chip period of the CDMA spreading signal, and combine them coherently to enhance the signal to noise ratio at the input to the demodulator. In this way, multipath results in diversity gain. The CDMA systems use multiple correlation techniques used in the Rake receivers [2] to detect and resolve individually the signals received at different delays. The number of signals that can be detected and coherently combined is related to the number of fingers in the Rake receiver. Any additional delayed signals act as interference to the receiver. In a non-CDMA system, the multiple delayed signals can result in inter-symbol interference if the relative delay between the signals is larger than or a significant fraction of the symbol time used in the digital communication system. In that case, sophisticated equalization techniques are required to cancel the inter-symbol interference effects [3].

Multipath signal propagation is characterized by the delay spread and coherence bandwidth parameters.

3.1.1 Delay Spread

The delay spread of a multipath channel is a measure of the signal time dispersion caused by the channel. The signal time dispersion is observed by looking at the impulse response of the channel. The impulse response of a multipath channel results in a series of pulses with different time delays and energies as shown in Figure 3.1. Delay spread is defined in two different ways: the maximum delay time spread and the root mean square (rms) delay spreads [4, 5]. The maximum delay time spread is the time interval during which reflections with significant energy arrive (about 10 to 20 dB below the strongest signal component).

The rms delay spread τ_{rms} is the standard deviation (or root-mean-square) of the power-delay profile. Mathematically, this is defined as

$$\tau_{rms} = \sqrt{(\tau^2)^- - (\tau^-)^2} \tag{1}$$

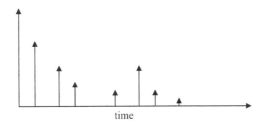

Figure 3.1 Multipath delay spread signal components.

where

$$(\tau^2)^- = \frac{\sum_j P_j \tau_j^2}{\sum_j P_j} \quad \text{and} \quad \tau^- = \frac{\sum_j P_j \tau_j}{\sum_j P_j}$$

and the summations are performed over the multipath signal reflections arrived at the receiver, τ_j is the delay of the jth version of the signal received, and P_j is the power of the jth received. For a digital signal with high bit rate, this dispersion is experienced as frequency selective fading or inter-symbol interference (ISI) as stated in the preceding section. No serious ISI is likely to occur if the delay spread is less than 10% of the symbol duration. The experimental measurements of power-delay profile using wideband signals are discussed in Reference 6.

Measurements made in the mobile communication environment indicate that delay spreads are usually in the range of 0.5 to 3 or 5 µs, with the higher end occurring in urban areas where most of the multipath occurs due to reflections from multiple tall buildings [7, 8]. This is considerably larger than the chip period of 0.26 µs of the WCDMA system, which uses a chip rate of 3.84 Mcps. This means that the multipath signals can be individually detected for the most part by a Rake receiver and then coherently combined to enhance the received signal to noise ratio. Thus CDMA systems can take advantage of the multipath phenomenon and turn it into positive effect. This benefit is achieved to the extent that sufficient numbers of Rake fingers are available in the receiver to detect and combine all the resolvable multipath signal components with significant energy. Thus a direct CDMA Rake receiver can achieve N-fold path diversity where N is the number of resolvable signal paths (that is, all signal paths for which the relative delay between any two is greater than the chip period).

3.1.2 Coherence Bandwidth

The coherence bandwidth is a frequency domain characterization of the multipath channel that is related to the delay spread and is inversely proportional to it. The coherence bandwidth defines the window of frequencies over which the signal frequency components of the communication signal fade coherently or in a correlated manner. Since the inter-arrival times of the multipath signal (reflections) fall within the delay spread, by definition, the smallest time scale at which the signal variations can occur due to multipath combining are of the order of the delay spread. In the frequency domain this means that the frequency at which the signal variations can occur due to changing multipath components (say when a user moves to a different location) is of the order of the inverse of the delay spread. Now, if this is smaller than the time scale of variations in the communication signal caused by modulation, the communication signal 'sees' a constant channel during a modulation period (i.e., symbol period). Then the channel is called frequency non-selective with respect to the communication signal, and no inter-symbol interference occurs. Therefore, all the frequencies within the signal band are treated similarly by the channel (fade coherently or in a correlated way). Hence the inverse relation between delay spread and coherence bandwidth. In the time-delay domain, the time-spreading phenomenon is characterized by the multipath delay spread, and in the frequency domain it is characterized by a channel coherence bandwidth.

To provide the exact relation between the coherence bandwidth and the inverse delay spread, the degree of signal correlation needs to be defined. Then the coherence bandwidth is technically defined to be the maximum frequency separations within the signal where the correlation between the fading on the two frequencies stays above a certain value. Depending on what value is selected for this correlation limit, different values result for the coherence bandwidth.

If the coherence bandwidth is defined as the range of frequencies over which the fade correlations is above 0.9, the following formula results [9]:

$$\text{Coherence BW} = \frac{1}{50\tau_{rms}} \quad (2)$$

If the coherence bandwidth is defined as the range of frequencies over which the fade correlations is above 0.5, the following formula results [4]:

$$\text{Coherence BW} = \frac{1}{5\tau_{rms}} \quad (3)$$

where τ_{rms} is the rms delay spread in seconds and coherence bandwidth is in Hz. Thus, the delay spread calculated from the delay profile of the received signal determines to what extent channel fading at two different frequencies is correlated.

For a delay spread of 2 μs, which is typical in urban cellular communication environments, the coherence bandwidths obtained from Equations (2) and (3) are 0.01 and 0.1 MHz, respectively. Even a delay spread of 0.5 s would result in a coherence BW of 0.04 MHz or 0.4 MHz depending on which equation is used. These are much smaller than the frequency separations between the uplink and downlink bands in the FDD mode of CDMA. Therefore, fading on the two links is correlated, which implies channel fade status detected on one link could not be used to compensate for fading on the other. That is why closed loop power control is implemented on each link in WCDMA to compensate for the multipath induced fades. Open loop power control is only used on the random access channels for initial network access and channel set up as just a rough means of estimating the initial powers needed to overcome channel attenuations.

3.1.3 Doppler Effect

The Doppler effect is the frequency shifts that are introduced in the signal as a result of relative motion between the transmitter and the receiver. The Doppler frequency shift in a signal is given by $f_D = \frac{V}{\lambda} Cos(\alpha)$, where V is the relative speed, λ is the signal wavelength, and α is the angle of arrival of the signal ray relative to the direction of motion. The Doppler frequency shift f_D is positive when the transmitter and receiver move toward each other, and negative when moving away from each other. The Doppler frequency shifts in the components of the multipath signal resulting from the relative motion between the transmitting and receiving antennas further contributes to variations in relative signal phases and hence in the signal envelope.

In the mobile communication environment with a vertical receive antenna and constant azimuth gain at the mobile terminal, dense objects in the vicinity can scatter the signal transmitted from the base station to arrive with a rather uniform distribution from all directions in the angular range $(0, \pi)$. This would result in a bowl-shaped Doppler frequency and power spectrum of an unmodulated CW (continuous wave) with a frequency spread in the range of $-V/\lambda$ to $+V/\lambda$ centered on the carrier frequency [5, 10], where V is the speed of the mobile terminal.

3.1.4 Small-scale Multipath Effects

When a mobile radio roams over a small distance, it will experience signal envelope variations (fading) caused by the changing multipath profile. The many irresolvable multipath signals resulting from reflections and scattering from objects in the vicinity of the mobile terminal interfere with each other and produce multipath fading. The multipaths, which differ by just half a wavelength, have opposite phases and cancel each other resulting in fades (they cannot be resolved by the Rake receiver because their path delay differences are less than a chip period of 0.26 µs). A half wavelength corresponds to a path difference of just 8.3 cm at 1800 MHz. Such small path differences can occur from the multipath geometry produced by the many signals reflecting and scattering objects in the vicinity of the mobile. Therefore, the received signal power changes rapidly in a random manner with a period of about half the signal wavelength when the mobile moves. These signal envelope variations that occur over small distances of the order of the signal wavelength translate into fast signal variations or fades in a mobile terminal scenario. These fast signal fluctuations can also occur due to the motion and vibrations of the signal reflectors and scattering objects. The word 'fast' here is meant in comparison to the signal fluctuations and path losses that occur over larger scale distances due to major changes in the terrain (buildings, hills, clutter, etc.) and environment. Since the changes in the channel multipath profile occur on the scale of the signal wavelength, the rate of change of these fast signal envelope variations or fades is obtained by the ratio V/λ, which is also the maximum Doppler frequency shift resulting from motion. These small-scale (in distance sense) variations in the signal envelope as a result of the changing multipath profiles are in fact confirmed by measurements. If the received signal power versus antenna displacement (in units of wavelength) is plotted for the case of a mobile radio, signal nulls on a small-scale distance (local signal strength variations) are observed at approximately half a wavelength [1]. Thus two antennas at the mobile terminal are likely to receive uncorrelated powers if their separation is more than a wavelength. Things are different in the case of a base station antenna, which is usually directional and mounted much higher with basically no scattering objects in the vicinity. The base station antenna receives signal reflections in a narrower range of angles. Therefore, spatial diversity at the base station requires much larger separation of the antennas to receive uncorrelated signal powers at the two antennas. That is also why antenna directivity results in different performances at the mobile and the base station.

The fast variations in the signal envelope, which result from the changing multipath in the mobile communication environment, are called *Rayleigh fading*. This name is based on the fact that the large number of randomly phased and attenuated components in the multipath signal with no line-of-sight signal present, produce a signal whose envelope is statistically described by a Rayleigh probability density function. When there is a dominant non-fading signal component present, such as a line-of-sight propagation path, the resulting signal envelope is described by a Rician pdf [4]. The Rician channels are used to model communications channels in mobile satellite communications networks, where most often a line-of-site signal component is also present between the mobile and the satellite.

The mathematical expression for the Rayleigh pdf is given by

$$P(r) = \frac{r}{\delta^2} e^{-\frac{r^2}{2\delta^2}} \quad \text{for} \quad r > 0$$
$$= 0 \quad \text{for} \quad r < 0 \quad (4)$$

where r is the envelope amplitude of the received signal and $2\sigma^2$ is the mean power of the received multipath signal. The Rayleigh faded component is sometimes called the *random*, *scatter*, or *diffuse* component. The Rayleigh pdf results from having no line-of-site component of the signal. Therefore, for a single link it represents the pdf associated with the worst case of fading per mean received signal power.

WCDMA uses closed loop fast power control on both the uplink and downlink to compensate and correct for these fast Rayleigh channel attenuations due to multipath fading. The power control takes place at the rate of 1500 times per second, which is many times faster than the rate of Rayleigh fades for slow and moderate mobile speeds.

3.1.5 Channel Coherence Time

The fast channel variations or Rayleigh fades caused by multipath occur at some rate that basically depends on the speed of the mobile terminal (if moving) or the mobility characteristics of the reflective and scattering objects in the area. This rate of the fast channel variations due to multipath effects is dominated by the mobile speed when the terminal is mobile and is given by V/λ. This is also the maximum Doppler spread resulting from the terminal motion. The inverse of the rate of channel variations due to multipath is referred to as the *coherence time* in the time domain [11]. The coherence time, T_0, is a measure of the expected time duration for which the channel's response characteristics are essentially invariant. The coherence time is reciprocally related to the Doppler spread, f_d, within a multiplicative constant. Therefore

$$T_0 \propto \frac{1}{f_d} \quad (5)$$

Intuitively, since the small-scale changes in the signal due to motion occur on the scale of a wavelength (changing multipath profile), the distance between a null and a peak in the faded signal is $\lambda/2$. This means the distance traveled must be half a wavelength before the signal changes significantly. This translates into a time interval of $\frac{\lambda/2}{V}$, which is equal to $\frac{0.5}{f_d}$, using the expression for the maximum Doppler frequency ($f_d = V/\lambda$). However, if T_0 is defined more precisely as the time duration for which the channel's response to a sinusoid has a correlation greater than 0.5, the relationship between T_0 and f_d is approximately [7]

$$T_0 = \frac{9}{16\pi f_d} \tag{6}$$

A popular general rule is to define T_0 as the geometric mean of Equations (5) and (6) [1]. This yields

$$T_0 = \sqrt{\frac{9}{16\pi f_d^2}} = \frac{0.423}{f_d} \tag{7}$$

which is close to the intuitively derived expression.

The channel coherence time notion is used to determine channel reciprocity in time division duplex communication (TDD). In the TDD mode of communication (implemented by the UMTS TDD mode), the same frequency band is used for uplink and downlink communication (mobile to base station and vice versa), but shared in the time domain. Now, if the spacing between the uplink and downlink time slots are within the coherence time of the frequency channel used, the channels are called reciprocal, meaning that they have the same fade characteristics. Therefore open loop power control can be effective on such channels. However, in the real world, there are other parameters that enter this scenario besides the coherence time parameter, which may not make the two links completely reciprocal. These factors would include human-made noise, which would be expected to be mostly present on the downlink side (at the mobile terminal locations), antenna diversities on the uplink (at base station), and perhaps even the different access protocols used on the two links.

Another area where the notion of channel coherence time is used is in the characterization of the Rayleigh fades as being slow or fast relative to the symbol rate used in a digital communication system. Fast Rayleigh fading describes a condition where the time duration in which the channel behaves in a correlated manner is short compared to the time duration of a symbol. Hence, fast fading can cause the baseband pulse to be distorted, resulting in a loss of SNR that often yields an irreducible error rate. Such distorted pulses cause synchronization problems (failure of phase-locked-loop receivers), in addition to difficulties in adequately defining a matched filter. A channel is generally referred to as introducing slow Rayleigh fading if the *coherence time > symbol time*. Here, the time duration for which the channel behaves in a correlated manner is long compared to the time duration of a transmission symbol. Thus, one can expect the channel state to remain virtually unchanged during the time in which a symbol is transmitted. The propagating symbols will likely not suffer from the pulse distortion described above. The primary degradation in a slow-fading channel, as with flat fading, is loss in SNR [1]. Regardless of this classification of the speed of multipath fading, Rayleigh or multipath fading is always referred to as *fast fading* in the mobile communication literature, when compared to the slow channel variations that occur over large-scale distances due to major changes in the terrain (hills, buildings, clutter, etc.). This will be discussed in a later section.

3.2 3GPP Multipath Channel Models

The design of radio communication links budgeting require the specification of the E_b/N_0 requirements for the base stations and the mobile terminals to achieve a certain performance such as BLER (block error rate) for each service. For this aspect of the system design the mobile radio channel is usually evaluated from 'statistical' propagation models: no specific terrain data is considered and channel parameters are modeled as stochastic variables. This is based on the consideration that, besides the receiver structure (modulation and coding used, etc.), it is the measures of signal variability (fade) and multipath effects that determine the E_b/N_0 requirements [3].

The 3GPP Standards specify certain performance tests for UE and base station receivers under the static white Gaussian noise with no fading as well as under a set of multipath propagation fading conditions. The 3GPP multipath propagation models are shown in Table 3.1. The taps for all the models have a classic Doppler spectrum. These models are used as reference measurement channels by 3GPP for specification and testing of UE and base station equipment performance requirements for speech (12.2 kbps codec rate) and for data channels with 64 kbps, 144 kbps, 384 kbps, and 2048 kbps with both uplink and downlink parameters defined. The uplink performance requirements are given for dedicated channels as the average required E_b/N_0 for a fixed BLER (block error rate) without power control being on. For the downlink side, 3GPP defines the link performance requirements as the average required E_b/N_0 for fixed BLER at assumed distances from the base station. Although the reference measurement channels may not be the most applicable in all practical cases, they are important, because the link performance requirements given in 3GPP specifications are for these channels.

Case 1 in Table 3.1 is a one-tap model that is heavily faded, and is mostly similar to the ITU Pedestrian S channel model at UE speed of 3 km/h. Case 2, which is similar to case 4, is

Table 3.1 3GPP propagation conditions for multipath fading environments [12].
[3GPP[TM] TSs and TRs are the property of ARIB, STIS, ETSI, CCSA, TTA and TTC who jointly own the copyright in them. They are subject to further modifications and are therefore provided to you 'as is' for information purposes only. Further use is strictly prohibited]

Case 1, speed 3 km/h		Case 2, speed 3 km/h		Case 3, speed 120 km/h		Case 4, speed 3 km/h		Case 5, speed 50 km/h		Case 6, speed 250 km/h	
Relative Delay (ns)	Average Power (dB)	Relative Delay (ns)	Average Power (dB)	Relative Delay (ns)	Average Power (dB)	Relative Delay (ns)	Average Power (dB)	Relative Delay (ns)	Average Power (dB)	Relative Delay (ns)	Average Power (dB)
0	0	0	0	0	0	0	0	0	0	0	0
976	−10	976	0	260	−3	976	0	976	−10	260	−3
		20000	0	521	−6					521	−6
				781	−9					781	−9

for low speed UE but with higher multipath diversity and less fading. Case 3 has four strong channel taps with UE speed of 120 km/h, which is much like the vehicular A channel at the same UE speed. Case 5 is the same as case 1 but with a UE speed of 50 km/h. Case 6 is similar to case 3 except that it is for a UE speed of 250 km/h.

3.3 ITU Multipath Channel Models

The 3GPP multipath models were proposed by 3GPP mainly as models under which to specify and verify equipment performance requirements for equipment manufacturers. The ITU multipath channel models are proposed by ITU [13], which are basically similar in structure to the 3GPP multipath channel models but are aimed more at network planners for system design and performance verification. The ITU models are used to capture essential multipath conditions in typical environments under which the average E_b/N_0 requirements of various services for certain performance levels are specified. Meanwhile the equipment manufacturers have to make sure that their equipment meets the ITU specified requirements.

Generally, the mobile user may operate under any possible environment with specific multipath and signal fading characteristics. Instead of constructing channel models for all possible operating environments, ITU-R M.1225 Recommendation [13] suggests a smaller set of test environments that adequately span the overall range of possible operating environments and user mobility. Possible operating environments include large and small cities, tropical, rural, and desert areas. The proposed ITU conceptual environments may not correspond to the actual mobile user's operating environment, but they should give a very good idea on how the mobile user performs under different typical environments. Specifically, ITU-R M.1225 Recommendation suggests three test environments for evaluating the performance of 3G systems: 1) indoor office test environment; 2) outdoor to indoor and pedestrian test environment; and 3) vehicular test environment. The indoor office test environment is characterized by small cells and low transmit power. Both base stations and pedestrian users are located indoors. The outdoor to indoor and pedestrian test environment is described by small cells and low transmit power. Base stations with low antenna heights are located indoors whereas pedestrian users are located on streets and inside buildings. The vehicular test environment is characterized by large cells, high transmit power, and fast moving terminals.

ITU recommends that the outdoor to indoor and pedestrian models (Table 3.2) be used to represent multipath conditions in micro-cells [13]. The indoor office channel models (Table 3.3) have been recommended for use in modeling indoor systems. The vehicular models (Table 3.4) are used to model multipath conditions in macro-cells, regardless of whether the user is inside the car or not. It is noted that the name 'vehicular' does not necessarily mean that the model is to be used for in-vehicle modeling. Likewise, for an in-vehicle user, Vehicular A is not necessarily on average the multipath channel. The same applies to the misleading terminology used for the 'outdoor to indoor and pedestrian' environment.

The operator may specify a certain percentage mix of the ITU representative models to capture the propagation characteristics specifics to a given environment. For instance a 'dense urban environment' may be modeled as consisting of 75% of a pedestrian channel model A at 3 km/h and 25% of pedestrian channel model at 50 km/h. Likewise a rural or suburban environment may be represented as a certain percentage mixture of the ITU

Table 3.2 ITU outdoor to indoor and pedestrian multipath channel models [Reproduced with the kind permission of ITU].

Tap	Channel A		Channel B		Doppler Spectrum
	Rel. Delay (ns)	Ave. Pwr (dB)	Rel. Delay (ns)	Ave. Pwr (dB)	
1	0	0	0	0	Classic
2	110	−9.7	200	−0.9	Classic
3	190	−19.2	800	−4.9	Classic
4	410	−22.8	1200	−8	Classic
5	NA	NA	2300	−7.8	Classic
6	NA	NA	3700	−23.9	Classic

Table 3.3 ITU indoor multipath channel model [Reproduced with the kind permission of ITU].

Tap	Channel A		Channel B		Doppler Spectrum
	Rel. Delay (ns)	Ave. Pwr (dB)	Rel. Delay (ns)	Ave. Pwr (dB)	
1	0	0	0	0	Classic
2	50	−3	100	−3.6	Classic
3	110	−10	200	−7.2	Classic
4	170	−18	300	−10.8	Classic
5	290	−26	500	−18	Classic
6	310	−32	700	−25.2	Classic

vehicular channel models at various speeds. With this construct, the E_b/N_0 requirement for the defined environment is obtained by a weighted averaging of the E_b/N_0 requirements documented by the vendors for the ITU channel models used to define the operator's environment.

Table 3.4 ITU vehicular, high antenna, multipath channel model [Reproduced with the kind permission of ITU].

Tap	Channel A		Channel B		Doppler Spectrum
	Rel. Delay (ns)	Ave. Pwr (dB)	Rel. Delay (ns)	Ave. Pwr (dB)	
1	0	0	0	−2.5	Classic
2	310	−1	300	0	Classic
3	710	−9	8900	−12.8	Classic
4	1090	−10	12900	−10	Classic
5	1730	−15	17100	−25.2	Classic
6	2510	−20	20000	−16	Classic

3.4 Large-Scale Distance Effects

When a terminal moves over large distances, it experiences signal variations and fade due to both distance and shadowing (path loss), and statistical variations at a given distance due to local terrain variations and shadowing effects. These large-scale fades are superimposed on the multipath fading that occurs over smaller-scale distances [14]. When estimating path loss for a link budget analysis in a cellular communication network, the signal variations and losses that occur on both the small- and large-scale distances should be accounted for. The former category and the critical parameters, and models typical of mobile communication environments, were discussed in the previous sections. The latter effects that occur over larger-scale distances are discussed in the next two sections.

3.4.1 Lognormal Fading

The lognormal fading is a statistical adjustment made to the distance dependent mean path losses that are obtained from a path loss model for a given environment. The path loss models are discussed in the next section (and subsections). Since most path loss models are statistical models that are derived from limited measurements carried out in the typical environments for which they are planned, they have limited accuracy as well as limited area resolutions. For the same typical environment of an urban area, not all locations at a given distance from a transmitter will have the same exact radio path characteristics. This consideration is in addition to the fact that the statistical path loss models predict path losses at a certain distance with limited area or distance resolutions.

Measurements have shown that at a given distance r from a transmitter, the path loss $L_p(r)$ is a random variable having a log-normal distribution about the mean distant-dependent value predicted by a distance dependent path loss model [1]. Thus, path loss $L_p(r)$ can be expressed in terms of $L_p^-(r)$ plus a random variable $X\sigma$:

$$L_p(r) = L_p^-(r) + X_\sigma \quad \text{all in dB} \tag{8}$$

where X_σ is normally distributed in the logarithm domain (in the dB scale) with zero mean. Models for predicting the $L_p^-(r)$ are discussed in the next section. The log-normally distributed variation to the mean path loss predicted by a path loss model, that is the X_σ component, is referred to as the slow fading, the shadowing, or the lognormal fading component. The standard deviation of this lognormal distribution depends on the environment and the accuracy of the path loss model used to predict the distant dependent mean path loss. In theory, this standard deviation can be reduced to zero if prediction methods using topographical databases with unlimited resolution and details are used to predict the signal mean path losses at each distance from the transmitter. In practice, the typical values of the lognormal fade standard distribution are about 8 dB in urban areas, 10 dB in dense urban areas, and 6 dB in suburban and rural areas with the usual Okumura-Hata path loss models (see Section 3.4.2.3). A simple path loss model based on only distance with no use of the terrain specific data will result in very large values for the standard deviation of the lognormal fade component.

The lognormal variations imposed on the mean path losses estimated by a path loss model require certain margins to be added to the radio link budgets to ensure a desired level of signal coverage. The required margin will depend on the desired *area coverage probability*, the standard deviation of the lognormal fade distribution (which will depend on the environment, such as typical urban or suburban area) for the area, and the path loss distant exponent. The area coverage probability is defined as the probability that the signal level in the area is above a certain threshold. The derivation of the formulation for the lognormal fade margin is given in Reference 5, and the formulas are discussed with examples in Section 5.3.1.3 of Chapter 5.

3.4.2 Path Loss Models

Path loss models are used to estimate the expected (mean) value of the signal path loss at a certain distance from the transmitter. The path loss depends on the signal frequency, the antenna heights of the transmitter and base station, and the specifics of the terrain and the morphostructure (buildings type and density, clutter such as vegetation, forestry, open areas, water, etc.). With this description, one can generally express the mean path loss $L_p^-(d)$ with reference to the path loss at a reference distance d_0 (located in the far field of the radiating antenna) and a path-loss distance exponent (which would depend on the environment) in the form [1]:

$$L_p^-(d) = L_p^-(d_0) + 10\,n\log_{10}(d/d_0), \quad \text{all in dB} \tag{9}$$

where n is the path-loss distance exponent (how the loss varies with distance), and $L_p^-(d_0)$ is the mean path loss at a reference distance d_0, which captures antenna heights and the frequency and environmental dependencies.

There are basically three different categories of path loss models, referred to as statistical (or empirical), deterministic, and semi-deterministic. The statistical models are formulas that describe the path loss versus distance on an average scale. They are derived from statistical analysis of a large number of measurements obtained in typically distinct environments such as urban, suburban, rural, which are incorporated in the form of tabulated data or best fit formulas for average path loss calculation versus distance in the particular environment. They do not require detailed site morphological information. Examples are the Okumura-Hata models (see Section 3.4.2.3). The deterministic models, such as those based on ray tracing, apply RF signal propagation techniques to a detailed site morphology description (using building and terrain databases) to estimate the signal strength resulting from multiple reflections, and line-of-site at various pixels in the area. The simple rather idealized free space and two-path ground reflection models may also be classified under the deterministic models. Because of the complexities involved in such models, they are only developed mainly for simple environments such as indoor and small micro-cell. Examples of these models are the ray tracing models used for modeling small micro-cells in urban and dense urban areas. The semi-deterministic (or semi-statistical) models are based on a mixture of a deterministic method of following individual signal propagation effects due to some site-specific morpho-data and a statistical generalization and calibration of model parameters based on collected path loss measurements. These models require more information than the statistical models but less than the deterministic models. Examples are the COST 231 Walfisch-Ikegami model and the generalized tuned Hata model.

3.4.2.1 The Free-space Path Loss Model

The free-space path loss model treats the region between the transmit and receive antennas as being free of all objects that might absorb or reflect radio frequency (RF) energy. It also treats the earth as being infinitely far away from the propagating signal (or, equivalently, as having a reflection coefficient that is negligible). In this idealized free-space model, the attenuation of RF energy between the transmitter and receiver behaves according to an inverse-square law. The path loss expressed as the ratio of the received power to the transmitted power in linear scale is given by

$$\frac{P_r}{P_t} = \frac{G_t G_r \lambda^2}{(4\pi d)^2} \tag{10}$$

where G_t and G_r are the transmitter and receiver antenna gains, λ is the signal wavelength, and d is the distance between the transmitter and the receiver.

3.4.2.2 The Two-ray Ground Reflection Path Loss Model

A single line-of-sight path assumed in the simple free-space path loss model is seldom the case in practice. The two-ray ground reflection model considers both the direct path and a ground reflection path. It is shown in Reference 9 that this model gives more accurate prediction at a long distance than the free-space model. The path loss expressed as the ratio of the received power to transmit power in linear scale is given by

$$\frac{P_r}{P_t} = \frac{G_t G_r (h_t h_r)^2}{d^4} \tag{11}$$

where G_t and G_r are the transmitter and receiver antenna gains, h_t and h_r are the transmit and receive antenna heights, and d is the distance between the transmitter and receiver.

Measurements have shown that the two-ray model resulting in a path loss exponent of 4 does not give a good result for a short distance close to the transmitter. This results in a breakpoint before which the path loss is not as steep as predicted by the two-ray model. Instead, the free-space model with a path loss exponent of 2 provides better approximation. And this leads to a two-slope path model where before the break point the free-space path loss model is used and after the breakpoint the two-ray path loss is used. The breakpoint is obtained where the linear envelope of the free-space path loss intersects with the two-path loss model, and is given by

$$D_{\text{breakpoint}} = \frac{2\pi h_t h_r}{\lambda} \tag{12}$$

For a carrier frequency of 1800 MHz, and a base station antenna height of 30 m, (assuming 1.5 m for mobile station antenna height), the breakpoint obtained from Equation (12) is 1702 m.

3.4.2.3 Okumura-Hata Path Loss Models

The Okumura-Hata model is a commonly used statistical model for calculating path losses in macro-cell environments. Okumura [13] made some of the earlier comprehensive path-loss measurements for a wide range of antenna heights and coverage distances in certain urban and suburban areas of Japan. Hata [15] later transformed Okumura's data into parametric formulas, now known as the Okumura-Hata path loss model. The model includes parametric correction factors for base station and mobile antenna heights, and for environmental effects such as for suburban, open areas, and irregular terrain. The Okumura-Hata model is highly referred to in the industry. Its range of usability in different environments and terrain types has made it very useful in many propagation and system simulation studies.

The spatial resolution of the Okumura-Hata model is about 20 m, because the original data on which the model was developed were averaged over 20 m intervals. The Okumura-Hata model is valid for frequency ranges of 150–1000 MHz, base station antenna heights of 30–200 m, mobile station antenna heights of 1–10 m, and distances of 1–20 km from the transmitting antenna. The model is based on overall statistical path loss measurements and form fitting and so it does not capture the detailed geometrical properties of buildings and the terrain. Therefore it is not expected to provide accurate results in microcellular environments where the size of the cells are small compared to buildings. The statistical based methods are not the best for modeling signal propagation behavior in cells where the antennas are placed below the roof-top level of surrounding buildings, making signal propagation more *localized* and site specific through sparse reflections from the buildings. The alternative for small micro-cells in dense urban areas is models based on ray tracing, which is discussed in Section 3.4.2.7.

Because of the frequency band limitation of the Hata model, the original model was later extended by COST 231 [16] to the frequency band of 1500–2000 MHz, which covers the bands allocated to the 3G networks.

The Okumura-Hata formula for the propagation loss has the following form:

$$L_p(d) = 69.55 + 26.16 \log_{10}(f) - 13.82 \log_{10} h_b + (44.9 - 6.55 \log_{10} h_b) \log_{10} d - a(h_m) - Q_r, \text{ dB} \tag{13}$$

where

$$150 \text{ MHz} < f < 1500 \text{ MHz} = \text{carrier frequency}$$
$$h_b = 30\text{–}200 \text{ m base station antenna height}$$
$$h_m = 1\text{–}10 \text{ m mobile antenna height}$$
$$d = 1\text{–}20 \text{ km distance from the transmitter}$$

and $a(h_m)$ depends on whether the model is used on the size scale of the city given as follows.

For a medium or small city:

$$a(h_m) = (1.1 \log_{10} f - 0.7) h m - (1.56 \log f - 0.8) \tag{14}$$

For a large city:

$$a(h_m) = 8.29[\log_{10}(1.54h_m)]^2 - 1.1 \quad f \leq 200\,\text{MHz} \tag{15a}$$

$$a(h_m) = 3.2[\log_{10}(11.75h_m)]^2 - 4.97 \quad f \geq 400\,\text{MHz} \tag{15b}$$

Q_r is a correction factor for open areas given as

$$Q_r = 4.78(\log 10f)2 - 18.33 \log 10f + 40.94 \quad \text{all frequencies (in MHz)} \tag{16}$$

and is zero for all other area types.

In practice, the height used for the mobile station antenna height is usually set at 1.5 m ($h_m = 1.5$ m). In that case, the expressions $a(h_m)$ will become very close to zero and not very sensitive to variations of the mobile antenna height.

3.4.2.4 COST 231 Hata Model

The Okumura-Hata model was extended in the COST 231 Project [16] to the frequency bands of 1500–2000 MHz, which includes the band allocated to 3G networks. The modified model is referred to as the COST 231 Hata model and is given below:

$$L_p(d) = 46.3 + 33.9\log_{10}(f) - 13.82\log_{10} h_b + (44.9 - 6.55\log_{10} h_b)\log_{10} d - a(h_m) + C_{\text{clutter}}, \text{dB} \tag{17}$$

where

$$1500\,\text{MHz} \geq f \geq 2000\,\text{MHz}$$

$$a(h_m) = (1.1\log 10f - 0.7)hm - (1.56\log f - 0.8) \tag{18}$$

and C_{clutter} is a clutter loss correction given by

$C_{\text{clutter}} = 0$ dB for a medium sized city and suburban centers with moderate tree density

$C_{\text{clutter}} = 3$ dB for metropolitan centers

The modeler may use measurements for his or her particular environment to find the best value for C_{clutter}. Model tuning is discussed in Section 3.4.3.

3.4.2.5 Two-Slope Extension to Hata Path Loss Models

Since the Hata models are valid only at distances larger than 1 km from the site, an extension to the formulas is usually used to address the path losses at distances closer than 1 km to the site. The Hata model does not apply at near to the site distances due to heavy fluctuations caused by combinations of LOS and NLOS signal paths. This makes the signal power increase over the values predicted by models such as Hata's, which are tuned for larger distances where signal variability due to LOS is less present. To take this effect into account, 'two-piece' models each with different distance path loss exponents are used. The two pieces join at a certain distance, normally the 1 km lower limit from where the Hata model becomes valid. An approach to modeling the near site path loss would be to construct a linear function by using the free-space path loss at a distance of 20 m and the path loss calculated with the Hata formula at a distance of 1 km. This then results in the following slope for path loss calculation within the near site:

$$\text{Slope-of-path loss Line}_{\text{near site}} = \frac{L_p^{Hata}(\log d_{BP}) - L_p^{FS}(\log 20)}{\log(d_{BP}) - \log 20} \quad (19)$$

where L_p^{Hata} is the COST 231 Hata path loss formula and L_p^{FS} is the free-space path loss formula given in Equation (10). The logs are all in base 10. The L_p^{FS} obtained from Equation (10) should be converted to dB before being used in Equation (19).

Then the path loss formula for the near site becomes

$$L_p^{\text{Near Site}}(d) = L_p^{Hata}(d_{BP}) + \text{Slope-of-path loss Line}_{\text{near site}} \cdot [\log(d) - \log(d_{BP})], \text{ dB} \quad (20)$$

for $0.020 \text{ km} << d \leq d_{BP}$
where d_{BP} is the breakpoint distance at which the Hata model becomes valid, that is $d_{BP} = 1$ km.

3.4.2.6 COST 231 Walfisch-Ikegami Path Loss Model

The COST 231 Walfisch-Ikegami model [17] uses more site-specific building clutter data to capture the propagation effects due to reflections and roof-top diffractions. The model therefore has deterministic elements rather than being purely statistical. It is more complex than the Okumura-Hata models and is more suited to smaller macro-cells in urban areas or micro-cells where the antenna is placed at roof-top levels. The model presents different formulas for cases when there is a LOS (line-of-site) signal component, and for when there is no LOS (NLOS) component. The formula for the case of LOS is simple and only depends on the frequency and the distance. In the NLOS case, the path loss is made up of a free-space path loss and two other components that are used to capture the effects due to signal diffractions and scattering from the roof-tops and multi-screen diffractions. These latter effects depend on the street width, the street direction relative to a direct line between the transmitter and receiver, roof heights, and building separations. The model therefore assumes that the signal propagates over roof-tops and is diffracted by the roof-top edges in reaching the receiver. If the roof-top diffraction effects are not present, for instance in small micro-cells where the antennas are placed below the average roof-top level of surrounding buildings, the model can overestimate the path losses. Therefore it should be used cautiously in application to small micro-cells, with results verified by measurements.

The model assumes uniform building spacing and identical dimensions with no terrain variations over the area. This is another aspect that should be kept under consideration when applying the model.

Multipath and Path Loss Modeling

The Walfisch-Ikegami formulas are given by:

$$Lp(d) = 42.6 + 26 \log_{10} d + 20 \log_{10} f \text{ dB}, \quad d \geq 0.020 \text{ km} \quad \text{when receiver is in LOS} \quad (21)$$

$$Lp(d) = 32.4 + 20 \log_{10} d + 20 \log_{10} f + L_{rts} + L_{msd} \text{ dB}, \quad \text{when receiver is in NLOS} \quad (22)$$

where L_{rts} is the roof-top to street diffraction and scatter loss effects, given by

$$L_{rts} = -16.9 - 10 \log_{10}(w) + 10 \log_{10}(f) + 20 \log_{10}(H_{roof} - h_m) + L_{cri}$$

where

$$L_{cri} = -10 + 0.354\varphi \quad \text{for} \quad 0^0 \leq \varphi \leq 35^0$$
$$L_{cri} = 2.5 + 0.075(\varphi - 35^0) \quad \text{for} \quad 35^0 \leq \varphi \leq 55^0$$
$$L_{cri} = 4.0 - 0.114(\varphi - 55^0) \quad \text{for} \quad 55^0 \leq \varphi \leq 90^0$$

L_{msd} is the multiscreen diffraction losses given by

$$L_{msd} = L_{bsh} + k_a + k_d \log_{10}(d) + k_f \log(f) - 9 \log_{10}(b)$$

where

$$L_{bsh} = -18 \log_{10}(1 + h_b - H_{roof}) \quad \text{for } h_b > H_{roof}$$
$$= 0 \quad \text{for } h_b \leq< H_{roof}$$
$$K_a = 54 \quad \text{for } h_b > H_{roof}$$
$$= 54 - 0.8 \times (h_b - H_{roof}) \quad \text{for } d \geq 0.5, \text{ and } h_b \leq H_{roof}$$
$$= 54 - 0.8 \times (h_b - H_{roof}) \times (d/0.5) \quad \text{for } d < 0.5, \text{ and } h_b \leq H_{roof}$$
$$K_d = 18 \quad \text{for } h_b > H_{roof}$$
$$= 18 - 15 \times (h_b - H_{roof})/H_{roof} \quad \text{for } h_b \leq H_{roof}$$

$K_f = -4 + 0.7 \times (f/925 - 1)$ for medium sized cities and suburban centers with moderate tree density and $K_f = -4 + 1.5 \times (f/925 - 1)$ for metropolitan centers.

The morphographic data H_{roof}, w, b, etc. given in Table 3.5 can be calculated by radio network planning tools using the vector building map data layer. Alternatively, the user may enter them for each cell independently of any digital map information. When the building heights and separations are not rather uniform, the results should be verified by measurements because the model assumes uniformity of the morphological parameters.

3.4.2.7 Ray Tracing Models

Ray tracing is a deterministic modeling approach based on geometrical optics. Ray tracing techniques were originally introduced in computer graphics applications to create photo-realistic pictures of 3-dimensional sceneries. In ray tracing, beams of light are

Table 3.5 The parameter ranges and units for the COST 231 Walfisch-Ikegami model.

Parameter	Valid ranges
frequency	f = 800–2000 MHz
Base station height	h_b = 04–50 m
Mobile station antenna height	h_m = 1–3 m
distance	d = 0.02–5 km
Building heights	H_{roof}, m
Street width	w, m
Building separation	b, m
Road orientation with respect to the direct radio path	Φ (degrees)
Environment	Urban

modeled as ray tubes with a fixed solid angle dQ and of identical shape and size at a distance r from the transmitter. The path of an individual ray is traced recursively until it strikes an object. If a ray strikes an edge or the surface of a building, hill, or any other obstacle indicated in the digital terrain map, diffraction or reflection of the ray takes place as appropriate. Reflections are modeled easily using the Fresnel equations of optics. In the case of diffraction, Huygens's and Kirchoff laws are used to compute the decline in the energy of the ray. In this way ray tracing tracks all possible major reflections and diffractions that end up at different locations. The sum of the received rays is formed to generate the resulting signal strength at each location in the modeled area.

Since ray tracing tracks all possible major reflections and diffractions, it requires detailed digital elevation models (DEM) and building heights and geometries [18]. Ray tracing models are therefore computationally intensive, but provide much better accuracy than statistical models, and estimate path losses at each specific location with high resolution. Because of their computational complexity they are only developed for small micro-cells where the rays may experience fewer reflections and diffractions due to the sparser medium of the smaller areas. However, they are well suited for predicting path losses for indoor applications, and for the small 3G micro-cells in dense urban environment. Examples of some of the available commercial ray tracing models include Volcan from Siradel and Wavesight from Wavecall, which are currently integrated in most 3G radio network planning tools. Ray tracing is discussed in more details in References 19 and 20.

3.4.2.8 Indoor Path Loss Modeling

Smaller indoor spaces lend themselves to manual path loss measurements for link budgeting and network planning. However, in the same way that the Okumura-Hata model has been developed semi-empirically for macro-cell coverage predictions, a model developed by Keenan-Motley is commonly used for indoor path loss predictions. This model has been accepted by COST 231 and contains 3D effects as well. In a simplified form, the Keenan-Motley model can be written in the following way for the 900 MHz band.

$$P_L = 31.5 + 20\log(d) + N_w \times W, \quad \text{for the 900 MHz} \tag{23}$$

where L is the path loss between isotropic antennas (dB), d is the transmitter-receiver separation (m), N_w is the number of walls passed by the direct ray, and W is the wall attenuation factor (dB).

The simplified model assumes that the path loss estimations are made on the same floor as the antennas are located. Thus the floor effects are neglected. This model calculates the loss for each wall in a straight line between the antenna and the prediction point, and adds the free space path loss. The following are rough estimates of wall loss parameters that can be used:

- 2 dB for plaster board walls in office buildings, etc.
- 5 dB for reinforced concrete walls in stairwells and car parks.

The free-space path loss increases with 6 dB for 1800 MHz in Equation (23), a figure that can also be used for the 1900 MHz.

A modified form of the Keenan-Motley model is

$$L_p(d) = 32.5 + 20\log(f) + 20\log(d) + kF(k) + PW(k) + D(d - d_{BP}) \tag{24}$$

where L = path loss (dB); F = frequency (MHz); D = distance between the transmitter and receiver (km); K = number of floors traversed by the direct wave; F = floor attenuation factor; P = number of walls traversed by the direct wave; W = wall attenuation factor (dB); D = linear attenuation factor (dBm) beyond the breakpoint, usually about 0.2 dBm/m; and d_{BP} = indoor breakpoint (m), typically 65 m.

These models do not consider the effects due to reflection and scattering, which are accounted for in models based on ray tracing.

3.4.3 Model Tuning and Generalized Propagation Models

The Hata formula is often generalized in propagation modeling tools to capture some of the deterministic elements that are specific to the propagation area. These elements can include correction terms for diffraction losses, building and clutter data such as height and separation, and environmental clutter type. We already noticed in the previous section that the alternative model provided by Walfisch Ikegami included some of these detailed deterministic elements related to specific building clutter data.

A generalized form of the Hata statistical prediction model is usually used in the following form:

$$\begin{aligned} P_{Rx} = P_{Tx} &+ K1 + K2\log(d) + K3\log(H_{eff}) + K4D + K5\log(H_{eff})\log(d) \\ &+ K6\log(h_{meff}) + K_{clutter} \end{aligned} \tag{25}$$

where K1, K2, K3, K5, and K6 are coefficients that are also present in the original Hata formula, but are left here unspecified for further tuning for the area. But their values should not be changed much from the original values used in the Hata formula to keep the model structure reliable. The coefficient K4, which multiplies a diffraction loss D, is included to adjust diffraction losses that are caused by buildings or terrain irregularities in the line-of-site of the receiver. The diffraction losses must be estimated separately for the area using diffraction algorithms such as the Epstein Peterson and included in the model tuning process. Alternatively this may be replaced by some parameters to capture data related to building heights and separations. The effective antenna heights for the base station and mobiles, H_{eff} and h_{meff}, are calculated by considering the terrain profile and the area to be covered. The effective antenna height values can vary, for instance, on whether the entire area is to be modeled or just a hilly road with antennas placed along. For an oscillatory hilly area, the effective antenna heights are normally calculated relative to the average terrain height as referenced to the sea level.

A new variable clutter correction parameter, $K_{clutter}$, is included to adapt the equation to each morphological class. This parameter allows the same formula structure to be used for each different environment and land usage class, such as urban, suburban, open, rural, etc., by simply shifting the basic curve to fit the particular clutter type.

3.4.3.1 The Model Tuning Process

Model tuning is a highly iterative and empirical process used to obtain the value of the model coefficients to minimize the mean and RMS errors. The RMS error (Root-Mean-Square) is a statistical measure of the spread of the error around the mean value calculated by examining the difference between the predicted and measured values. This process is often done manually, and then adjustments made by computerized tools. The mean error should be driven to as close to zero as possible. The RMS error results range typically from 7.4–10.0 dB for the dense urban environment, with a value of 8 dB commonly achieved in good tuning.

Data preprocessing and filtering is an important step in using measurements to tune model coefficients,. First the signal strength measurement should be corrected for receiver antenna gains, G_{Rx}, and any receiver antenna cable or feeder losses, L_{Rx}. The path loss is then obtained by

$$L_P = EIRP - G_{Rx} + L_{Rx} - R_{xlev} \tag{26}$$

where EIRP is the equivalent isotropic radiated power of the transmit antenna, and R_{xlev} is the indicated (detected) signal strength by the receiver. Then the local mean signal levels should be obtained from the measurements by averaging out the small-scale distance variations due to multipath. This is usually achieved by averaging the measurement samples (at least 50) collected over a distance of 40λ. This can ensure an averaging accuracy of within of $+/-1$ dB as verified by tests [9].

It is important to consider that the measurements collected are not necessarily all informative for the model predictions. For example, data that appear exceptionally different from the rest and not characteristics of the clutter and morphology should be excluded and filtered out. Such data would include for instance measurements that show exceptionally weak signal strength from blocking by bridges or collected under tunnels, which would not be representative of the typical environment for which the model is intended. Data collected over bins that are much higher than the site antenna should also be cautiously treated because they are likely to come from remote interfering sites. Another category of data samples that could mislead the tuning for the larger area includes measurements collected very close to the site such as within 100–200 m. Also noise-like data that are below the receiver sensitivity levels should be filtered out and excluded from use in the model tuning.

Once the necessary preprocessing and filtering is performed on the collected measurements, the model parameters are iteratively tuned one at a time. This makes it possible to analyze the effect of each parameter and identify the trends for successive parameter changes. Relevant data should be used to tune each parameter and minimize the RMS error between the measurements and the predicted values. For instance, to tune the diffraction coefficient K4, the measurements collected in the LOS locations should be filtered out because diffraction effects are caused by NLOS propagations. One important aspect of the model tuning is the minimization of the regression error in the determination of the model path loss slope (the parameter K2). The path loss slope is estimated by a plot of the measurements collected in the radial directions from the site versus the logarithm of the distance. This parameter affects the path loss attenuation with distance and should be accurately tuned by using measurements collected in several different locations on radial directions from the site. An accurate path loss slope results in cell dimensioning with small error margins when the model is used for link budgeting. In consequence, the number of predicted cells necessary to cover an area will be more reliable. There may be several different clutter types for the area being modeled. In each iteration round, the dominant clutter is singled out and filtered. Then using only this clutter type, the model is modified to minimize the mean and RMS errors by modifying the base model parameters (K1–K6). Then, the clutter correction or offset, C_{corr}, is used as an endpoint for final adjusting of the model predictions for the environment.

For improved near and far performance, dual slope attenuation can be introduced by specifying both near-site and far-site values for K1 and K2 and the crossover point. The measurements collected in radial directions from the site should be graphed and inspected to determine where the breakpoint (d_{BP}) occurs. When a breakpoint distance has been found, the relevant subset of the plotted measurements should be used to calculate the values for the near-site and far-site coefficients K1, and K2.

3.4.3.2 Map Data Requirement

Most network planning tools contain sophisticated radio propagation prediction tools that can be readily used by the radio network planners. However, the proper utilization of these tools require the feeding of adequate digital map data that capture the essential terrain height data and morphology specifics that are critical to the analysis. Detailed morphological data and building height data are essential in modeling micro-cells in urban and dense urban areas where the planning must be more detailed. The morphology maps, which are also referred to as the clutter maps, contain information about land usage such as showing if a particular area is classified as a lake, forest, park, or an industrial area. Usually dense urban areas may contain a larger number of clutter types. An adequate representative number of clutters should be used to obtain a balance between model accuracy and the computation time. The clutter maps should also be updated and recent because area usage can change from time to time.

The map resolution chosen should achieve a good balance between the expected planning accuracy, cost involved, and computation time. The map resolution has a major impact on the simulation time and the speed requirements of the computer. However, high-density areas such as the urban and dense urban areas require higher resolution maps, in the order of 12–20 m, to result in more accurate planning. For 3G planning, accurate representation of traffic density and distribution over the area is required to obtain realistic models of the interference and hence its impact on network coverage and capacity. It is important to use vector maps that allow preparation of the traffic maps with clutter maps. Vector maps make it possible to associate specific traffic densities to specific clutters to obtain accurate local distributions of the traffic in the area.

3.4.3.3 Model Resolution Requirement

The model resolution is the smallest bin size over which RF propagation predictions and signal coverage validations can be performed. The 3G networks will require small cell sizes of the order of just a few hundred meters in dense urban areas to handle the traffic demand of the high rate data services, as well as heavy voice traffic. This calls for higher resolution radio coverage validation and prediction. However, choosing arbitrarily small resolutions for modeling RF predictions can end up consuming a large amount of computational resources and much time for planning and doing 'what if' studies. On the other hand choosing coarse resolutions introduces uncertainties about the quality of RF coverage predictions and path losses for properly estimating the interference geometries. Therefore a tradeoff is needed in the estimation of an adequate prediction model resolution. The work reported in Reference 21 concludes that the optimum model prediction resolution depends on the intended cell coverage radius in the following way:

$$\text{Optimum RF prediction resolution} = R/40 \qquad (27)$$

where R is the intended cell coverage radius. This means that for a cell of the size of 400 meters in a dense urban area, a model prediction resolution of 10 meters is required. Therefore, the area map data should also be obtained with this level of resolution. The ray tracing models are capable of providing RF path loss predictions within just a few meters of resolution.

3.5 Far-Reach Propagation Through Ducting

Radio propagation can be enhanced by tropospheric ducting phenomena as the atmosphere warms and cools throughout the day. On clear cloudless days with little or no wind to stir up the atmosphere, the upper air cools down at sunset as does the earth surface temperature but at a different rate. The varying cooling rate of air at different altitudes sets up temperature inversions. This results in an abrupt change in the refractive index at the interface between the air layers at different temperature. The refractive index changes from a smaller value in the cool layer to a larger value in the warm layer. By analogy with the optical Snell's Law, this causes significant reflections of radio waves back towards the earth's surface where they are further reflected, thus causing a ducting effect [22]. Ducting is a wave guiding effect formed by multiple reflections between layers of air with different refractive indexes. This helps to prevent large-scale signal dispersions resulting in the propagation of the signal over large distances with significant power. During tropospheric ducting, the signal is trapped inside of an atmospheric signal duct (*waveguide*) that bounces the signal through the atmosphere to locations past 60 miles. The end result is that radio waves will propagate well beyond their intended service area with less than normal attenuation. This effect has been observed most often in the frequency ranges from 450 MHz down to 100 MHz (TV signals), but it can generally happen in the UHF frequency ranges up to 3GHz. In fact the author has observed the ducting effect at 900 MHz through his work on GSM in certain parts of Asia, where signals were received in the −80 dBm ranges over NLOS distances from high altitude GSM sites located 20–30 km away in rural areas. Under normal conditions, the signal that is not blocked or obstructed simply travels in a straight line out into space, never to return to earth. However, the ducting effect causes the normal path of the signals to be bent downward, returning the signal to the surface of the earth and at great distances from its point of origin with unexpected power levels. The reader may refer to References 23 and 24 for terrestrial signal strength measurements in the 30 MHz–3 GHz range and to Reference 25 for measurements of tropospheric ducting effects on radio paths over the sea at 2 GHz. Ducting and the anomalous radio propagation effect can occur at most latitudes but they are more common in tropical climates usually associated with high pressure areas (anticyclones).

The ducting effect is something that is exploited by amateur radio enthusiasts to achieve communications over abnormally long distances for the frequencies used. This effect can not be exploited reliably for commercial communication services because the conditions can form and disperse in minutes. Furthermore, it can cause interference well outside of the normal service area. Ducting can result in troublesome communication on GSM networks by enhancing co-channel and adjacent channel interferences from far sites, particularly ones that are at elevated heights and with little down-tilting to provide large coverage in rural areas. Therefore a simple solution around ducting is to increase the number of sites by reducing antenna heights and providing sufficient down-tilting to avoid bouncing the signal into the atmosphere. The new handset technology based on interference rejection cancellation can also help to mitigate the effects of ducting, which occurs mostly on the downlink when it does occur.

References

[1] Sklar, B., 'Rayleigh fading channels in mobile digital communication systems part I: Characterization', *IEEE Communications Magazine*, **35**(7), July 1997, pp. 90–100.
[2] Noneaker, D.L. and Pursley, M.B., 'Rake Reception for a CDMA Mobile Communication System with Multipath Fading', in Glistic, S.G. and Leppanen, P.A., Code Division Multiple Access Communications, Kluwer Academic, Dordrecht, 1995, pp. 183–201.
[3] Proakis, J.G., *Digital Communications*, 3rd Edition. McGraw-Hill, New York, 1995.
[4] Rappaport, T., *Wireless Communications*, Prentice Hall, Upper Saddle River, NJ, 1996.
[5] Jakes, W.C., Ed., *Microwave Mobile Communications*, John Wiley & Sons, Inc, New York, 1974.
[6] Pahlavan, K. and Levesque, A.H., *Wireless Information Networks*, John Wiley & Sons, Inc, New York, 1995, Chapters 3 and 4.
[7] Greenwood, D. and Hanzo, L., 'Characterisation of Mobile Radio Channels', in Steele, R., Ed., *Mobile Radio Communications*, Pentech Press, London, 1994, Chapter 2.
[8] Parsons, D., *The Mobile Radio Propagation Channels*, John Wiley & Sons, Ltd, Chichester, 1996.
[9] Lee, W.Y.C., *Mobile Cellular Communications*, McGraw-Hill, New York, 1989.

[10] Clarke, R.H., 'A Statistical Theory of Mobile Radio Reception', *Bell Sys. Tech. J.*, **47**(6), July–August 1968, 957–1000.

[11] Amoroso, F., 'Use of DS/SS Signaling to Mitigate Rayleigh Fading in a Dense Scatterer Environment', *IEEE Pers. Commun.*, **3**(2), April 1996, 52–61.

[12] GSM 03.30 version 8.3.0, 'Radio Network Planning Aspects', ETSI, 1999.

[13] ITU-RM. 1225, 1997, 'Guidelines for Evaluation of Radio Transmission Technologies for IMT-2000', 1997.

[14] Lee, W.C.Y., 'Elements of Cellular Mobile Radio Systems', *IEEE Trans. Vehic. Tech.*, **VT-35**(2), May 1986, 48–56.

[15] Hata, M., 'Empirical Formulae for Propagation Loss in Land Mobile Radio Services', *IEEE Trans. Vehic. Tech.*, **VT-29**(3), 1980, 317–25.

[16] COST 231, (Damasso, E. and Correia, L.M., Eds), 'Digital Mobile Radio Towards Future Generation Systems', Final Report, COST Telecom Secretariat, Brussels, Belgium, 1999.

[17] Walfisch, J. and Bertoni, H.L., 'A Theoretical Model of UHF Propagation in Urban Environment', *IEEE Transactions on Antennas and Propagation*, **36**(12), December 1988, 1788–1796.

[18] Kiirner, T., Cichon, D.J. and Eicsback, W., 'Concepts and Results for 3D Digital Terrain Based Wave Propagation: An Overview', *IEEE Journal on Selected Areas in Communications*, **11**(7), September 1993, 1002–1012.

[19] Mckown, J. and Hamilton, R., 'Ray Tracing as a Design Tool for Radio Networks', *IEEE Networks Magazine*, **5**(6), November 1991, 27–30.

[20] Rossi, J-P. and Gabillet, Y., 'A Mixed Ray Launching/tracing Method for Full 3-D UHF Propagation Modeling and Comparison with Wide-band Measurements', *IEEE Transaction on Antennas and Propagation*, **50**(4), April 2002, 617–523.

[21] Bernardin, P. and Manoj, K., 'The Post-processing Resolution Required for Accurate RF Coverage Validation and Prediction', *IEEE Trans. Vehic. Tech.*, **49**(5), September 2000, 1516–1521.

[22] Hitney, H.V., Richter, J.H., Pappert, R.A., Anderson, K.D. and Baumgartner, G.B. Jr., 'Tropospheric Radio Propagation Assessment', *Proceedings of the IEEE*, **73**(2), 1985, 265–283.

[23] ITU-R Recommendation P.1546, 'Method for Point-to-area Predictions for Terrestrial Services in the Frequency Range 30 MHz to 3000 MHz', International Telecommunication Union, 2003.

[24] ITU-R Recommendation P.452, 'Prediction Procedure for the Evaluation of Microwave Interference between Stations on the Surface of the Earth at Frequencies above about 0.7 GHz', International Telecommunication Union.

[25] Gunashekar, S.D., Siddle, D.R. and Warrington, E.M., 'Trans-horizon UHF Radio Wave Propagation on Over-the sea Paths in the British Channel Islands', Radio Systems Research Group, Department of Engineering, University of Leicester, Leicester LE1 7RH, United Kingdom. http://www.ursi.org/Proceedings/ProcGA05/pdf/F01P.10(0086).pdf.

4

Formulation and Analysis of the Coverage-capacity and Multi-user Interference Parameters in UMTS

It is well known that capacity and coverage are interlinked through interference in CDMA networks. This chapter derives and presents the detailed analysis and formulations that will illustrate the tradeoffs between coverage, capacity, and base station power as influenced by the multi-user interference. For the uplink, the coverage-capacity formula will be derived for a homogeneous service environment, to illustrate the interaction between coverage and capacity. On the downlink side, which is the most complicated link in 3G radio modeling, the generic formulations for a mixed service environment illustrating the driving factors behind radio capacity in UMTS are derived and presented. The mathematical analysis and formulations given here are used to make numerous conclusions regarding the implications and hints for network planning and optimization. It is therefore intended to provide the mathematical basis and the guidelines for the network planning process, which we will specifically cover in Chapter 5.

4.1 The Multi-user Interference

Interference is inherent to 3G networks, which are based on CDMA technology. CDMA separates each user's voice or data information by multiplying it by pseudo random bit sequences, which are mutually orthogonal to each other. The resulting bit streams are referred to as chips and have a rate of 3.84 Mcps, which spreads the narrow band information bits of the user across a much wider bandwidth of approximately 5 MHz in UMTS. Theoretically, there should be no mutual interference among the users communicating simultaneously with the network due to code orthogonality. However, due to the fact that users on the uplink communicate asynchronously with the base station (their signals are received at the base station asynchronously), code orthogonality is lost at the base station on the uplink, resulting in some mutual interference between the users.

On the downlink side, it is the multipath environment that causes multiple delayed versions of the users' signals to arrive at each mobile station, effectively resulting in some loss of the code orthogonality among the different users' signals as received at the mobile [1–2]. As UMTS uses a frequency re-use of 1, every cell in the network operates on the same frequency band for a given carrier, making each site interfere with other neighboring sites. The loss of code orthogonality on the downlink in a given cell due to multipath is usually quantified through an *orthogonality factor*. The orthogonality factor (expressed in the range 0 to 1) represents a measure of residue code orthogonality left in the presence of multipath, and when subtracted from 1 represents the fractional effective intracell interference.

The multi-user interference will influence both the overage and capacity in UMTS (or any CDMA based network). As a result, the cell size in UMTS is not fixed. As the number of users communicating simultaneously in a cell rises, so will the multi-user interference, which then deteriorates the SIR (the signal to interference ratio) of the communication links. This will then affect the capability of the users to communicate with the base station, resulting in reduced coverage [1, 3]. It is therefore critical, particularly for green field operators, to plan site locations and site densities based on a reasonably accurate forecast of the traffic and services for at least the near term future. That would help to save much time and cost in re-planning and adjustments as the traffic grows.

The following notations are defined for convenience and used throughout the analysis.

- E_b: signal energy per information bit as arrived at the receiver
- N_0: effective thermal noise power spectral density
- I_0: multi-user interference power spectral density at the receiver
- R_b: service information bit rate (user's bit rate)
- W: UMTS spreading bandwidth (i.e., the chip rate of 3.84 Mcps in UMTS)
- G: processing gain (i.e., W/R_b)
- ρ: required effective signal to noise (+ interference) ratio at the BTS receiver input (for a homogeneous service environment)
- M: number of connections in the cell (uplink or downlink as is the case)
- P_s: signal power as received at the receiver
- I_{hc}: multi-user interference power due to own cell users
- I_{oc}: multi-user interference power arising from users in other cells
- f_{UL}: ratio between other cells to own cell multi-user interference (i.e., I_{oc}/I_{hc}) on uplink (at Node B)
- f_{DL}: effective cell's averaged 'other-cells-to-own cell multi-user received power' on the downlink
- g_{mk}: net radio downlink channel gain and loss (net channel effect) on the connection k in cell m
- υ_k: service activity ratio on connection k (for homogeneous service environment, $\upsilon_k = \upsilon$, $k = 1, 2, \ldots$)
- $\alpha_k =$: orthogonality factor on connection k in the downlink
- P_m: power transmitted from the base station to user m for communication with that user. This will depend on the user E_b/N_0 requirements, and the net channel gain g_{mk} between the BTS and the user
- γ_{cmm}: percentage BTS power allocated to the pilot and common control channels
- ρ_k: required signal to noise (+ interference) ratio at the mobile on connection k

- P_N: effective thermal noise power at the receiver side (includes receiver's noise figure effect)
- $P^{(n)}{}_T$: total transmit power required from base station n to support the users and the common control and signaling channels in the cell (this may be set to a value lower than the full BTS transmit power capability by the operator or the RRM algorithm)
- f_k: ratio of interference from other BTSs (other cells) to own BTS at user location k within the cell

4.2 Interference Representation

There are a few quantities that are defined to characterize and measure the multi-user interference in 3G networks. There are defined in the following sections.

4.2.1 Noise Rise

The interference contributions from users within and outside a given cell result in a 'noise rise' relative to the total thermal noise at the base station. This effect is quantified by what is called the noise raise ratio, defined as the ratio of the multi-user interference plus the effective thermal noise power to the effective thermal noise power at the receiver:

$$\text{Noise Rise} = (I_{hc} + I_{oc} + P_N)/P_N \text{ in linear scale} \quad (1)$$

or, in dB scale,

$$\text{Noise rise} = 10 \times \log[(I_{hc} + I_{oc} + P_N)/P_N], \text{ dBs} \quad (2)$$

in which I_{hc} is the multi-user interference power due to own cell users, I_{oc} is the multi-user interference power arising from users in other surrounding cells, and P_N represents the effective thermal noise power at the receiver (background noise corrected by the receiver noise figure) within the channel band.

The noise rise represents the ratio of the total interference power from noise and multi-user communication to the effective thermal noise power. If there is no multi-user interference, this ratio becomes 1 in linear scale, and therefore 0 dB in the decibel units, indicating no noise rise above the effective thermal noise floor.

4.2.2 Load Factor

The load factor is defined as the ratio of the multi-user interference power to the multi-user interference plus noise power at the receiver (MS or BTS) given by

$$\text{Load factor (denoted as } \eta) : \eta = (I_{hc} + I_{oc})/(I_{hc} + I_{oc} + P_N) \text{ in linear scale} \quad (3)$$

The load factor represents the ratio between the interference (from multi-user effects) to the total interference plus the thermal noise power, and is therefore always less than one and is positive (as expressed in linear scale). If there is no multi-user interference, this ratio becomes 0, meaning no load due to interference is placed on the receiver. If the multi-user interference is much larger than the effective thermal noise power, this ratio will approach to 1 as seen from Equation (3), meaning 100% loading due to interference (the multi-user interference far exceeding the thermal background noise).

By comparing Equations (1) and (3), the following relation between the noise rise and load factor is derived:

$$\text{Noise rise} = \frac{1}{1-\eta} \qquad (4)$$

which shows that the noise rise grows unboundedly as the load factor approaches 1.

4.2.3 Geometric Factor

The multi-user interference experienced by a user on the downlink is a mixture of the interferences from the own cell and from surrounding cells. The make-up of this interference will depend on user location, radio environment, and neighboring cell layout and design. The geometric factor is used to quantify this geometry of interference on downlinks by the ratio between the own cell interference to other cell interference plus the thermal noise power expressed as

$$G_e = \frac{I_{hc}}{I_{oc} + P_N} \qquad (5)$$

If the network is interference is large, that is if $I_{oc} \gg P_n$, then $G_e = I_{hc}/I_{oc}$. The geometric factor depends on the distance of the mobile station from the Node B antenna. A typical range is from -3 dB to 20 dB where -3 dB is typically for the cell edge [4].

4.2.4 The f Factor

The ratio between the own cell interference power and the total multi-user interference power is usually referred to as the f factor given by

$$f = \frac{I_{hc}}{I_{hc} + I_{oc}} \qquad (6)$$

This ratio is between 0 and 1 (in linear scale) and attains its maximum value of 1 when the multi-user interference arriving from other cells drops to 0. Therefore, it can be used as a measure of spectral efficiency or virtual 'frequency re-use' in CDMA networks.

Of all the above parameters the load factor (from which the noise rise can also be derived, as was shown) is the most commonly used parameter. This parameter is used in the link budging process for 3G networks.

4.3 Dynamics of the Uplink Capacity

This section analyzes the effect of multi-user interference on uplink capacity, and derives what is referred to as the 'pole capacity' for the uplink. The conceptual formula derived can be used to study the tradeoffs between capacity and coverage on the uplink.

We will assume perfect power control, implying that all users signals will be received at the base station (i.e., Node B) at the same power level, P_s.

Power control on the uplink is expected to ensure that all user signals are received at the required signal to noise ratio, ρ. Therefore

$$E_b/(I_0 + N_0) = \rho \qquad (7)$$

where I_0 and N_0 denote the power spectral densities for the multi-user interference and the effective thermal noise at the receiver, respectively, and E_b represents the received energy per information bit.

Therefore, defining W as the effective noise equivalent channel bandwidth (that is, the WCDMA chip rate of 3.84 Mcps or MHz), we can express I_0 as

$$\begin{aligned} I_0 &= (I_{hc} + I_{oc})/W \\ &= (1+F)I_{hc}/W, \text{(using the definition for the f factor)} \end{aligned} \qquad (8)$$

From the definition of I_{hc}, we can write

$$I_{hc} = v(M-1)P_s \qquad (9)$$

where v denotes the service activity ratio (in the case of conversational voice, this is about 50%), M represents the number of simultaneous users using the service in the cell, and P_s is the signal power at the receiver input.

But the energy per bit, E_b, can be expressed as

$$E_b = P_s/R_b \qquad (10)$$

where R_b is the service information bit rate. Substituting Equations (8) through (10) into Equation (7), using the definitions for the processing gain G ($G = W/R_b$), and expressing the noise power in the band, P_N, as $P_N = N_0.W$), we obtain:

$$\rho = \frac{G}{(M-1)(1+f_{Ul})v + P_N/P_s} \qquad (11)$$

Solving for M, gives

$$M = 1 + \frac{G/\rho - P_N/P_s}{v.(1+f_{UL})} \qquad (12)$$

And from the definition for the load factor as given by Equation (3), we obtain:

$$\text{Load factor } \eta = \frac{(1+f_{UL})(M-1)v.P_s}{(1+f_{UL})(M-1)v.P_s + p_N} \qquad (13)$$

We can see from Equations (12) and (13) that both the number of active users M in the cell, as well as the load factor, will increase as the received power P_s is increased. This may be

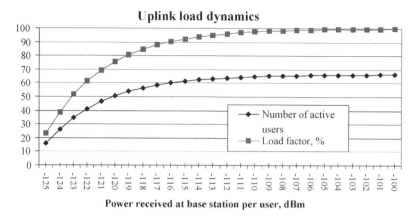

Figure 4.1 Illustrating the dynamics of the uplink capacity for a sample case.

seen more tangibly in Figure 4.1, which shows the trend for a sample case where we have set $\rho = 7\,\text{dB}$, $R_b = 12.2\,\text{kbps}$, $\nu = 0.6$, $f_{UL} = 0.6$, and $W = 3.84\,\text{Mcps}$ (as in UMTS).

Figure 4.1 shows that as the power is increased, the number of active users approaches a limiting value of about 66 users in this example, while the load factor approaches 1 (100%).

Equation (12), the capacity equation, shows that generally as the power P_s is increased, the number of active users increases but stays limited to an upper bound obtained by setting P_s to infinity (practically to a value that would make the interference terms in the equations much larger compared to the thermal noise power), which is then:

$$M_{\text{pole}} = 1 + \frac{G}{\rho.\nu(1+f_{UL})} \qquad (14)$$

This capacity limit is called the pole capacity on the uplink side [4]. Likewise Equation (13), the load equation, shows that the load factor will approach its limit of 1 (100%) as the pole capacity is reached (i.e., as the required power becomes higher and higher). Once the pole capacity limit is reached, further increase of mobile station power will not help to accommodate further users. And before this limit is reached, the power required to maintain the link budget, particularly for users at the cell fringe (who will require to transmit more power to overcome the normally higher path loss to the base station) may have already exceeded the mobile station transmit power limit depending on the load in the cell. This implies a shrinking of the cell coverage range.

In practice, when the load factor approaches to even a value of 60%, coverage is lost as the users at the cell fringe who will require the largest power (to overcome the normally larger path loss) will not be able to maintain the link budget, and hence lose coverage. Therefore, the effective range of the base station or the cell size will depend on the number of active users in the cell. This interaction between the cell size and the cell capacity on the uplink is called cell breathing in CDMA networks [3]. Thus coverage in UMTS is normally uplink limited.

Since higher bit rate services will require a higher signal to noise ratio, and hence a higher transmit power, they would be the first to suffer as higher loads are placed on the cell. This means that we can expect to have different coverage ranges for different services depending on their bit rates, and also the traffic distribution in the cell. This must be considered in the network planning process by examining the coverage required for each service [4]. One way to increase the overall system capacity at the desired coverage level is to introduce additional carriers, particularly for accommodating the high bit rate services. The use of multi-carriers will help to reduce the uplink load on each carrier, and hence extend the service coverage ranges. This is discussed in more detail in Chapter 6.

For practical design purposes, the network planning should aim for a design to limit the load factor (on each carrier) to no more than about 55–65%. The RRM algorithms within the RNC dealing with admission and load (interference) control should also be set to ensure this limit.

4.4 Downlink Power-capacity Interaction

The analysis of capacity and interference on the downlink side is significantly more complicated due to the fact that the relevant coverage and interference parameters would vary depending on the location of the mobile within the cell. However, from a transmit power requirement for an individual service at a general location within the cell, an overall power-capacity equation can be derived based on a cell 'averaged' effective overall parameter representation as is shown in the following section.

4.4.1 The General Power-capacity Formula on Downlink

This section derives the general power-capacity formulation for a mixed service environment on the downlink side. To start the derivation process, it is considered that signals aimed at other users in the cell are also received at each mobile station in the cell. Theoretically, however, these other users' signals are not supposed to interfere with the desired user due to the fact that they are all transmitted synchronously by the base station and should therefore maintain code orthogonality. However, the path delay spreads introduced by the multipath radio propagation causes some loss of the orthogonality, and hence mutual interference among users. The measure of orthogonality left in the signals will be modeled by a parameter defined as α, which will be between 0 and 1, with the value of 1 representing complete orthogonality. The value of α will depend on the radio propagation environment in the area and typically is in the range of 0.4–0.9 [4].

Power control on the downlink is expected to ensure that the signal to noise (+ interference) ratio received at mobile m, on the link with the base station, meets the required value, ρ_k, on the connection k to the base station.

$$E_b/N_T = \rho_\kappa \qquad (15)$$

where N_T represents the total thermal noise + multi-user interference power spectral density at the location k in cell m, and ρ_k is the required signal to noise (+ interference) ratio on connection k.

But, from the definition for N_T given above, this is equal to the power spectral density of the thermal noise (i.e., P_N/W) + the power spectral density of the in-cell interference and the contributions from the other cells received by the desired mobile station. It is therefore easy to see that it can be approximately expressed by

$$N_T = \frac{1}{W}\left[P_T^{(m)} g_{mk}(1-\alpha_k) + \sum_{n=1,n\neq m}^{N} P_T^{(n)} g_{nk} + P_N\right]$$

in which the first term represents the in-cell multi-user interference, and the second term (the summation) represents the multi-user interference from other base stations. $P^{(m)}T$ is the total power transmitted by base station m on all channels. The parameters g_{mk} represent the path gain (inverse of path loss) for the radio link between base station m and the mobile k, and N is the number of interfering surrounding base stations.

Now, E_b, the received energy per bit is calculated from

$$E_b = P_k R_k g_{mk}$$

where P_k is the power transmitted on the connection to mobile k, and R_k is the service bit rate. Substituting the above expressions for N_T, and E_b into Equation (15), and solving for P_k, gives

$$P_k = \frac{\rho_k R_k}{W}\left[P_T^{(m)} g_{mk}(1-\alpha_k) + \sum_{n=1,n\neq m}^{N} P_T^{(n)} g_{nk} + \frac{P_N}{g_{mk}}\right] \qquad (16)$$

According to the definition for the other-cells-to-own-cell interference ratio at location k in the cell, f_k, we have

$$f_k = \frac{\sum_{n=1,n\neq m}^{N} P_T^{(n)} g_{nk}}{P_T^m \cdot g_{mk}} \qquad (17)$$

Solving for $\sum_{n=1,n\neq m}^{N} P_T^{(n)} g_{nk}$, and substituting into Equation (16), gives

$$P_k = \frac{\rho_k R_k}{W}\left[P_T^{(m)} g_{mk}(1-\alpha_k) + f_k \cdot P_T^{(m)} + \frac{P_N}{g_{mk}}\right] \qquad (18)$$

as the transmit power required to support user k at location k in the cell (cell m).

Now, from a transmit power balance at BTS m, we have

$$\sum_{k=1}^{M} P_k \upsilon_k = (1-\gamma_{cmm}) P_T^{(m)} \qquad (19)$$

where γ_{cmm} is the percent BTS power allocated to the pilot and common control channels. Substituting for P_k from Equation (18), and solving for $P_T^{(m)}$, gives

$$P_T^{(m)} = \frac{P_N \sum_{k=1}^{M} \frac{\rho_k R_k \upsilon_k}{W g_{mk}}}{1 - \left\{\gamma_{cmm} + \sum_{k=1}^{M} \frac{\rho_k R_k \upsilon_k}{W}[(1-\alpha_k)+f_k]\right\}} \qquad (20)$$

as the total transmit power required from BTS m to support the M connections in the cell. As can be seen from the above formula, the transmit power required will depend on the distribution of the users in the cell, the RF propagation environment (as determined by the parameters g_{km} and α_k), as well as the inter-cell interference geometry and the cell plan as determined by the parameters f_i. This complexity between the service capacity and required power on the downlink side is what leads to the need for recourse to Monte Carlo simulations in studying the various power-capacity tradeoffs and services support capabilities in 3G networks [5–7].

Notice that the degree of the power required will increase as the channel gain factor g_{mk} becomes smaller, which happens most normally as the user moves away from the site and gets closer to the cell edge coverage areas. The technologies to increase the channel gain include providing spatial diversity at the user mobile station, which is not always practical, or transmit diversity at the BTS, which is practical. Another means of helping to offset the effect of reduced channel gains on the downlink, such as at the cell edge areas, is to reduce the receivers noise figure, which then lowers the required signal to noise ratio requirement for the same service quality.

4.4.2 Downlink Effective Load Factor and Pole Capacity

The form of Equation (20) allows us to define a 'cell effective overall load factor', η_{DL}, for the M service connections as

$$\eta_{DL} = \gamma_{cmm} + \sum_{k=1}^{M} \frac{\rho_k R_k \upsilon_k}{W}[(1 - \alpha_k) + f_k] \qquad (21)$$

Substituting into Equation (20), gives

$$P_T^{(m)} = \frac{P_N \cdot \sum_{k=1}^{M} \frac{\rho_k R_k \upsilon_k}{W \cdot g_{mk}}}{1 - \eta_{DL}} \qquad (22)$$

Since the power is always positive (in linear scale), and the numerator of Equation (22) is always a positive quantity (all its constituents are positive), the denominator in Equation (22) must always stay positive. This means that we must have $\eta_{DL} \leq 1$. Now, when the 'load factor' approaches 1, Equation (22) clearly indicates that the required transmit power from the BTS will grow larger and larger. When this happens, the cell is said to have reached its pole capacity, which will be finite. Therefore, the downlink capacity limit is achieved for an integer value of M (for a given set of the geometric and radio environmental parameters associated with the user distributions in the cell, that is α, f_k, etc.) that results in the largest value for η_{DL}, below 1. Beyond that, any increases in the base station power will not help to support more users. At this capacity limit, more base station power to support more users will simply increase the downlink noise floors to a limit that cannot achieve the required signal to noise ratio (a positive feedback effect). Therefore, any more available base station power should be directed to a second carrier to get the capacity benefits.

This phenomenon is further illustrated in Figure 4.2, which plots the required base station power P_T (in dBm) from Equation (6) and the effective 'load factor' η_{DL} from Equation (21)

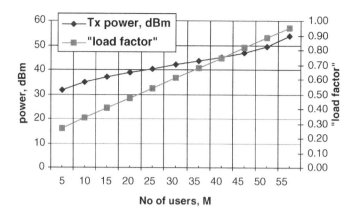

Figure 4.2 The required base station power, and the 'load factor', versus the number of users, M (parameters used are: f = 0.7, α = 0.6, $\gamma_{cmm} = 0.20$, $\upsilon = 0.65$, and an average path loss ($1/G_{mk}$) of 145 dB).

versus the number of users M, for a special case of using AMR 12.2 kbps voice (with an activity factor υ of 0.5), assuming a signal to noise ratio requirement of 7 dB. The data plotted in Figure 4.2 assume average values of 0.6 and 0.7 for the orthogonality factor α, and the inter-cell interference geometric factor f, respectively. The effective thermal noise floor power assumed for the mobile receiver is assumed to be -103 dBm. The plot shown for this special case shows that the required power approaches a value of 53.8 dBm as the number of users M attains its maximum value right before the 'load factor' exceeds its allowed limit value of 1. Any more base station power beyond this limit should be directed to a second carrier to achieve additional capacity.

As seen from Equation (21), this limit capacity on the downlink will depend on the distribution of the users in the cell, the RF propagation characteristics for the area (as determined by the parameters g_{mk} and α_k), as well as the inter-cell interference geometry and cell plan, which are reflected in the parameters f_k.

When capacity in a cell becomes downlink limited, sharing the base station power across multi-carriers can increase the system capacity. The sharing of the base station across multi-carriers reduces the interference load on each carrier, and thus helps to reduce the power required to achieve certain capacity (i.e., increases the limit pole capacity on each carrier).

4.4.3 Single Service Case and Generalization to Multi-service Classes

The downlink power-capacity equation, Equation (20), can be simplified for the case with all the M connections in the cell using the same service with the same bit rate and service quality (i.e., $R_k = R$, $\rho_k = \rho$, $\upsilon_k = \upsilon$, $k = 1, 2, \ldots M$). Then,

$$P_T^{(m)} = \frac{P_N \rho.\upsilon.R.M/(W.g_m)}{1 - \left\{\gamma_{cmm} + \frac{\rho R \upsilon}{W} M[(1-\alpha) + f_{DL}]\right\}} \quad (23)$$

in which α and f_{DL} are defined as

$$\alpha = \frac{1}{M}\sum_{k=1}^{M}\alpha_k \qquad (24)$$

$$\frac{1}{g_m} = \frac{1}{M}\sum_{k=1}^{M}\frac{1}{g_{mk}} \qquad (25)$$

$$f_{DL} = \frac{1}{M}\sum_{k=1}^{M}f_k \qquad (26)$$

and we can think of them as representing the effective overall orthogonality factor, the overall channel gain, and the other-cells-to-own-cell interference factor for the given distribution of the M users in cell m, respectively. Notice that, the cell capacity is also affected by the interference from neighboring base stations as determined by the parameter f_{DL}. This is referred to as the soft capacity effect. This means that it is important to try to minimize mutual interference between the sites in order to maximize the overall capacity of the network.

For the general practical case of dimensioning for a service mix distribution, the terms of the power-capacity equation, Equation (20), can be grouped according to different user profiles: that is, separating different services according to bit rate, activity ratio, E_b/N_0 (i.e., ρ) requirements, and mobility (location distribution) within the cell. Then, by analogy to Equations (20) and (23), it is easy to see that the power-capacity equation for M_J service-mobility classes in cell m takes the form [8]:

$$P_T^m = \frac{P_N \sum_{j=1}^{M_J}\frac{\rho^{(j)}v^{(j)}R^{(j)}M^{(j)}}{W \cdot g^{(j)}}}{1 - \left\{\gamma_{cmm} + \sum_{j=1}^{M_J}\frac{\rho^{(j)}v^{(j)}R^{(j)}M^{(j)}}{W}[1-\alpha^{(j)}+f_{DL}^{(j)}]\right\}} \qquad (27)$$

in which M_J is the number of service classes, and $M^{(j)}$ denotes the number of connections in each service class j, $j = 1, 2, \ldots, M_J$.

4.4.4 Implications of Downlink Power-capacity Analysis

The analysis given in the above sections for the downlink power-capacity interactions shows that capacity on the downlink is limited normally by the total base station transmit power, and the loading placed on the cell by the specifics of the traffic distribution, service power requirements (SNR), radio environment, etc. Since the power required for each service will generally depend on its bit rate (the higher the bit rate, the higher the power required), how far it is situated from the site, and the radio propagation environment in the area, the cell capacity for a given base station transmit power and load factor will depend on the service mix, the traffic distribution over the area, and the cell morphology (the g_m parameter). Thus cells in UMTS will not have a fixed capacity contrary to the case in GSM. As the traffic distribution changes in the cell and its neighborhood, the cell traffic supporting capacity will change. This must be considered in the process of network planning by carefully examining the services mix and distribution.

For practical design purposes, the network planning should aim for a design to limit the load factor to no more than about 75–85%. The downlink can tolerate higher loading due to the fact that there is some code orthogonality left in the signals (as represented by the parameter α).

4.5 Capacity Improvement Techniques

The analysis in the previous sections shows that it is the multi-user interference, coupled with multipath effects, that is the driving force in the determination of coverage and capacity in CDMA systems. The multipath not only results in fading of the received signals, but also results effectively in partial loss of code orthogonality, thus leading to multi-use interference on the downlink as well. Therefore mechanisms to improve capacity and coverage will have to deal with reducing the multi-user interference, and also either mitigating the adverse effects of multipath or turning it into beneficial use. The Rake receivers [2] within both the mobile and base station receivers are what put the scattering of a given user's signal from multipath to beneficial use. The chip duration at 3.64 Mchips/s in WCDMA is at 0.26 μs. The Rake receiver can separate those multipath components that are at least 0.26 μs apart and combine them coherently to obtain multipath diversity gains. This helps to increase the effective signal to noise ratio seen by the receiver.

The loading from multi-user interference can be reduced by cell sectorisation. Base station sectorisation helps to confine the other user's signal to a small spatial area, and hence protect interference to users in other areas of the cell [9]. By sectorising a cell to three and six sectors, capacity gains of almost 3–6 fold can be achieved, respectively, considering the frequency re-use of 1 used in WCDMA.

The multi-user interference can be cancelled by various mechanisms, such as the basic well known equalization techniques [10], as well as the CDMA specific multi-user detection technique. Since the base station knows about the codes assigned to all users in the cell, it can use the knowledge to estimate and cancel the interference placed by other users on a given user's signal. This technique of interference estimation and cancellation has been referred to as multi-user detection in CDMA systems and has been researched extensively by researchers (see for instance References 11–13). It does, however, require extensive processing and is more practical for implementation on the uplink (within the base station). Vendors have just started implementing the multi-user base station receivers.

Signal fading at the base station receiver due to multipath fading can be compensated for by receiver spatial diversity mechanisms. Receiver spatial diversity can be implemented through antenna diversity or through using x-polarized antennas. For instance, a 2-branch receiver diversity can help to increase the receiver sensitivity by about 3 dB. The cross polarized antennas are effective in more dense urban areas where the existing of multitude reflecting obstacles can result in dual polarization of the received signals.

On the downlink side, because it is not feasible to use more than one antenna in small inexpensive mobile stations, the WCDMA standard supports the use of base station transmit diversity. With downlink transmit diversity, the signal is transmitted via two base station antennas, where a receive diversity antenna can be used as the second transmitter as well.

Both the receive and transmit diversity provide gain against fast fading, where this gain is larger when there is less multipath diversity [14] such as in low mobility environments. Transmit diversity helps to keep the codes orthogonal on the downlink in flat fading channels and hence results in reduced multi-user interference. Transmit diversity can be implemented in either closed or open loop mode. In the closed loop mode, feedback from the mobile station is used to adjust the transmit phases at the base stations and make the signals combine coherently at the mobile.

The use of additional carriers, as stated before, is also a simple and effective way to help reduce the cell loading from multi-user interference, and increases the overall system capacity. The use of multi-carriers divides the users among more than one carrier, which helps to reduce the mutual interference between users, and hence increases the coverage range for each carrier and the total system capacity.

Finally, capacity improvement far beyond what has been discussed above is to be achieved through the evolving future generation of wireless networking technology based on MIMO (multiple input, multiple output) systems. In the MIMO systems, each user's data are split into N multiple streams that are transmitted and received through M antennas. Then the spatial correlation (signature) induced by the rich multipath environment along with space-time coding is exploited to separate the different user data streams at the receiver. In this way, MIMO usefully exploits the multipath effects rather than mitigating them. Thus the simultaneous transmission and reception of multiple data streams over the same channel bandwidth helps to achieve very high capacity systems. The reader interested in this evolving technology can consult References 15–18.

4.6 Remarks in Conclusion

Derivation of the capacity equations for both the uplink and the downlink in UMTS cells has shown that capacity and coverage are tightly coupled through the multi-user interference in the cell. The multi-user interference can be represented through the load factors defined on both the uplink and downlink, and will be used for link budgeting purposes against interference. The load factor presents the overall interference, keeping in mind that a different traffic mix (different rates and E_b/N_0 requirements) and its distributions in the cell can result in the same values for the load factors. This is to still not mention the influence of the radio propagation environment in this area. Thus, designing to a certain load factor would not mean designing to a specific coverage or capacity. For the same load factor setting, a different scenario of the traffic mix and distribution may be supported. Thus coverage and capacity will not be static as in GSM, because they will be dependent on the service mix and area distribution. As services and their requirements change, so will the coverage provided to each and the capacity. These must be considered carefully in the network planning process.

The conclusions and implications drawn from this study are summarized below and these should be carefully noted and used as the guiding rules when planning and optimizing the UMTS networks.

1. Coverage and capacity are intertwined in UMTS through the multi-user interference effect. As one is stretched, the other shrinks (i.e., the so called cell breathing phenomenon).

2. Services with different bit rates are expected to have different uplink coverage range from the base stations (higher bit rates will have smaller coverage range particularly in a highly loaded cell).
3. Cell coverage range will depend on the service mix, the traffic location distribution, and the radio propagation environment in the area.
4. Cell performance is affected by interference arising not only from the users communicating from the cell itself, but also from its neighboring cells.
5. Cell capacity is increased if neighboring cells are lightly loaded or induce less interference to the cell. This leads to the concept of borrowed capacity from neighboring cells.
6. Cell capacity on the downlink is limited by the total available base station transmit power and the downlink load factor.
7. Cell loading will vary according to the service mix and volume, traffic distribution over the area, the path loss geometry, and interference arising from neighboring cells.
8. Cell performance on the downlink is affected by interference arising not only from the own base station communicating with other users in the cell, but also from base stations in communication with users in neighboring cells.
9. High sites (sites overlooking remote area), as potential sources of interference and performance degradation, should be carefully checked for generating undesired remote fragmented coverage.
10. Multipath can help through space diversity achieved in the Rake receiver, but can also hurt by causing loss of code orthogonality on the downlink side. Transmit diversity can help to prevent code orthogonality loss.
11. Multi-carriers can be used to reduce interference and increase system capacity and coverage.
12. Use of multi-user detection in the base stations can help to reduce the uplink interference and increase coverage and capacity on the uplink fading.
13. Both receiver and transmitter diversity can help to mitigate fast fading effects.
14. The concept of 'load factor' is a good criterion for checking the maximum acceptable traffic in the cell to ensure adequate coverage and performance. On the downlink, a good general rule is to make sure the plan will not place more than about 80% load on the cell. This should be reduced in severe multipath conditions. On the uplink, a good general rule is to make sure the plan will not place more than about 60% load on the cell.

References

[1] Gilhousen, K.L., Jacobe, I.M., Padovani, R., Viterbi, A.J., Weaver, L.A. and Weatley, C.E., 'On the Capacity of a Cellular CDMA System', *IEEE Trans. Vehic. Tech.*, **40**(2), May 1991, 303–312.
[2] Jakes, W.C., *Microwave Mobile Communications*, Wiley & Sons, Inc., New York, 1974.
[3] Dehghan, S., Lister, D., Owen, R. and Jones, P. 'W-CDMA Capacity and Planning Issues', *Electronics and Communication Engineering Journal*, June 2000, 101–118.
[4] Laiho, J., Wacker, A. and Novosad, T., *Radio Network Planning and Optimization for UMTS*, John Wiley & Sons, Ltd, Chichester, 2002.
[5] Wacker, A., Laiho-Steffens, J., Sipila, K. and Jasberg, M., 'Static Simulator for Studying WCDMA Radio Network Planning Issues, *Proceedings VTC 1999 Spring Conference*, Houston, TX, May 1999, pp. 2436–2440.
[6] Lee, J.S., and Miller, L.E., *CDMA Systems Engineering Handbook*, Artcech House, 1998.
[7] Laiho-Steffens, J., Sipila, K. and Wacker, A., 'Verification of 3G Radio Network Dimensioning Rules with Static Network Simulations, *Proceedings VTC 2000*, Spring Conference, Tokyo, Japan, May 2000, pp. 478–482.

[8] Siplila K., Honkasalo. Z.C., Laiho-Steffens. J. and Wacker. A., 'Estimation of Capacity and Required Transmission Power of WCDMA Downlink Based on a Downlink Pole Equation', *Proceedings VTC 2000 Spring Conference*, Tokyo, Japan, May 2000, pp. 1002–1005.

[9] Wacker, A., Laiho-Steffens, J., Sipila, K., and Heiska, K., 'The Impact of Base Station Sectorisation on WCDMA Radio Network Performance, *Proceedings of VTC 1999*, Houston, TX, May 1999, pp. 2611–2615.

[10] Silva, J. C., Jesus, S.J., Souto, N.S., Cercas, F.C., Correia, A. and Rodrigues, A.J., 'Partitioned MMSE Receiver for Wideband CDMA Systems', *Proceedings International Symp. on Wireless Personal Multimedia Communications WPMC*, Aalborg, Denmark, 2005, **1**.

[11] Andrews. J.G., 'Interference Cancellation for Cellular Systems: A Contemporary Overview', *IEEE Wireless Communications*, April 2005, 19–29.

[12] Juntti, M. and Latva-aho, M., 'Multiuser receivers for CDMA Systems in Rayleigh Fading Channels', *IEEE Trans. Vehic. Tech.*, 2000, 126.

[13] Silva, J. C., Souto, N.S., Cercas, F.C., Rodrigues, A.J. and Correia, A., 'Multipath Interference Canceller for High Speed Downlink Packet Access in Enhanced UMTS Networks', *Proceedings IEEE International Symposium on Spread Spectrum and Techniques Applications (ISSSTA)*, Sydney, Australia, September, 2004, Vol. 1, pp. 609–612.

[14] Holma, H. and Toskala, A. *WCDMA for UMTS*, John Wiley & Sons, Ltd, Chichester, 2000

[15] Giangaspero, L. and Agaross, L. 'Co-channel Interference Cancellation based on MIMO Systems', *IEEE Wireless Communications*, Dec 2002, 8–17.

[16] Vieira, P. and Rodrigues, A.J., 'Multiuser MIMO Performance Applied to UMTS HSDPA', *Proceedings IEEE Vehicular Technology Conference*, Milan, Italy, May 2004, Vol. 2, pp. 833–837.

[17] Gesbert, D., Shafi, M., Shiu, D. and Smith, P., 'From Theory to Practice: An Overview of Space-time Coded MIMO Wireless Systems', *IEEE Journal on Selected Areas on Communications*, **21**(3), 2003, 281–302.

[18] Adjoudani, A., Beck, E.C., Burg, A.P. *et al* 'Prototype Experience for MIMO BLAST over Third-generation Wireless System', *IEEE Journal on Selected Areas in Communications*, **21**(3), 2003, 440–451.

5

Radio Site Planning, Dimensioning, and Optimization

In 2G systems, interference analysis is required for frequency allocation. In WCDMA systems, interference analysis is critical for the determination of coverage and capacity, which are interlinked through the multi-user interference as discussed in Chapter 4. In 2G systems, the initial network planning is centered on estimating the number of sites required to provide the RF signal coverage over the area, and this can be done for the most part independent of user distribution and traffic volumes. Once coverage is provided, new capacity can most often be added by simply integrating more hardware into the sites.

In WCDMA, coverage depends on the user traffic distribution, the services, and the traffic intensity. Once the latter changes, so can the coverage range of the cell, and that with respect to each service E_b/N_0 requirement. Moreover, unlike in GSM where coverage balance on the uplink and downlink can usually be achieved, either link can become the limitation in UMTS. Most often capacity is limited by the base station (Node B) total transmit power due to the power sharing aspect on downlink. This means that normally the cells in WCDMA are capacity limited on downlink, particularly in congested small cells in the dense urban area. The cells are expected to be coverage limited on the uplink in areas with light traffic density such as in rural areas, given the limited transmit power capability of the mobile stations. Therefore, it is important to have more a realistic picture of the traffic mix and its distribution, as well as its growth and possible variations over the different time periods, as the basis for robust network planning and deployment in WCDMA. Alternatively, the network configuration and link budgeting should take into account all practically possible variations in the traffic mix with different service quality requirements and distribution patterns to generate robust network designs.

A major aspect of radio network planning is the determination of radio site locations and the configuration and dimensioning of them. Various approaches have been presented in the literature to handle this problem, some of which involve rather complex mathematical formulation and analysis. The simplest approaches are based on a mixture of heuristics and link budgeting. We will discuss the former category first and draw on the conclusions of their analysis for providing new useful insights in the use of the more simple routine practical

---

UMTS Network Planning, Optimization, and Inter-Operation with GSM   Moe Rahnema
© 2008 John Wiley & Sons, (Asia) Pte Ltd

methods, such as those based on link budgeting and iterative simulations, which will be discussed subsequently.

5.1 Radio Site Locating

Practically, the optimal base station locations must either be decided from a set of feasible allowed potential site locations such as in green field planning or from a set of existing GSM sites for co-location plans. In either case, appropriate models and methodologies need to be used to decide on the optimal or near optimal radio site locations.

The classical coverage models, adopted for 2G GSM systems, are not well suited for planning 3G base station locations. They are based on radio signal propagation to ensure RF coverage without taking into account much of the specifics on traffic distribution and required capacity. Capacity enters the picture most often after the sites are planned and deployed. The conventional planning methods [1] simplify the radio network planning process into an RF coverage planning, and a frequency/capacity planning phase. The frequency planning phase is driven by capacity requirements. The coverage planning phase uses radio propagation models for the areas to decide where to place base stations to provide sufficient RF signal coverage in the areas desired. In the frequency/capacity planning phase, estimated traffic requirements are used to decide how many channels/frequencies to assign to each site. The grouping and assignment of frequencies to various sites is decided by taking into account geographically based cell interactions with respect to RF visibility (interference) so that the required signal to interference + noise ratios are met within each site.

This two-phase approach is not applicable to CDMA systems, where the coverage area of a cell depends on both radio propagation environment and the user traffic densities and distributions over the area. Basically no frequency planning is needed (except possible frequency coordination with adjacent operators in the case of an operator with multi-carriers). But the complexity of the radio planning lies in the coupling and the interaction between coverage and capacity through the multi-user interference as discussed in Chapter 4. In WCDMA systems, the radio access mechanism allows for a more flexible use of the radio bandwidth and the capacity of each cell is not hard limited by a fixed channel assignment as in GSM systems. But it depends on the actual interference levels that determine the achievable SIR values. Since the interference levels will depend on the base station locations, and the actual traffic mix and distributions, BS location in UMTS cannot be based on only RF coverage considerations, but must also take into account the specific of traffic densities and different services quality requirements.

In Reference 2, an integrated approach to cellular network planning has been proposed using discrete population models for the traffic characterization based on the demand node concept (DNC). The DNC concept is used to formulate the base station locating task as a Maximal Coverage Location Problem (MCLP). This reduces the problem of radio site locating to a set coverage problem (SCP) well known in economics for modeling and solving facility location problems. The approach presented [2] was originally developed for the base station locating problem in 2G networks by placing sites where the traffic demand is concentrated and thus helping to save on site costs. However, this approach is also well applicable to WCDMA systems, as it considers the traffic distribution and densities in the coverage planning phase. In fact in 3G networks, the best location for radio sites is where the traffic demands are concentrated, or basically at the centroids of the traffic hot spots, as concluded in Reference 2 for saving on 2G site costs. This is justified by considering that the closer the site is to the traffic centers, the less will be the transmit power requirements (which are set by the closed loop power controls). The reduced transmit powers in turn results in less multi-user interference levels, which translate into more capacity. In fact, a number of optimization models and algorithms have been presented [3, 4] for finding the best radio site locations in WCDMA based on considerations of traffic distributions and the minimum service SIR requirements. The algorithms presented consider the effect of the closed loop power control loops which try to adjust the transmit powers so that the minimum required SIR is achieved for each service at the receiving end of the radio links. The algorithms presented [3, 4] consider the optimal site location calculation based on using either the uplink or the downlink in the considerations of traffic requirements. The models presented result in a set of constrained linear programming mathematical equations [2], which are solved through application of Greedy search procedures and Tabu algorithms. It is shown that the uplink based site calculations result in a more stringent requirement for the number of radio sites compared to a downlink based model in a balanced traffic scenario such as voice calls. This is not unexpected considering the lower intra-cell interferences experienced on the downlink (as a result of leftover code orthogonality), and the more limited transmit powers from the mobile stations on the uplink.

Based on the results presented [3, 4], it is expected that the uplink based models for site calculations provide good estimates of the number of sites that are actually needed, whereas the models based on downlink provide better estimates of the amount of traffic that can be covered. Therefore, based on the work presented [2–4], it is concluded that the best site locations for 3G planning are at the centroids of the traffic hot spots, and that the number of sites for coverage determination can be based on uplink considerations, whereas the downlink considerations will more drive the capacity requirements. These insights can be utilized in the more simple practical dimensioning based on link budgeting.

5.2 Site Engineering

Site engineering is concerned with the basic RF parameters related to power settings on the pilot and common channels, and antenna configuration such as the antenna height, number and orientation of sectors.

5.2.1 Pilot and Common Channel Power Settings

The powers on the downlink common channels are all set relative to the P-CPICH power. Therefore, if the P-CPICH power changes, the power on all other channels changes accordingly. Typical settings are given in Table 5.1. Typically, 5–10% of the total cell transmit power capability is allocated to P-CPICH depending on the neighboring site and RF geometries. Pilot channel power setting is an important task in WCDMA radio access design. The optimum pilot power results in coverage in intended areas while minimum interference to neighboring cells. Too low a value for the pilot power results in insufficient pilot coverage and poor cell dominance. On the other hand excessive pilot powers reduces the total power available for the traffic channels, and extends the cell coverage beyond the planned service coverage. This collects traffic from areas where the mobiles do not have

Table 5.1 Typical downlink common channels power allocation.
[from *Radio Network Planning and Optimisation for UMTS*, J. Laiho, A. Wacker and T. Novosad © 2002. Copyright John Wiley & Sons, Ltd. Reproduced with permission]

DL common channel	Typical power level	Note
P-CPICH	30–33 dBm	5–10% of maximum cell Tx power
P-SCH and S-SCH	−3 dB	Relative to P-CPICH power
P-CCPCH	−5 dB	Relative to P-CPICH power
PICH	−8 dB	Relative to P-CPICH power, and for Np = 72
AICH	−8 dB	Relative to P-CPICH power
S-CCPCH	−5 dB	Relative to P-CPICH power, and for SF = 256 (15 ksps)

sufficient link budget to connect to the cell. Thus the increasing or decreasing the power of the P-CPICH signal makes the cell larger or smaller relative to the traffic channel coverage. Therefore the tuning of the pilot powers can help to balance the cell load among the neighboring cells, in addition to making sure that adequate signal is received by the user terminals for pilot decoding. The work presented in Reference 5 discusses procedures for tuning the pilot power to achieve good balance between coverage and load.

Generally, the power setting on the pilot channel is based on the criteria of 1) making sure that adequate pilot signal power is received in the areas of intended service coverage to allow call set up, and 2) pilot pollution is avoided. The minimum pilot power required for a terminal to be able to decode the signal information is specific to the receiver electronics. The 3GPP Specifications [6] require that the terminal be able to decode the pilot signal with an E_c/I_0 of −20 dB, where E_c is the energy per chip in a pilot signal, and I_0 is the total interference power spectral density resulting from all other pilots, the traffic channels from the own and other cells, and the thermal noise. Receivers with high quality sensitivities can handle signals with several dBs lower than that. It is not easy to determine the minimum E_c/I_0 required for proper pilot signal decoding because that would also depend on several factors, including the multipath profile. However the limit chosen should not be lower than that of the specifications, and a margin for the soft handover area must be provided. A value in the range of −18 to −16 dB under loaded condition is usually considered acceptable in practice. It is however important to prevent excessive pilot powers as this would consume the limited power capability of the base station, considering that the power of other common channels are set relative to the pilot channel. Furthermore excessive pilot power can result in pilot pollution.

Pilot pollution occurs in an area when multiple pilot signals are received within about 3 dB of each other. If none of the pilots is dominant enough, the mobile may not be able to initiate a call, and ping-pong handovers also result. To accommodate handovers, the pilot signals from neighboring cells should overlap in cell border areas. However, each pilot signal in the overlap areas contributes to interference to other pilots, and the lowering of the E_c/I_0 as seen by each. Therefore receiving too many pilots in an area can result in reduced network performance and should be avoided through proper planning. As a guide, the maximum number of pilots received with adequate signal strength in soft handover overlap areas should not exceed the number of design soft handover branches. This should

Table 5.2 Pilot coverage verification.

Coverage level	RSCP, dBm	E_c/I_0, dB, (unloaded condition)
Sufficient	RSCP = −100	$E_c/I_0 \geq -14$
Poor	$-115 \leq$ RSCP < -100	$-16 \leq E_c/I_0 < -14$
No coverage	RSCP < -115	$E_c/I_0 < -16$

be about 2–3 in any case to limit the soft handover overhead in the network. The total number of pilot signals with adequate signal strength in an area should also be less than the maximum number of the Rake receiver fingers in the mobile base stations. The Rake receiver is used to both detect the resolvable multipath components from the same signal as well as resolving pilot signals from neighboring cells. Usually up to six fingers are provided in the mobile base stations Rake receiver [7].

5.2.2 Pilot Coverage Verification

The indicators used in measuring pilot coverage are the E_c/I_0 and the E_c, where the E_c is the received code power of the CPICH channel as measured by UE. Since E_c/I_0 contains the effect of thermal noise and interference from own cell traffic channels, and other cells channels (traffic and pilot), the E_c/I_0 alone is not the only indicator of RF coverage provided by a cell. For instance, E_c/I_0 can be low not because the RF coverage from the cell is not good, but because there is too much interference from neighboring cells (I_0 is too high). On the other hand, E_c can be used mainly to estimate the path loss since the power of the CPICH is either known or can be read by the system information. Therefore, the cell nominal RF coverage classification is normally based on the combination of RSCP and E_c/I_0. The coverage classifications given in Table 5.2 are usually used by the industry based on the two stated variables. The E_c/I_0 values given in Table 5.2 are under no load conditions, that is, it excludes the interference from the traffic channels. A 98% coverage would then mean that 98% of the measurements fall in the acceptable ranges.

These criteria may be used to verify discreetly degrees of coverage level on an area basis.

The transmission powers of AICH and PICH are common traffic channel configuration parameters that are set by the network planner relative to the P-CPICH in order to have the same coverage over the same area. These parameters are sent to the base station whenever the corresponding common traffic channel is set up or reconfigured.

The PICH transmission power depends on the number of paging indicators (PI) per frame, Np, which is always provided to the base station together with the power of the S-CCPCH carrying the PCH. More paging indicators per frame results in fewer bit repetitions per frame and hence the higher is the PICH power needed relative to the P-CPICH. Typical offsets are −10 dB for Np = 18 or 36, −8 dB for Np = 72, and −5 dB for Np = 144 [8].

To increase the reliability in the transmission of the pilot and TFCI control symbols, these symbols are transmitted with higher power relative to the rest of the data in the S-CCPCH. The TFCI and the pilot symbol transmission powers are specified with the power offset parameters PO1 and PO2, respectively, as shown in Figure 5.1. These parameters, as well as the FACH parameters, maximum FACH power, PCH parameters, and the PCH power are given to the Node B when the S-CCPCH, that is the FACH and PCH, are set up or

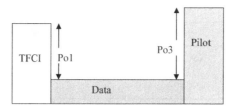

Figure 5.1 Downlink transmit powers on S-CCPCH: the pilot and TFCI symbols are transmitted with power offsets of Po3 and Po1, respectively

reconfigured. Typical power offsets of the pilot/TFCI symbols relative to the power of the S-CCPCH data filed are 2 dB for 15 ksps, 3 dB for 30 ksps, and 4 dB for 60 ksps. The power offsets may vary during the communication according to the bit rate used. Typical power values for the data filed of S-CCPCH relative to P-CPICH are +1 dB for SF of 64 (60 ksps), −1 dB for SF of 128 (30 ksps), and −5 dB for SF of 256 (15 ksps).

5.2.3 RACH Coverage Planning

The uplink coverage on the random access channel, RACH, should be planned adequately to allow connection start-up signaling. The RACH coverage should match the coverage of the uplink dedicated channels to be supported in the cell.

The standards support both a 10 ms and a 20 ms interleaving on RACH, which result in 16 kbps and 10 kbps data rate on RACH, respectively, with a 20 byte RACH message size. The interleaving depth used in the cell is broadcast to the mobile on the BCCH channel. Since the RACH channel bit rates are small compared to most dedicated channels used for high data rate transmission, the RACH coverage should be checked against the coverage provided for the lower data rates (on dedicated channels), and particularly on the AMR voice channels whose bit rates are in the same range as for the RACH. But the RACH coverage cannot be expected to match the coverage on dedicated channels with comparable bit rates. This is due to the fact that the RACH channel does not benefit from any macrodiversity gains (no handover are supported on common channels). Moreover, the short burst transmissions on RACH channels make the reception more difficult than receiving the dedicated channels. However, a higher FER can be allowed on the RACH, with the drawback of increasing the call set up delays. The average FER on the RACH channel can be controlled by system parameters that are broadcast on the BCCH.

The 20 ms RACH option has a higher E_b/N_0 requirement due to the longer time span that increases the possibility of radio channel changes during the transmission. The RACH power is set only one time during the RACH preamble, with no subsequent power control. Therefore it is reasonable to use the 20 ms RACH option on large cells that are uplink coverage limited. The 10 ms RACH option is then a better option for uplink capacity limited cells. Suitable RACH options to match the coverage of dedicated channels are presented in Table 5.3 by accounting for all the differences just explained. These options are based on the results from Reference 9, which are also presented in Reference 10.

Planning carefully for RACH coverage is critical when the network is designed to handle the low bit rate services whose bit rates fall within the same range as for the RACH.

Table 5.3 RACH options for matching the coverage of dedicated channels.
[*WCDMA for UMTS*, H. Holma and A. Toskala, © 2000. Copyright John Wiley & Sons, Ltd. Reproduced with permission]

DCH bit rate	RACH option
AMR $\leq$ 7.95 kbps	20 ms, FER $>$ 10%;
AMR 12.2 kbps	20 ms, FER $\leq$ 10%
Bit rate $>$ 20 kbps	10 ms, FER $\leq$ 10%

5.2.4 Site Sectorisation

Sectorisation helps to achieve much higher capacities compared to omni-sites as a result of confining the radiation more to the desired users, therefore reducing the interference to others. This reduces the base station power required to achieve a certain C/I. Also the higher gains of sector antenna helps to reduce the base station transmit power for each sector, allowing more users to be supported for a given maximum amount of base station power. The gains achieved from sectorisation in WCDMA are much higher than in the GSM (TDMA) systems because of the frequency re-use of 1, which results in the entire re-use of the allocated band in each sector.

However, the biggest problem of sectorisation, and especially higher levels of sectorisation, is the increased cell overlaps. The overlapping sectors increase interference leakage between adjacent sectors, which results in reduced capacity. Furthermore, excess adjacent sector overlaps result in increased soft (softer) handover probabilities and overhead, which in turn reduces system capacity. Therefore the degree of overlap between sectors must be controlled so that it reduces to an acceptable level. The degree of sector overlaps is influenced by the number of sectors used in a site. In fact, the results of simulation confirmed by analytical models presented in Reference 11 show that the capacity gains with higher order sectorisations (beyond 3) are not proportional to the number of sectors. Furthermore, the degree of overlap between sectors is influenced the specific antenna radiation patterns and beamwidth, which can be affected by the radio propagation environment. For instance obstacles in the vicinity of the antenna can increase the sidelobe radiation levels and the beamwidth. These effects, though, can be made small by careful choice of the antenna beamwidths. The simulations presented in Reference 11 and confirmed by theory provide the sectorisation gains achieved on the uplink, and downlink of sectored sites, with sectorisation gain defined as the ratio of the number of users supported per sectored site to the number of users supported per omni-site. The results for various sectored configurations were taken from Reference 11 and are given in Table 5.4 to provide the network planners with some insight into the site configuration designs. Notice that with 3-sectored sites, the maximum gains are achieved with antennas of 65 degree horizontal beamwidth.

Simulation results have confirmed the expectations that a 3-sector configuration using two carriers in each would give better capacity (due to the lower inter-cell interference) compared to a 6-sector site configuration using one carrier in each (for multi-carrier design, see Chapter 6).

5.2.5 Controlling Site Overlap and Interference

A certain degree of cell coverage overlap is required for the smooth functioning of soft and softer handovers to provide ubiquitous service coverage. However, if cell-overlap in the network exceeds certain limits, it can result in excess handover overheads. Pilot pollution

Table 5.4 Sectorisation gains [11].
['The Impact of the Base Station Sectorization on WCDMA Radio Network performance', A. Wacker, J. Laiho-Steffens, K. Sipila and K. Heiska, *Proc. VTC 1999, Fall Conference* © 1999 IEEE. Reproduced with permission

Number of sectors	Antenna beamwidth	Sectorisation gain	
		Uplink	Downlink
1	omni	1	1
3	90 deg	2.57	2.47
3	65 deg	2.87	2.97
3	33 deg	2.82	2.81
4	90 deg	3.11	2.85
4	65 deg	3.59	3.71
4	33 deg	3.44	3.32
4	90 deg	4	3.42
6	65 deg	4.7	4.51
6	33 deg	5.02	5.07

and interference can also result if the pilot signals from too many cells (more than 2 to 3 is considered high in UMTS) are received in one area (many overlapping cells). Pilot pollution is detectable through test pilot scanners.

A measure of cell overlap is obtained by estimating the soft handover overheads. Likewise a measure of sector overlaps (sectors from same sites) is obtained by estimating the softer handover overhead. Normally based on experience and simulation results, the soft and softer handover overheads should be limited to a range of 30–40%, and 5 to about 15% for efficient radio designs, respectively.

The following measures can be taken to limit the cell overlap and interference:

- Avoid high sites as much as possible, because very high sites are a source of pilot pollution and cannot always be effectively controlled with antenna tilting.
- Use proper down–tilting. Antenna tilting is the most important CDMA site optimization technique because it allows control of cell boundaries and interference. However, antenna tilting must not be over-rated. One scenario where antenna tilting may prove to be of limited use is on high sites where radical down-tilting can affect antenna patterns and result in coverage gaps and interference. Furthermore, down-tilting of sites in urban environments can result in signal diffraction and limit its effectiveness.
- Use antennas with well-controlled side lobe suppressions (more than say 20 dB suppression for the upper side lobes).
- Examine and use proper adjustment for the antenna azimuths. Proper 'azimuthing' of neighboring sites is important for the effectiveness of antenna tilting.
- Use antennas with no more than 65 degrees horizontal beam width for 3-sectored sites, and no more than 33 degrees for 6-sectored sites if possible.
- Use proper setting of the P-CPICH power based on area and coverage desired. Effective power planning together with antenna tilting provides a powerful combination for interference management, particularly in cases where antenna tilting on its own is not very effective [12].

It is important to avoid high-elevation sites in WCDMA, that is, sites which are considerably above the average radio site elevation in the area. High sites are often a cause of pilot pollution due to their large RF coverage footprints. The larger than desired footprint of high elevated sites can be reduced by lowering the elevation of the offending antennas, reducing the transmit powers, and/or introducing sufficient amount of down-tilt.

5.3 Link Budgeting for Dimensioning

Link budgeting is part of the network planning process, which helps to dimension the required coverage, capacity, and quality of service requirement in the network. The outcome of link budgeting is normally the maximum allowed path loss for each service. This is then used with radio path loss models appropriate for the area to estimate the maximum cell coverage range, and then used to estimate the required site density. Since different services will end up with different coverage ranges, the service resulting in the smallest coverage range will determine the nominal cell radius for an all-service uniform coverage objective in a single carrier design scenario. Because in the end the service profile and traffic densities will impact the load factor and hence the coverage range for each service, link budgeting based dimensioning is an iterative process. This means that after a nominal cell coverage radius is determined based on an initially assumed load factor, the service profile and densities within the found cell coverage range should be used in calculating the resulting load factor. If the resulting load factor is larger than the assumed design value, then the cell coverage range should be reduced (or the service with the strictest requirements moved to a second carrier in a multi-carrier site design, see Chapter 6) and the load factor recalculated. If the new load factor still exceeds the design value, the iterative process of reducing the cell coverage range should be repeated until the calculated load factor no longer exceeds the design value.

Alternatively Monte Carlo simulation on a UMTS modeling tool can be used to determine the cell coverage range for a given design load factor, service profile, and traffic densities. The simulation tools take into account the cumulative effects from individual services on the interference floor at the base stations, and at each user location in the downlink. In this way, the tools generate as output some of the assumptions, such as the load factors, inter-cell interference, etc., which are used as basic inputs to a simple link budgeting approach. However, simple link budgeting exercises on the uplink in WCDMA provide a quick means for system dimensioning estimates, and an indication of cell coverage range and hence site count requirements. The link budgeting also provides insight into the effects of different parameters on service coverage.

There are two main limiting factors in UMTS link budgeting: the mobile station power in the uplink and the base station transmit power in the downlink. Simple link budgeting can be done for quick dimensioning estimations under specified criteria and presumptions such as:

- Type of service, specified with bit rate, E_b/N_0 requirement, expected mobility speed, and under a certain load condition (interference floor)
- Type of radio propagation environment (terrain, car/building penetration)
- Behavior and type of mobile (speed, max power level)
- Site and hardware configuration (BTS antennas, BTS power, cable losses, equipment reference sensitivities)
- Required area coverage probability

Then the site dimensioning is based on the service with most stringent requirement in a non-hierarchical single carrier radio access architecture. In multiple layered radio architectures discussed in Chapter 6, different service categories may be assigned to different layers implemented on dedicated carriers (frequency bands), which would help to relieve the number of site requirements.

In GSM systems, link budgeting can be done for each user irrespective of other users in the cell. Link budgeting in WCDMA is much more complicated as the power transmitted by or to each user impacts the link budget parameters of other users. For instance, the transmit power allocated to each user on the downlink or used by each user on the uplink translates to a degree of interference to the other users in the cell. Thus each user influences the transmit power requirements for other users, resulting in a positive feedback effect in the power requirement loop for each user. This means that the coverage-capacity prediction has to be done iteratively until the transmit power requirements for the various users in the cell stabilize. This mutual user impact is more complicated to model on the downlink for two reasons: 1) it depends on the mobile location dependent parameters, as discussed in detail in Chapter 4; and 2) due to the base station power sharing among the users. On the uplink, the multi-user interference raised noise floor (the mutual user effect) at the base station receiver can be modeled through a single lumped parameter, which we have called so far the load factor at the base station. This would easily allow the calculation of the raised noise floor at the base station receiver. What this argument is leading to is that it is possible to design a rather simplified form of link budgeting on the uplink for quick estimation of various services coverage ranges. This can then be verified later through a semi-analytic method of verifying the assumptions used, such as the assumed load factor resulting from the traffic mix, or through the iterative simulation process. The iterative simulation process is a necessity for verifying the detailed service coverage and capacity on the downlink as will be discussed in Section 5.3.2.

5.3.1 Uplink Link Budgeting and Static Analysis

As was stated in the previous section, the uplink link budgeting is a rather quick simple way of estimating the coverage ranges that are possibly achievable for different services under certain assumed load conditions. The outcome is then used as initial cell range estimates for calculating site count requirements. The uplink link budgeting is simplified by modeling the multi-user effect through a single lumped parameter called the uplink load factor. The load factor is used to model the cumulative multi-user interference at the base station. The load factor is a measure of the design capacity for the cell and is used in the calculation of the base station sensitivity for each service. Since it is the traffic mix and intensity in the cell (and in fact the surrounding cells as well) that determine the final value of the load factor, any assumed value in the initial uplink link budgeting should in principle be verified subsequently by recalculating the assumed initial value considering the traffic profile within the found cell nominal coverage range. The cell nominal coverage range is decided based on a coverage probability for a given service as decided by the operator. If the re-calculated uplink load factor exceeds the target design value, then the calculated nominal cell coverage range should be reduced and the iterations repeated until the resulting load factor no longer exceeds the design value as explained in the previous section. Alternatively some of the traffic from certain services, for instance the higher bit rate RABs, may be moved to a second carrier and the calculations repeated for checking of the resulting new load factors against target values. Once the uplink link budgeting iterations are completed, the resulting downlink load factor and total required base station transmit power should also be calculated and checked against the allowed target values. If the latter also check out, the link budgeting process is completed, and the final cell range is estimated from the maximum allowed path loss from the uplink and an appropriate path loss model for the area. Otherwise further iterations are required until convergence is achieved on the downlink as well. This analysis based on link budgeting is deterministic in nature, and does not require knowledge of the user location, mobility pattern, or traffic raster (distribution) over the area. The complete link budgeting iterations based on this deterministic analysis for uplink and downlink are illustrated in the flowchart given in Figure 5.2.

5.3.1.1 Uplink Load Factor Formulation

The load factor lumps the interference contributions from all the users with their different services into a single factor. This is used as a common link budget parameter to calculate the cumulative noise floor at the receiver for all the services. For a given cell load (capacity) as determined by the load factor, the coverage range of each service is then estimated by its E_b/N_0 requirement and bit rate. The load factor for the uplink was formulated and calculated for the special simple case of a homogeneous user environment in Chapter 4. In this chapter, we will derive the formula for the uplink load factor for the more general case of mixed services. The uplink load factor defines the ratio of the multi-user interference to the multi-user interference plus the effective thermal noise at the base station.
Mathematically

$$\eta = \frac{I_{hc} + I_{oc}}{N_T} \qquad (1)$$

where η is the uplink load factor, I_{hc} is the own-cell interference (interference caused by users transmitting from within the cell), I_{oc} is the other-cells interference (interference caused by the transmission of the users from surrounding cells), and N_T is the total effective thermal noise + multi-user interference power at the base station.

Taking the engineering approach and expressing the other-cells interference, I_{oc}, as a fraction of the own-cell interference, I_{hc}, by the factor f (which was defined in Chapter 4), we can write

$$I_{oc} = f.I_{hc} \qquad (2)$$

Now, the own-cell interference at the base station can be calculated by the sum of the signal power received from each connection j, which is $E_j.R_j.v_j$, assuming perfect power control resulting in all connections fully compensated for the path losses, where E_j and R_j are the energy per bit and the connection bit rate, respectively, and v_j is the activity factor for the connection.
Therefore

$$I_{hc} = \sum_{j=1}^{M} E_j.R_j v_j \qquad (3)$$

where M denotes the number of uplink connections in the cell.

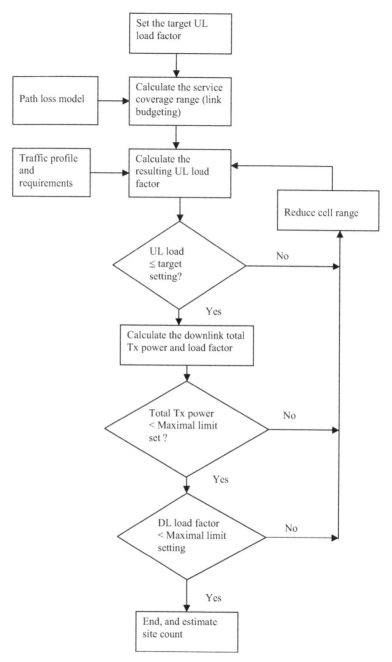

Figure 5.2 Static analysis based on iterative link budgeting.

Substituting into Equations (2) and (1), and combining, we obtain

$$\eta = (1+f) \sum_{j=1}^{M} E_j . R_j . v_j / N_T \quad (4)$$

from the definition of signal to noise ratio for the jth connection,

$$\rho_j = W \frac{E_j}{N_T - E_j} \quad (5)$$

where W is the WCDMA channel band, that is the chip rate (3.84 Mcps).
For the case of many connections leading to the approximation that $N_T \gg E_j$, we have

$$\rho_j \cong W \frac{E_j}{N_T}$$

Solving for E_j, and substituting in Equation (4), gives the following expression for the uplink load factor:

$$\eta \cong (1+f) \frac{1}{W} \sum_{j=1}^{M} \rho_j . R_j . v_j \quad (6)$$

which is also the formula given in Reference 13. This equation can be used to check if a certain service mix can meet a specified design value for the uplink load factor, or for estimating the amount of traffic that can be supported for a specified load factor. For example, for a specified design load setting of 70%, it can be used to determine how many voice connections can be supported in the cell based on uplink considerations within a cell coverage range as determined by the uplink budgeting presented in the next three sections. It is therefore an equation to guide in site dimensioning based on uplink considerations.

5.3.1.2 Base Station Sensitivity Estimation

The sensitivity is defined generally as the minimum signal power required at the input to the receiver to meet the quality requirement for the service (such as the FER limit for voice, for instance). Since the service quality for a given receiver structure and design is determined by the signal to noise ratio [14], the sensitivity will depend on the noise floor present at the receiver input. In GSM or TDMA/FDMA networks, the base station sensitivity is normally assumed to be constant implying the same coverage threshold for all base stations. In WCDMA, the base station sensitivity varies from cell to cell, and with the service distribution and bit rates used. This is the case as the number of users and the bit rates used by each will impact the noise floor at the base station. Therefore, we need to investigate how to calculate the dependency of receiver sensitivity on the cell load for use in the link budget.

In link budget considerations, the base station noise level over one CDMA carrier is influenced by the multi-user interference. The multi-user interference raises the noise floor at the receiver, and this increase is calculated by the $1/(1 - \eta)$ factor, in which η is the load factor. Now the E_b/N_0 requirement, denoted by ρ, can be written as

$$\rho = E_b/N_0 = \frac{S/R_b}{\dfrac{P_n}{(1-\eta).W}} \tag{7}$$

in which R_b is the service bit rate, P_n is the total effective thermal noise power within the carrier band at the receiver (and is dependent on the receiver noise figure), S is the receiver signal power at input to the receiver, η is the uplink load factor, W is the WCDMA effective channel band (the chip rate). Solving for S, gives the following formula for the receiver sensitivity:

$$\text{Receiver sensitivity} = \frac{1}{1-\eta}\rho.P_n/\left(\frac{W}{R_b}\right) \quad \text{(in linear scale)} \tag{8}$$

in which the signal to noise ratio requirement, ρ, will depend on the receiver structure, the service quality requirement and bit rate, as well as the mobile speed. The effective thermal noise power P_n should be corrected for any external environmental pollution such as human-made noise. The ratio W/R is the processing gain as discussed in Chapter 2, and it is seen to improve the sensitivity of the base station. Therefore the effect of the processing gain is incorporated in the receiver sensitivity calculation, and should not be included in the link budget formula anymore. The $1/(1 - \eta)$ factor is the noise rise due to multi-user interference at the base station, and incorporates the necessary *interference margin*. Thus Equation (8) combines the effect of the processing gain and the load (interference margin) to generate an updated base station sensitivity for each service.

The effective thermal noise power within the band, P_n is simply given by

$$P_n = N_t + N_f + W \quad \text{(in linear scale)} \tag{9}$$

where N_t is the thermal noise power spectral density at the base station, N_f is the base station receiver noise figure, and W is the WCDMA chip rate (3.84 Mcps).

5.3.1.3 Soft Handover Gain Estimation

Soft handover contributes to the gains in the link budget in two ways. The most common is the gain provided against the fast fading, which results from the multipath effects. This is achieved due to the macro diversity combining gain, which reduces the required E_b/N_0 relative to a single link. The second benefit of soft handover is that it contributes to a margin against lognormal shadowing effects (fading) [15]. This happens because the lognormal fading is partly uncorrelated between different cells (or sectors), and the handover process also results in the selection of the best server. In the following, we provide a means to estimate this gain by the influence it places on the area coverage probability. The area coverage probability is defined as the probability that the signal level over the whole cell area is above a certain threshold.

Since soft handover requires some degree of overlapping between adjacent cells, more than one cell can provide coverage in the overlapped areas. This means that on average the cell area coverage probability will increase relative to that for a single isolated cell. This translates into a smaller number of required base stations to meet the same overall coverage probability. Therefore, the location coverage probability should be modified to account for the stated multi-server effect due to soft handover process. The cell edge coverage probability, which can be translated into area coverage probability, has been analyzed [15] with the following result:

$$P_{out} = \frac{1}{\sqrt{2\Pi}} \int_{-\infty}^{\infty} e^{-\frac{x^2}{2}} \left[Q\left(\frac{\gamma_{SHO} - c.\sigma.x}{d.\sigma}\right)\right]^2 dx \qquad (10)$$

in which P_{out} is the coverage outage at the cell edge, γ_{SHO} is the required fading margin when the SHO gain is included, σ is the standard deviation of the lognormal fading, and for a 50% correlation of the lognormal fading between two links to two base stations, $c = d = 1/\sqrt{2}$. This probability at the cell edge can be converted to the area coverage probability [16]. Then, the SHO gain against slow lognormal fading is calculated from

$$G_{SHO} = \gamma_{Single} - \gamma_{SHO} \qquad (11)$$

where the lognormal fade margin required for a single link, γ_{Single}, is dependent on the area coverage probability desired, the path loss exponent, and the standard deviation of the lognormal fading for the environment. The derivation of the formula to calculate γ_{Single} is given below [16]:

$$P_{area\ cov\ prob} = \frac{1}{2}\left[1 - erf(a) + \exp\left(\frac{1 - 2ab}{b^2}\right)\left(1 - erf\frac{1 - ab}{b}\right)\right] \qquad (12)$$

in which

$$erf(x) = \frac{2}{\sqrt{\pi}} \int_0^x e^{-t^2} dt$$

$$A = \frac{\gamma_{single_r}}{\sigma\sqrt{2}} \quad \text{and} \quad b = \frac{10.n}{\sigma\sqrt{2}} Log_{10}e$$

where n is the path loss distant exponent, σ is the standard deviation of the log normal fading in the area, and $P_{area\ cov\ prob}$ is the desired area coverage probability.

The tabulated values of the above equation are available in the RF planning literature. For a 95% area coverage probability, a path loss exponent of 3.5, and a standard deviation of 7 dB for lognormal fading, γ_{Single} is obtained to be 7.3 dB from Equation (12). And from the integral Equation (10) for a cell edge probability of 99% (corresponding to a 95% area coverage probability), the value of the required fade margin with soft handover gain included, γ_{SHO}, is obtained to be 4 dB. Therefore the soft handover gain is $7.3 - 4 = 3.3$ dB, using Equation (11). Because of the complexity of Equation (10), in practice a soft handover gain in the range of 2–3 or so is normally assumed, in which case the margin required against slow fading (under soft handover conditions), that is, the quantity γ_{SHO}, is simply obtained from Equation (11) and the tabulated values of γ_{Single} given in the literature.

5.3.1.4 The Uplink Link Budgeting Formulation

There are a few parameters that are specific to only WCDMA link budgeting as compared to GSM and these are listed and explained below before we proceed.

Processing Gain
Processing gain gives the gain achieved by spreading a narrow band signal over a wideband spectrum. In UMTS, it is given by the ratio of the chip rate, which is 3.84 Mcps, to the service bit rate, Rb. In dB, it is $10 \times \log 10$ (3840 kbps/Rb kbps). The processing gain was already accounted for in the calculation of the receiver sensitivity, and so will not be included in the link budget formula here.

Interference Margin
Interference margin is calculated from the UL/DL loading values. The interference margin on the uplink compensates for the degradation of Node B sensitivity due to increased network load (noise rise at Node B). The interference margin on the DL compensates for the network loading on the DL at the mobile station. The interference margin was used in the receiver sensitivity calculation through the $1/(1-\eta)$ factor and hence should not be included in the link budget formula again.

Soft Handover MDC Gain
Soft handover macro-diversity combining (MDC) gain is explained in detail in Chapter 8. The soft handover gain is realized through different combining techniques used to process the frames from different paths in soft handover. This gain is around 1–1.2 dB on the downlink (using maximal ratio combining) and zero on average on the uplink due to frame selection combining being used. This gain is in addition to handover gains resulting from protection against shadow fading as explained in Chapter 8.

Power Control Headroom
Power control headroom provides a margin in the uplink link budget to allow power control to work effectively when the mobile approaches the cell boundary. It ensures that power is left for the uplink power control to follow the fast fading, and is effective at slow mobile speeds. Therefore it is important to include this margin at slow mobile speeds. More on this is given in Chapter 8.

Rayleigh or Fast Fading Margin
Rayleigh or fast fading margin is the margin required to accommodate the increases in the transmit powers by the fast power control loops to compensate against the deep Rayleigh fades. This margin on the uplink is usually combined with the power control headroom, and the combined margins are then referred to as the power control headroom, where a value of about 1.2–4 dB is used. On the downlink side, a smaller value of around 0.7 dB is usually used. On the downlink side, the margins against fast fading benefit from an averaging effect

in the base station power sharing over the connections. Therefore large margins are not required.

The uplink link budget formula can be expressed (in dB scale) as

$$L_{Pmax} = P_{UE} - S_{UL} - M_{LNF} - M_{Ray} - L_{\text{feeder and Jumper}} + G_{HO} + G_{\text{diversity}} - L_{Body} - L_{oth} + G_{oth}, \text{ in dB} \tag{13}$$

where P_{UE} is the mobile station transmit power, S_{UL} is the interference compensated sensitivity of the base station for the service (which was formulated in Section 5.3.1.2), M_{LNF} is the required log normal fading margin for a specified signal coverage probability, M_{Ray} is the Rayleigh fade margin, $L_{\text{feeder and Jumper}}$ are the losses from the receive antenna feeders and jumpers, G_{HO} is the total gains due to handover, $G_{\text{diversity}}$ is the receive antenna diversity gain, L_{Body} is the body loss effects (3–5 dB), and L_{oth} and G_{oth} are used to capture any other losses or gains, respectively. Note that a required interference margin of $1/(1 - \eta)$ was incorporated in the calculation of the base station sensitivity given in Equation (8). If this margin is taken out from the sensitivity calculations, then it should be added to the link budget equation.

The uplink link budget equation shows that the maximum allowed path loss, L_{pmax}, and hence the coverage range, is interlinked with capacity through the base station sensitivity S_{UL}, which incorporated the noise rise factor $1/(1 - \eta)$. This noise rise factor represents the noise enhancement through the interferences from the other simultaneous connections to the base station. It is clear now that uplink coverage range estimation requires multiple link budgets with service dependent parameters such as target E_b/N_0 and the bit rate.

Using the above procedures, a template for calculating the uplink link budget was designed and is given in Table 5.5. This template was used to perform link budgeting for AMR voice at 12.2 kbps, and for data services at 144 kbps, and 384 kbps for a load factor of 50% the results of which are also shown in Table 5.5. The link budgeting was performed for a 95% area coverage probability, assuming a path loss exponent of 3.5 and a lognormal standard fade duration of 7 dB. This resulted in the requirement for a lognormal fade margin (slow fade margin) of 7.3 dB as shown in Table 5.5. It is seen that the maximum allowed path loss (and hence the service coverage range) for the higher data rate services gets smaller compared to that of voice or a lower data rate connection.

The results of the link budgeting for 144 kbps data can be compared to GSM voice at the 1800 band. Assuming a GSM terminal transmit power of 30 dBm, a base station sensitivity of -110 with antenna diversity (which is typical), and all the other assumptions being the same as for the WCDMA case, we have

GSM voice maximum allowed path loss = MS TX power-base station sensitivity
 − body loss − lognormal fade margin + Rx antenna gain − cable loss
 = 30 − (−110) − 3 − 7.3 − 2 + 18 − 3 = 144.7 dB

which is the same as for WCDMA 144 kbps data within 0.1 dB. Therefore we can expect with moderate uplink load factors of 50%, the coverage range for 144 kbps in UMTS to be about the same as for GSM voice at 1800 band.

Table 5.5 Link budgeting template for WCDMA and results for AMR voice 12.2 kbps, 144 kbps data, and 384 kbps data.

Parameter	Voice AMR 12.2 kbps	Data, 144 kbps	Data, 384 kbps	Unit	Formula used when a value is calculated
Enter service bit rate: a	12.2	144	384	kbps	
Enter chip rate: b	3,840,000	3,840,000	3,840,000	chips/s	
Enter Tx power: c	21	21	21	dBm	
Enter Tx antenna gain: d	0	2	2	dB	
Enter transmit antenna cable/body loss: e	3	0	0	dB	
Calculated Tx EIRP (incl. Losses): f	18	23	23	dBm	'= c + d − e'
Enter thermal noise density: g	−174	−174	−174	dBm/Hz	
Enter receiver noise figure: h	4	4	4	dB	
Calculated receiver effective noise density: i	−170	−170	−170	dBm/Hz	'= h + g'
Enter target load factor (set): j	50	50	50	%	
Calculated interference margin: k	3.0103	3.0103	3.0103	dB	'= 10 × log10 (1/(1 − j/100))'
Calculated interference corrected receiver noise density floor (calculated): l	−166.99	−166.99	−166.99	dBm/Hz	'= I + k'
Enter required service E_b/N_0: m	5	1.5	1.5	dB	
Calculated receiver sensitivity (required signal power): n	−121.126	−113.906	−109.646	dBm	'= L + m + 10 × log10(1000 × a)'
Enter receiver antenna gain: o	18	18	18	dB	
Enter receiver cable losses/body: p	3	3	3	dB	
Enter antenna slant loss: q	0	0	0	dB	only on downlink
Enter slow fading margin (for the desired area coverage probability for the service): r	7.3	7.3	7.3	dB	
Enter HO gain (incl. any macrodiversity combining gain): s	2	2	2	dB	
Enter space diversity gain: t	0	0	0	dB	
Enter any other gain: u		0	0	dB	
Enter indoor penetration loss: v	0	0	0	dB	
Enter TPC headroom (fast side fade margin): w	2	2	2	dB	only on uplink
Calculated Maximum Allowed Path Loss: x	146.826	144.6061	140.346	dB	'= f − n + o − p − q − r + s + t + u − v − w'

Radio Site Planning, Dimensioning, and Optimization

Table 5.6 Coefficient values (K) to estimate coverage area from cell range.

value of K	Site configuration			
	Omni	2-sector	3-sector	6-sector
	2.6	1.3	1.95	2.6

The maximum allowed path losses obtained from the link budgeting process, after the iterations given in Figure 5.2 are completed, are then converted to cell coverage ranges for each service using a radio propagation model suited to the area (see Chapter 4). The coverage range obtained can then be converted to coverage area using the formula

$$\text{Coverage area} = K \cdot r^2 \tag{14}$$

where the value of K would depend on the sectorisation, that is the number of sectors used in the site. Approximate values for k are given in Table 5.6. The number of sites is then obtained by dividing the area to be covered by the coverage area of one cell as estimated by Equation (14).

5.3.2 Downlink Load and Transmit Power Checking

On the downlink side, simple link budgeting is not possible due to base station power sharing and the location dependent interference parameters. The base station power required per connection will also depend on the requirements of other users as they would influence the noise floor within the cell. Besides, it is expected as was discussed in Section 5.1, that the service coverage range is more dominated by the uplink in 3G, whereas the downlink drives the capacity due to the base station power sharing among the connections. Therefore the simple link budgeting for the downlink amounts to a determination of the total base station transmit power required for supporting a number of connections from a number of different services (a service mix) within the service coverage ranges determined by the uplink link budget and the traffic model. The formula for the base stations transmit power was derived in Chapter 4 (Equation (20)) and is repeated below:

$$P_T^{(m)} = \frac{P_N \cdot \sum_{k=1}^{M} \frac{\rho_k R_k \upsilon_k}{W \cdot g_{mk}}}{1 - \left\{ \gamma_{cmm} + \sum_{k=1}^{M} \frac{\rho_k R_k \upsilon_k}{W} [(1 - \alpha_k) + f_k] \right\}}$$

in which the bracketed expression in the denominator is the effective downlink load factor, η_{DL} for the cell, that is (Equation (21) from Chapter 4):

$$\eta_{DL} = \gamma_{cmm} + \sum_{k=1}^{M} \frac{\rho_k R_k \upsilon_k}{W} [(1 - \alpha_k) + f_k]$$

For the location dependent parameters, α_k and f_k, typical average estimates over the whole (as given in Section 5.3.5) are used for this quick link budgeting. For the channel gain parameter, g_{mk}, the following formula can be used that generates a constant value for the whole cell:

$$g_{mk} = M_{dist}/Lp_{max} \tag{15}$$

where Lp_{max} is the maximum path loss obtained from the uplink link budgeting (obtained after the iterations given in Figure 5.2 are completed) and M_{dist} is a factor related to the path loss distribution of the UEs within the cell. That is, it scales down the path loss given by Lp_{max} by considering that not all the traffic is concentrated at the cell edge. In practice a value of around 6 dB (3.98 in linear scale) is normally assumed for M_{dist}. However its precise value would depend on the path loss distribution of users within the cell.

Equation (20) from Chapter 4 (repeated above) can be used to guide the dimensioning of the base station for a service mix, whose uplink coverage is already met by the uplink link budgeting. If the required base station power is above the maximum base station transmit power limit, and/or the downlink load factor, η_{DL}, exceeds the target design value, some of the services are moved to a second carrier if available or the cell coverage range is reduced and the link budgeting iterations repeated as illustrated in the iterative link budgeting flowchart given in Figure 5.2. A reasonable threshold for the maximum downlink load factor is 75%.

Thus the final cell range is determined not just by the link budgeting done on the uplink, but also by the capacity limitations as determined by the downlink transmit power requirements and the downlink target load factor. For the location dependent parameters g_{mk}, α_k, and f_k, typical average estimates as given in Section 5.3.5 are used for this quick link budgeting. It is noted that Equation (20) from Chapter 4 does not incorporate the soft handover connections explicitly. However, the reduced power required on connections with soft handover branches to neighboring cells (due to macro diversity gain of soft handovers) is assumed to compensate for the power transmitted on soft handover branches from the neighboring cells in this quick approximate link budgeting process.

5.3.3 Downlink Link Budgeting for the Pilot Channel (P-CPICH)

It is important to ensure a sufficient link budget for the primary pilot channel P-CPICH for a required coverage. Likewise, the coverage of the pilot channel for an acceptable power may need to be estimated. In either case, the downlink cell range dimensioning for the CPICH channel can be performed through the following link budgeting formula:

$$L_{Pmax} = P_{CPICH} - S_{UE} - M_{LNF} - L_{BL} - L_{CPL} - L_{BPL} + G_{Tx} + G_{Rx} - L_{\text{feeder and jumpers}}, \text{ in dBs} \tag{16}$$

where P_{CPICH} is the power allocated to the pilot channel, S_{UE} is the user equipment sensitivity for the pilot channel detection (which is derived below), M_{LNF} is the required lognormal fading margin, L_{BL} is the body losses, L_{CPL} is the car penetration losses, L_{BPL} is any building penetration losses (for extending coverage indoor), G_{Tx} and G_{Rx} are the base station and the user equipment antenna gains, respectively, and $L_{\text{feeder and jumpers}}$ is the combined feeder and jumper losses. Since the pilot channel is not power controlled, there are no power control related margins involved.

The user equipment sensitivity for the pilot channel, S_{UE}, is derived in analogy with the derivation of the base station sensitivity for a traffic channel (Equations (8) and (9)), except that here we have $R_b = R_c$, and hence there is no spreading gains, and the noise rise due to

Radio Site Planning, Dimensioning, and Optimization

interference should be calculated at the extreme end of the cell boundary. It is easy to see that the UE sensitivity for pilot channel detection then takes the form

$$S_{UE,CPICH} = \frac{E_c}{I_0} \cdot \left\{ P_n + \frac{P^T}{L_P}[(1-\alpha) + f] \right\} \quad \text{in linear scale} \tag{17}$$

in which E_c/I_0 is the required pilot signal to noise ratio for the coverage criteria (see Section 5.2.1), P_n is the total thermal noise power within the band (calculated by the same formula as given in Equation (9), but at the UE), P^T is the base station (Node B) transmit power allocated to all channels other than the pilot and the synchronization channels, and α and f represent the orthogonality factor and the other-cells to own-cell interference ratio at the cell boundary, respectively. The L_p is the path loss from the base station to the cell boundary (in linear scale, and larger than 1) and is used to down scale the P^T for estimating the received interference power from the own cell at the cell border. A typical reasonable value to assume for α is about 0.35 for the urban environments and 0.85 for rural areas. A suggested value to assume for f at the cell border for use in Equation (16) is around 2.

5.3.4 HS-PDSCH Link Budget Analysis

This section presents a link budgeting analysis for the downlink high-speed shared packet channel, HS-PDSCH, under assumed base station transmit power and interference geometry parameters. The latter are used to calculate the interference noise floor on the downlink. Since this channel uses adaptive modulation depending on the radio conditions of the channel, the bit rate can change from time to time. Therefore, the analysis of link budgeting and power requirements is based on a symbol level. Furthermore, due to packet retransmissions, the actual average symbol rate (actual number of symbols transmitted per time period on average) is m times the effective average symbol rate, where m is the average number of times a packet is transmitted due to the hybrid ARQ mechanism [17].

The following notation is used for this analysis:

- C = total Node B transmitted power
- P_{hs} = the portion (fraction) of Node B power available for HS-PDSCH
- N = the number of simultaneous HS-PDSCH code channels in use
- S = the power assigned per HS-PDSCH code channel
- m = average number of transmissions per packet (accounting for packet re-transmissions)
- R_s = the average symbol rate per code channel on HS-PDSCHP (includes re-transmissions)
- R_e = the average effective symbol rate per code channel on HS-PDSCH
- ρ_s = the required signal to noise ratio (on a symbol basis), E_s/N_0 (accounting for re-transmissions)
- S_{sen} = the minimum required energy per symbol per code channel at the input to the UE to satisfy the required ρ_s, with retransmissions accounted for
- N_0 = the effective thermal noise power spectral density at the input to the mobile station antenna
- W = the WCDMA chip rate (3.84 Mcps)
- α = the orthogonality factor at the user terminal location

- f = the ratio of the multi-user interference received from other cells to that from the home cell at the terminal location

From the above definitions, the hybrid ARWQ re-transmissions imply:

$$R_s = m.R_e \tag{18}$$

And from the definitions given above,

$$S = C.P_{hs}/N \tag{19}$$

From the definition of the symbol power to noise power spectral density ratio,

$$<E\Theta> \rho_s = \frac{S_{sen}.W/R_s}{[(1-\alpha)+f](C-P_{hs}.C)+N_0..W} \tag{20}$$

where the first term in the denominator represents the multi-user interference power and the second term the thermal noise power. Note that no interference is assumed between multiple codes of HS-PDSCH, since they are assigned to the same user at a time, and similarly detected by the same Rake receiver.

Solving for S_{sen}, gives the minimum received symbol power required per code channel at the user terminal, that is the UE sensitivity per code channel:

$$S_{sen} = \frac{R_s}{W}\{[(1-\alpha)+f)]C(1-P_{hs})+N_0W\}\rho_s \tag{21}$$

Then the maximum allowed path loss is calculated as

$$\begin{aligned}L_{max,pathloss} = S - S_{sen} + G_{ant,NB} - L_{feeder,NB} + G_{ant,UE} \\ - L_{feeder,UE} + G_{HO} - M_{shadowing} - L_{penetration} + G_{oth}\end{aligned} \tag{22}$$

where $G_{ant,NB}$ is the transmit antenna gain, $L_{feeder,NB}$ is the transmit antenna feeder losses, $G_{ant,UE}$ is the receiver antenna gain, $L_{feeder,UE}$ is the UE antenna feeder losses, G_{HO} is the gains from any soft handovers, $M_{shadowing}$ is the required margin against slow shadow fading (lognormal fading), $L_{penetration}$ is any building or car penetration losses, and G_{oth} is used to capture any other gains or losses, such as those stemming from the packet scheduling algorithms implemented, etc. Note that the ratio W/R_s represents the spreading gain, which is fixed at 16 on the HS-PDSCH channel [17].

5.3.5 Setting Interference Parameters

In this section, typical reasonable values for interference parameters f (other-cells to own-cell interference ratio), α (the orthogonality factor on downlink), and design values for uplink and downlink load factors (see Chapter 4) are given based on previous simulation results, but can be updated with any new results concluded through simulation or measurements for the area and the antennas used.

The f Parameter
Suggested values for the f parameter on the uplink are shown in Table 5.7.

Table 5.7 Suggested f values on uplink.

Environment	Site-configuration	
	3-Sector	6-Sector
Typical Urban, TU-3	0.79	0.99
Typical Urban, TU-50	0.77	0.96
Typical Rural, RA 50	0.87	1.1

The notation in Table 5.7 is as follows:

- ITU-3: the ITU defined model for typical urban environment with 'vehicular speed' of 3 km/hr
- ITU-50: the ITU defined model for typical urban environment with 'vehicular speed' of 50 km/hr
- RA-50: the ITU defined model for typical rural environment with 'vehicular speed' of 50 km/hr

For the downlink side, the value of f would vary from location to location within the cell with typical values between 0.4–1.1 or more (at cell edge). However, mean values over the cell (i.e., sector) of 0.72 for the 3-sector site and 0.84 for the 6-sector site are typical numbers reported by simulations carried out by vendors. The planner may use these values in the absence of any more updated data.

The Downlink Orthogonality Factor, α

The downlink orthogonality factor is a measure of channelization code orthogonality left in the downlink signals after experiencing multipath. In the case of zero multipath, the code orthogonality is 1 (all received users signals are orthogonal to each other). However in practice, there is always some multipath in the mobile communication environment. The amount of multipath is the largest in the macro-cells where the signals can experience many reflections and diffractions through the larger path distances traveled, etc. Therefore a value of around 0.5 is normally used for the orthogonality factor. In the small micro-cell areas, where there are usually more line-of-sight signals with relatively small amounts of delay spread, an orthogonality value of from 0.7–0.9 is used.

Setting Load Factors

The load factors determine the multi-user interference resulting at the base station and on the downlinks. The higher a load for which the network is design, the heavier the interference. As was shown in Chapter 4, when the load factors approach the limit of 1, the pole capacities are reached, at which point infinite power is required from both the base station and the mobile station to cope with the interference floors. Therefore the network plan should target to leave some margin to the limiting value to avoid performance degradation.

Normally, in urban areas where there is high traffic density and demand for capacity, the network is planned for larger load values. In rural areas the traffic density is sparser and therefore the network is planned for smaller values of load to increase the cell coverage

Table 5.8 Recommended target load guidelines.

Environment	Suggested load factor setting guidelines	
	Uplink	Downlink
Urban	not to exceed about 70%	not to exceed about 80–85%
Suburban, etc.	not to exceed about 60%	not to exceed about 70–75%

ranges and thus save on site counts. However, in both cases, carefully chosen levels of uplink loading should be planned to handle both the short term and medium term projected traffic. After the network is deployed, it is rather difficult to increase the cell uplink load limit without having to interleave additional sites to maintain service coverage performance. On the downlink side, a higher load due to increased traffic can be handled simply by adding more base station power amplifier modules. The guidelines given in Table 5.8 are based on vendor recommendations, and it is best to ensure that they are not exceeded in most circumstances.

The exact technology that is used in implementing the WCDMA the receiving systems will also have a bearing on the limit of the load factors that can be used. For instance, if multi-user detection is used in the base station (which tries to estimate and cancel the multi-user interference), even higher loads can be planned for than those given in Table 5.8.

5.4 Simulation-based Detailed Planning

In the previous sections on uplink link budgeting for the determination of nominal service coverage, we assumed an aggregate load factor at the base station. This helped to calculate the total interference noise floor at the base station receiver, and hence use it to estimate the link budgeting and coverage range for each specific service. However, because the aggregate load factor will depend on the service mix and traffic within the area, an iterative process was required until the final cell coverage range calculated based on the strictest service resulted in an uplink load that did not exceed the initially assumed target design value in the static iterative link budgeting. This analysis did not require traffic rasters, i.e., the distribution of the traffic over the area. On the downlink, simple link budgeting was not even proposed except for the pilot channel and a specific service (the HS-DPSCH), the latter by assuming certain aggregate interference parameters resulting from other connections in the cell. The downlink link budgeting is complicated as users at different locations within the cell see a different level of interference (noise rise). For instance, users nearer the cell edge will experience weaker signals from the own cell, but stronger interfering signals from the neighboring cells than users closer to the site. Also, the orthogonality factor, which affects the multi-user interference, can vary from location to location within the cell on the downlink. Furthermore, there is power sharing involved on the downlink, which implies further service interaction on the downlink in the determination of coverage for each service. These considerations mean that coverage for a given service will not only depend on the requirements for that service but also on the service mix (service profile) and traffic densities within the cell.

As an alternative to the deterministic static link budgeting analysis proposed in Figure 5.2, a simulation process can be implemented on digital computers to determine the convergence between service coverage, and system resources and service requirements, for a specific mobile location list and expected traffic at each location. That is, to determine the convergence between the users that can be served from a specific mobiles location distribution list and the design/system parameters. This is referred to as static simulation [18–20] because in each run the locations of the mobiles are fixed within the cell. In the iterative steps for the UL and DL, the required transmission powers for each mobile and the base station are calculated, respectively. Based on that, the mobile sets that can be served at each step are determined until certain iterated parameters such as the changes in the re-calculated noise floor (load factor) do not change above a certain preset value.

Each mobile service-location list will represent a particular snapshot of the mobile locations under unchanging traffic situation; that is, the location of the mobile users along with the bearer used by each at a particular time. A single snapshot may be used to determine the service coverage picture for a particular mobile service-location distribution. Sometimes, a particular snap shot can be quite revealing as certain phenomena of the networks operating characteristics can show up in a specific iteration, that is, with a certain distribution of users locations. Also, a representative mobile service-location distribution list may be used to perform 'what-if' studies, such as the change on network performance when a cell is reconfigured or a site is relocated and so on.

Multiple snapshots, each representing possible users locations within the area, can be used to simulate the impact of user location randomness on network coverage-capacity. The results from each snapshot can then be combined to provide statistically reliable and relevant results by averaging out the dependency of the results from mobile locations. This is particularly more critical for scenarios in which there are just few users but with very high bit rate data. This method is referred to as the so-called Monte Carlo simulation based analysis [18–20].

Aggregate link performance analysis based on iterative simulations as just discussed forms an essential part of WCDMA radio network planning (RNP) tools. The RNP tools should be able to use and combine the results of any number of iterations specified by the network planner as well as store the results pertaining to individual iterations for later review. Advanced RNP tools are also capable of generating mobile location distributions, inputting user specified lists, and setting and checking of pilot power, as well as accounting for adjacent channel interferences from a second carrier or a different operator.

In the following two sections, the iteration steps for the uplink and the downlink are discussed, and illustrated by flowcharts.

5.4.1 Uplink Simulation Iterations

The objective in the uplink iterations is to determine a subset of users, services, and locations from a given user population, service mix, and distribution, which can provide a convergence between the final base station sensitivity and the available power from each mobile. Each initial service mix and distribution (user locations, services for each, and service E_b/N_0 requirements) forms one snapshot of a potential system traffic configuration, which is generated randomly according to some desired statistics (expected user locations, services, etc.). The simulator must be able to read in a user-defined data file, which specifies the geographic traffic weighting and service to be provided at the location. For each Monte Carlo snapshot then, the mobiles and services for each are assigned random positions observing the specified weightings and service.

For each snapshot, the power control process is run at each step of the iteration by estimating the transmit power required from each user at the location assumed to fulfill the E_b/N_0 requirement of the base station for the service. The received power at the base station for each user and hence the transmit power required from the user will depend on the sensitivity level of the base station at each iteration step, the service bit rate, mobile speed, and the link losses. Adjustments should be made for the service activity factor, the soft handover gains, and the power control headroom needed to compensate for fast fading. The sensitivity of the base station at each step is updated by the noise rise obtained by the $1/(1-\eta)$ factor where η, the load factor, was defined by Equation (1). If the transmit power required from a mobile exceeds the maximum allowed or feasible value, it is either moved to a second carrier, if available (a multi carrier design case), or put to outage. This iterative process is continued until the changes in the power requirements from each mobile fall below a specified convergence threshold. In the meantime, if the uplink loading of the cell exceeds a specified design limit, mobile stations are discreetly moved to another carrier or otherwise put to outage. Once convergence is achieved, the number of mobiles failing to achieve their targets can be determined. This number and its constituents (services used by each mobile) will vary from snapshot to snapshot. Therefore it is important to perform sufficient number of snapshots to obtain a statistically significant representation of service mix and distributions that can be supported. Figure 5.3 shows a typical flowchart for the mobile uplink iterations. The detailed results from implementing the uplink static simulations just discussed are presented in References 8 and 21. The simulation runs presented have shown that overall the uplink dimensioning results (in terms of the number of connections covered from single as well as mix services within the cell coverage range estimated by link budgeting) are more or less in line with the simulation results. In the case of speech services, the simulations underestimated the number of users supported by about 16% compared to the results from dimensioning. The difference is attributed to the large standard deviations in the value of the other-to-own cell interference factor (the f factor in Equation (6)), which is assigned just a single fixed value in the dimensioning efforts for all cells.

5.4.2 Downlink Simulation Iterations

The objective in the downlink iteration is to allocate the base station transmit powers for each link in use by the mobile station (including SHO connections) until the effective C/I received at each mobile meets the required carrier-to-interference defined by

$$\text{Target C/I} = \frac{E_b/N_0}{W/R} \qquad (23)$$

where E_b/N_0 is the required received E_b/N_0 for the MS, which depends on the terminal speed, service bit rate, and quality requirements. The effective received (C/I)m for mobile m is calculated using maximal ratio combining according to the following equation, by summing the C/I values of all links k in use by mobile m:

$$(C/I)_m = \sum_{k=1}^{K} \frac{P_{km}/L_{km}}{(1-\alpha_k).P_k/L_{km} * I_{oth,k} * N_m} \qquad (24)$$

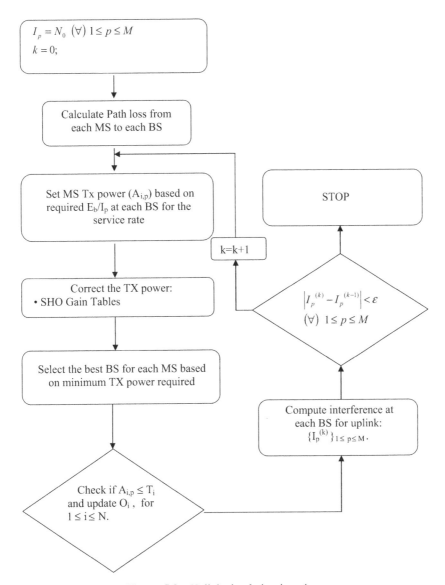

Figure 5.3 Uplink simulation iterations.

where P_k is the total transmit power of the base station to which link k is established, L_{km} is the link loss from cell k to mobile station m, α_k is the cell specific orthogonality factor, P_{km} is the power allocated to the link from base station k to mobile m, $I_{oth,k}$ is the other cell interference, and N_m is the background thermal noise in the receiver of mobile m.

Since the required transmit powers to each mobile will depend on the powers allocated to other mobiles, an iterative process is necessary until convergence between the target C/I and the received effective C/I at each mobile station is achieved. Thus at each step of the iteration, the transmit powers to each mobile are adjusted according to the difference between the target and the effective received C/I at the mobile. If, at each step, certain link power limits or the total base station transmit power are exceeded, mobile stations are taken out randomly (or perhaps according to the ones requiring the largest transmit power), or moved to a different carrier if available. In the meantime, the received primary CPICH E_c/I_0 is checked for each mobile, and if it is below a certain threshold necessary for proper synchronization, the mobile station is put to outage or moved to the second carrier if available. Figure 5.4 shows a typical flowchart for the downlink iteration.

The initial transmit power in the iteration process is estimated based on the initial sensitivity value calculated for the mobile assuming a reasonably acceptable value for the local downlink load factor. This is used to estimate the transmit power required for the best server. After the transmit power of the dominant server is allocated, the transmit power of the other SHO connections for each mobile are determined by making adjustments according to the difference between the P-CPICH power of the base stations where the link is located and that of the best server. Mathematically, the transmit power for the soft handover link j to mobile m is calculated from

$$\text{TxPower}_{m,j} = \text{TxPower}_m + \text{P\_CPICHPower}_j - \text{P\_CPICHPower}_{(bestserverDL)} \quad (25)$$

where TxPower_m is the power transmitted from the best server to the mobile.

Estimating the C/Is and Adjusting for the Tx powers

To estimate the C/I at each iteration step in the downlink for each mobile, we need to calculate the resulting C and I at each step for each link to the mobile. Since there are soft handover connections involved, we need to estimate the contribution of each to the carrier-to-noise ratio and sum up in linear scale to get the effective C/I. The power received at the mobile station m over a link to base station j is

$$C_{m,j} = P_{m,j}/L_{m,j} \quad (26)$$

in which $P_{m,j}$ is the transmitted power over link j to mobile m, and $L_{m,j}$ is the path loss over the link.

The own cell interference is

$$\text{Iown}_{m,j} = (1 - \alpha_m) \times (P_{T,m} - P_{m,j})/L_{m,j} \quad (27)$$

where α_m is the orthogonality factor as seen by mobile m, P_{Tj} is the total transmitted power of base station j (including over the common signaling channels).

The other-cell interference for mobile m in base station j can be written as

$$\text{Ioth}_{m,j} = \sum_{\substack{\text{Other BSs, excluding BSJ,} \\ \text{and the SHOBSs}}} P_{T,i}/L_{i,m} \quad (28)$$

where $P_{T,i}$, the total transmitted power from a base station i, should be adjusted for the adjacent channel leakage ratio (ACLR) of the base station transmitter, and the adjacent channel protection (ACP) of the receiver [22] in mobile station m, if the base station is

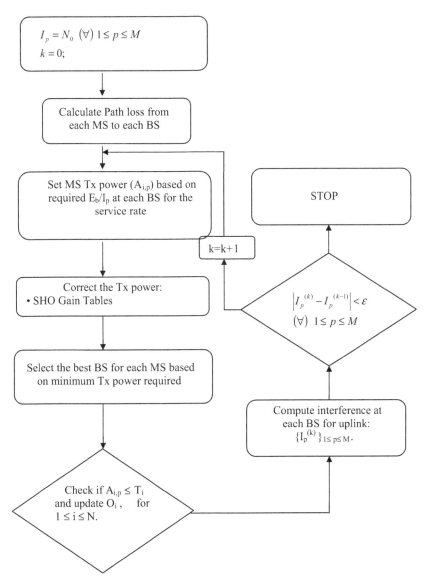

Figure 5.4 Downlink iteration flowchart.

transmitting on a different carrier frequency (from the same operator or a different operator). The $L_{i,m}$ is the path loss ratio from a base station i to the mobile station m.

Then, the contribution to C/I on the downlink to mobile m, from base station j is calculated by

$$(C/I)m, j = \frac{P_{m,j}/L_{m,j}}{Iown_{m,j} + Ioth_{m,j} + N_T} \qquad (29)$$

in which N_T is the total effective thermal noise in linear scale of mobile m (and should include the effect of the receiver noise figure)

The total effective C/I for mobile m is obtained by summing over all the connections to that mobile (include SHO connections), therefore

$$(C/I)_m = \sum_{i=1}^{SHOConnections} (C/I)_{m,i} \qquad (30)$$

in which the summation is done in linear scale.

Now, the SHO gain with respect to base station j (the best server), is simply

$$\text{SHO Gain} = (C/I)_m - (C/I)_j \qquad (31)$$

The (C/I)m is compared to the target (C/I)m for the service, from which the SHO gain is subtracted:

$$\text{Delta}(C/I)_m = [\text{Target}(C/I)_m - \text{SHO Gain}] - (C/I)_m \qquad (32)$$

This is used to correct the transmit power from base station j to mobile m (in dB):

$$\text{BaseStation TxPower}_{m,j} = \text{BaseStation TXPower}_{m,j} + \text{Delta}(C/I)_m \qquad (33)$$

The iterations for Delta (C/I) and the transmit power calculations are repeated until the maximum value of Delta (C/I) is less than a specified threshold.

The results of downlink simulations carried out [21] have confirmed the results from dimensioning, in that the downlink is the limiting factor for capacity. The results show that the capacity estimates based on just the uplink are a bit too optimistic, and therefore the downlink analysis should be carried out for capacity estimations.

5.4.3 Area Coverage Probabilities

Generally, coverage for a service is dependent on the reception of a desired signal level against the total effective noise floor at the receiver. In WCDMA, the effective noise floor is dependent on the traffic distribution, traffic mix, and their E_b/N_0 requirements. As these change, so does the network coverage, and the coverage itself under a given traffic mix and distribution will also be service dependent. In other words, under a given network configuration, traffic distribution, and mix, the coverage extent for a given service will depend on its E_b/N_0 requirement. While one service may be supported with high coverage probability, another with a higher E_b/N_0 requirement may not. And furthermore coverage for a service may be uplink or downlink limited depending on the effective noise floors at the base station and at the mobile location. Therefore, unlike in GSM, it is not possible to quantify a certain area coverage probability in UMTS independent of the service and independent of the traffic mix and distribution in the area. Thus each service category is expected to have a different area coverage probability in a given dimensioning result.

In simulating a given traffic load distribution on the dedicated channels, it is also important to ensure that sufficient coverage is provided on the cell's common control and signaling channels. Otherwise, convergence on the uplink and downlink traffic channels alone in the iterative simulation processes discussed in the previous sections will be practically useless, if the users cannot detect the pilot channels, or receive paging messages, etc. We will discuss coverage analysis on the signaling channels in the following two sections.

5.5 Primary CPICH Coverage Analysis

One of the important optimization parameters is the percentage base station power that is allocated to the primary pilot channel, P-CPICH. Normally 5–10% of the total BS power is assigned. The power assigned to the P-CPICH channel should provide a tradeoff between providing adequate signal for synchronization and measurement by users in the cell, for adequate soft handover areas with neighboring cells, and prevention of pilot pollution. Pilot pollution occurs when users hear too many equally strong pilot signals in the same area (normally not to exceed 3).

The E_c/I_0 is a key indicator used in handover decisions, cell reselections and other physical layer procedures, and is therefore often used as a coverage indicator for the cell. In E_c/I_0, the E_c is the energy contained in one chip of the signal and I_0 is the total effective noise + interference power spectral density in the band. Usually a threshold in the range -14 to -12 is considered adequate. The E_c/I_0 can be calculated from the following expression

$$E_c c/I_0 = \frac{P_{CPICH}/L}{\sum_{i=1}^{N} P_{Tx,i}/L_i + I_{ACI} + N_t} \tag{34}$$

where P_{CPICH} is the power of the P-CPICH for the cell, L is the link loss to the base station, $P_{Tx,i}$ is the total transmit power of BS i, Li is the link loss to BS i, I_{ACI} is interference from adjacent channels, N_t is the total effective thermal noise of a mobile station, and N is the total number of base stations in the RF visible range of the cell for which the pilot channel is being analyzed. The E_c/I_0 is then compared to a design level threshold, and the P-CPICH coverage area ('area coverage probability') is defined as the ratio of pixels in which the threshold is exceeded. For adequate soft handover overlap with neighboring cells, we should have about 2–3 neighboring cells whose E_c/I_0 will fall within roughly less than 5–6 dB from the E_c/I_0 of the best serving cell.

5.6 Primary and Secondary CCPCH Coverage Analysis

The most important common control channels are the Broadcast Common Channel (BCCH) and the Paging Channel (PCH). The BCCH can be carried either by the broadcast Channel (BCH) or (rarely) by the Forward Access Channel (FACH). BCH is then mapped on the Primary Common Control Physical Channel (P-CCPCH), while FACH and PCH are mapped to either the same or a different Secondary Common Control Physical Channel (S-CCPCH). For mobiles to be able to decode properly the information transmitted on the BCCH and PCH channels, a certain threshold must be met for the E_b/N_T on these channels. Generically, we can write the following expression:

$$E_b/N_T = \frac{P_{CCPCH}}{(1-\alpha)\cdot\dfrac{P_{tot,BS}}{L_{BS}} + \sum_{k,k\#BS} P_{tot,k} + I_{ACI} + N} \cdot \frac{W}{R_{CCPCH}} \quad (35)$$

where P_{CCPCH} is the transmit power and R_{CCPCH} is the bit rate of either primary or secondary channel CCPCH, $P_{tot,BS}$ is the total transmit power, L_{BS} is the link loss of the best server, α is the effective orthogonality factor in the area, I_{ACI} is adjacent channel interference, N is the effective background noise, and W is the chip rate. The summation in the equation is taken over all the network base stations in the area except the cell whose E_b/N_T is being calculated. Equation (35) is calculated for all the pixels in the cell (best serving cell), and the area coverage probability for the selected channel is estimated by the fraction of pixels in which E_b/N_T exceeds the required threshold. By constructing the CDF using different thresholds for E_b/N_T, the necessary E_b/N_T to achieve a certain given coverage probability and hence the transmit power can be estimated.

5.7 Uplink DCH Coverage Analysis

Since coverage is mostly uplink limited for the traffic channels, the coverage on a dedicated channel (DCH) can be estimated and verified for a given service with a certain bit rate by checking in each pixel whether the maximum allowed transmit power of the MS can fulfill the E_b/N_0 requirement at the base station. The signal to (noise + interference) received at the base station from a virtual mobile (i.e., passive terminal) at a pixel i within the coverage area determined for the service (based on either a static iterative link budgeting analysis or the Monte Carlo detailed simulation snapshots) is as follows:

$$\frac{P_{Tx,MS}/R}{\dfrac{N_0}{1-\eta}\cdot\dfrac{1}{W}\cdot L_P} = \rho \quad (36)$$

where $P_{Tx,MS}$ is the power transmitted from the mobile, R is the service bit rate, Lp is the path loss factor on the radio link between the mobile and the base station, N_0 is the effective thermal noise at the base station receiver input, η is the design uplink load factor, ρ is the required E_b/N_0 for the service, and W is the WCDMA chip rate.

Solving Equation (36) for $P_{Tx,MS}$, gives

$$P_{Tx,MS} = \frac{L_P\cdot N_0\cdot\rho}{(1-\eta)\cdot\dfrac{W}{R}} \quad (37)$$

Then the coverage probability for the service is defined as the ratio of pixels within the area for which the required transmission power from the mobile, $P_{Tx,MS}$, as given by Equation (36), will not exceed:

> MIN (MAX transmission power capability of the mobile station, MAX allowed transmission power into the cell).

5.8 Pre-launch Optimization

Pre-launch optimization is carried out when a cluster of sites is on air (in barred mode with no commercial traffic). The objective of pre-launch optimization is to achieve the best parameter settings and configuration for the planned network. The main objectives of pre-launch optimization are as follows:

- Verification and adjustment of neighbor lists (CDMA systems are very sensitive to neighbor lists)
- Verification of actual pilot scrambling codes versus planned
- Minimization of pilot polluters
- Evaluation of bearer throughput and performance (FER (frame error rate), BLER (block error rate), BER (bit error rate), RxQual (Receive quality), etc.)
- Measurement of end-to-end delay
- Optimization of soft, softer, and inter-frequency and inter-system handover

Drive testing is the main approach in the pre-launch optimization phase to collect the data needed to achieve the above objectives. For stress testing the network throughput in this phase, orthogonal channel noise simulators (OCNS) are used to load the sites on the downlink side and thus raise the effective load in a cell and its neighborhood.

5.9 Defining the Service Strategy

Before planning and deploying the network, the operator must define and develop a service strategy that fits its business objectives. The service strategy then forms a basis for the network planning criteria. WCDMA offers a wide range of services with bit rate and QoS characteristics fitted to different types of applications as discussed in Chapter 15. The service strategy should define what services and applications are going to be provided and with what level of quality in different areas. The characteristics of the user populations for whom the services are targeted and the competition in the planned service areas should also be factored in. Agreements must be reached within the operator's organization and with service providers as to a set of reasonable QoS levels for different services. The service area demography should also be considered by taking into account user educational and income levels, because these items affect expectations and perception of the service. The QoS levels and threshold should also take into account whether the service is going to be extended in buildings and inside vehicles.

5.10 Defining Service Requirements and Traffic Modeling

Advanced network planning requires the specification of detailed and realistic service characteristics and requirements such as bit rate and E_b/N_0 requirements for real time services. For non-real time services, the delay tolerance limits and bit rate or throughput characteristics (such as minimum, average, maximum, etc.) should also be specified. Due to the inter-linking of capacity and coverage in UMTS, it is also critical to have a reasonably accurate forecast of the traffic and different services for at least the near future. This would help the operator to plan site locations, densities, and the layering architecture properly to provide the required coverage while being able to handle the required capacity. Projecting the realistic trends in future traffic and services can help to develop flexible network plans that can save much time and cost in re-planning and adjustments as the traffic grows. Existing operators can make use of the knowledge on some of the traffic patterns on their GSM networks such as the traffic measurements collected in the location of traffic hot spots, peak hour traffic, etc. This information can be combined with knowledge about new services in 3G to project what may be expected in the future. Green field operators can obtain demographic data, population densities, income levels, area business developments, etc. to estimate the level of traffic and range of services they may expect their networks to handle. The operator can assign different weighting methods when assigning traffic densities to different areas. For example, equal distribution or weighting based on clutter or other criteria can be used. Speed targets can also be assigned to mobiles based on clutter and area type.

Traffic modeling in 3G systems also requires that the traffic densities be estimated within raster resolutions usually in the range 1–200 meters depending on the environment. Typically the minimum acceptable resolution is 5 meters for dense urban areas, due to the small geographical cell sizes in such areas. For urban areas, a resolution of 10–15 meters is considered to be reasonable. The finer the resolution of course, the more accurate will be the results.

Different traffic forecasts and distributions can be used as snapshots into the aggregate link performance simulator (RNP TOOL). The results can then be combined to average out the effect of uncertainties and randomness in the traffic distribution and forecasts. This will result in statistically more reliable network designs.

The E_b/N_0 requirements for a service depend generally on the receiver structure, detection used, coding, and channel [14]. This is collectively reflected in the vendor's receiving equipment (base station and mobile) reference sensitivity figures. ITU has specified E_b/N_0 requirements for certain standard voice and data services through extensive link level simulations carried out under standard ITU multipath channel profiles. Vendors also provide reference sensitivity figures based on extensive link level simulations carried out with their equipment, which may exceed the ITU reference sensitivities. The receiver reference sensitivities used in the ITU link level simulations are the 3GPP minimum specified requirements for base station and mobile receiver sensitivities. The ITU's Vehicular A model at 3 km/h is usually used for specification of the E_b/N_0 requirements for each service in urban and suburban environments. The ITU's Vehicular A model at 50 km/h is then used for specification of the E_b/N_0 requirement for each service in rural type areas.

The E_b/N_0 requirements for higher bit rate services are usually lower for the same quality. The higher bit rate services require higher transmission power to result in the same energy per bit (and hence the same E_b/N_0 for the same quality). But the higher transmission power result in improved channel estimation, which then helps to reduce the required E_b/N_0. The results of link level simulations carried out [10] for packet data under the ITU Pedestrian A channel model at 3 km/h for various bit rates are given in Figure 5.5, which shows the typical variations of E_b/N_0 required with the data rate.

The E_b/N_0 requirement for a service also depends slightly on the speed of the mobile user. Higher speeds require somewhat higher E_b/N_0 (by a few tenths of a dB) for channel estimation and tracking. Typical trends in the variations of the E_b/N_0 requirements for AMR 12.2 speech under different mobile speeds are shown in Figure 5.6 [10].

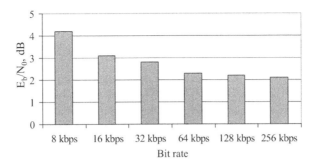

Figure 5.5 Packet data link level simulation results [10] with different bit rates under the ITU Pedestrian channel model A at 3 km/h.
[*WCDMA for UMTS*, H. Holma and A. Toskala, © 2000. Copyright John Wiley & Sons, Ltd. Reproduced with permission]

5.11 Scrambling Codes and Planning Requirements

The scrambling or pseudo-noise (PN) codes generated by mostly a linear feedback shift register have low cross-correlation. This makes them suitable for cell and call separation. They have very low cross-correlation, which make them suitable for cell and call separation in a non-synchronized system. The scrambling codes are used on the downlink to identify a whole cell. On the uplink a scrambling code is assigned to each call or transaction by the system. The scrambling codes are divided into short and long code classes. The short code is a complex code constructed from two 256 chip codes of the very large Kasami set [23]. The long code is a 38,400 chip segment (10 ms) of Gold code of length $2^{41}-1$. Limiting the code length to 38,400 chips was an attempt to facilitate and speed up the code phase search and acquisition process for the terminals (on the downlink side). The uplink scrambling codes can be either the short or the long codes and are assigned by the system.

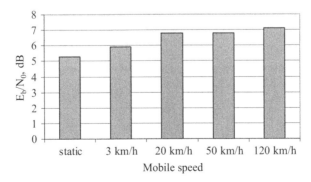

Figure 5.6 Link level simulations results for AMR 12.2 speech under ITU channel model A [10].
[*WCDMA for UMTS*, H. Holma and A. Toskala, © 2000. Copyright John Wiley & Sons, Ltd. Reproduced with permission]

The downlink uses the long Gold codes that are also used on the uplink. The complex scrambling code on the downlink is formed by simply a delay between the I and the Q branches. No short codes are used in the downlink. A primary set of codes limited to 512 has been assigned for downlink scrambling in UMTS to avoid elongating the cell search time. The 512 primary codes are considered to be sufficient from the network planning perspective. Codes from a secondary set of codes can be assigned for beam steering on dedicated channels in case of adaptive antenna deployment without disturbing the downlink code plan. The scrambling codes must be assigned to sectors in the network planning process. Because of the large number of scrambling codes, the code planning is a very trivial task. The codes are organized in 64 code groups each containing 16 different scrambling codes. All cells visible to a mobile in an area should be assigned different scrambling codes. The suggested code assignment strategy would be to assign codes from different scrambling code groups to neighboring cells, resulting in a code re-use of 64.

It is recommended that only one scrambling code be used on the downlink per sector in order to maintain orthogonality between different downlink code channels. If under certain circumstances, for capacity reasons for instance, a secondary code is needed to be added to a cell, then only the additional users that could not be accommodated under the first code should be moved to the second one. This would help to reduce the number of code channels interfering with each other compared to an even distribution of users between the two codes.

5.12 Inter-operator Interference Protection Measures

Inter-operator interference from operators using adjacent frequencies can have a considerable impact on the performance of WCDMA networks. An operator has some control over the interference generated from its own network, but has no effective control over interference generated by other operators. This section addresses the issue of inter-operator interference. The sources of this interference based on the WCDMA system are considered and it is shown how likely it is to affect the system performance and create dead zones, around base stations of other operators.

Interference between operators occurs when their frequency bands are sufficiently close to each other. The 3GPP specifications TS 25.101 and TS 25.104 provide details of the filter requirements intended to eliminate much of this problem. The filters should have a response that will pass the desired signal and cut-off quickly within the guard-band between adjacent frequency bands. The specifications have been designed with cost and design constraints taken into account. Consequently there are scenarios where inter-operator interference can cause problems unless the filters exceed these specifications.

5.12.1 The Characterizing Parameters

Consider the mechanisms for the leakage of interference between two operators A and B. First, operator A will transmit some of its power into the adjacent carrier bandwidth of operator B due to an imperfect transmit filter. This is measured as the Adjacent Channel Leakage Ratio (ACLR). Second, the receiver filter of operator B is not able to receive the desired signal perfectly while completely rejecting the adjacent carrier signal. This is known as Adjacent Channel Selectivity (ACS) and is illustrated in Figure 5.7.

Radio Site Planning, Dimensioning, and Optimization

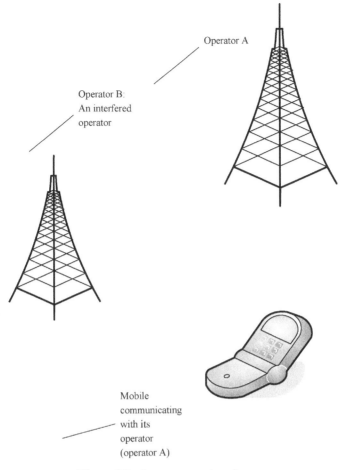

Figure 5.7 Inter-operator interference.

Therefore, operator B will be subject to interference from Operator A via these two mechanisms. Generally, the total inter-operator interference between any two operators is measured by the Adjacent Channel Interference Ratio (ACIR) and determined by both ACLR and ACS according to Equation (38) given below:

$$\text{ACIR} = \frac{1}{\frac{1}{ACLR} + \frac{1}{ACS}} \quad (38)$$

These mechanisms occur on both the uplink and downlink. The 3GPP specifications indicate that the user equipment (UE) will have poorer ACLR and ACS values compared with the base station. Consequently the UE ACLR is dominant for the uplink and the UE ACS is dominant for the downlink, as easily concluded from Equation (38) written for uplink and downlink:

$$\text{ACIR}_{UL} = \frac{1}{\frac{1}{ACLR_{UE}} + \frac{1}{ACS_{BS}}} \qquad (39)$$

$$\cong ACLR_{UE}$$

$$\text{ACIR}_{DL} = \frac{1}{\frac{1}{ACLR_{BS}} + \frac{1}{ACS_{UE}}} \qquad (40)$$

$$\cong ACS_{UE}$$

5.12.2 Effects on Downlink and Uplink

The effect of the adjacent channel interference on the downlink is to create dead zones around the interfered operator's base station. Dead zones are regions where communication between a mobile and a Node B cannot take place due to overwhelming ACL (adjacent channel leakage) from cells that operate on the adjacent frequencies. This can be experienced in both the uplink and more severely in the downlink. In the downlink side dead zones are formed when the base station of one operator is located at the cell edge of the other operator. In the case of the WCDMA downlink, as the terminal moves away from its own operator's base station and towards that of a different operator (on an adjacent carrier), the received signal strength generally diminishes and the interference signal from the other operator will rise. Inter-cell interference from the own operator may also rise under these conditions. This affects the region around the other operator's base station creating a potential dead zone, where terminals suffer dropped calls due to insufficient link budget to satisfy the required quality of service.

In the case of the WCDMA uplink, as the terminal moves away from its own operator's base station and towards that of the competing operator, the transmitted signal is increased due to power control. Therefore the interference received by the other operator's base station will rise resulting in a loss of capacity (due to higher interference). In effect, the interference can act to restrict where traffic can be supported, which might be more significant than absolute capacity to the operator. Therefore, any loss of capacity in this case is typically characterized by cell shrinkage. Thus the uplink may not exhibit the type of dead zones seen on the downlink, but will still suffer cell shrinkage due to the inter-operator interference.

The inter-operator interference will affect a larger area around the interfered operator's base station in the case of high bit rate data services due to their higher demanding link budget.

5.12.3 The Avoidance Measures

The 3GPP specifications set the limits shown in Table 5.9 on adjacent power interference in UTRA.

The effects of these leakages can be very significant if adequate isolation is not provided between systems operating on adjacent bands. The scenarios that can hurt the most are when a user from another system is located at its cell edge, which happens to be close to the base

Table 5.9 UTRA specifications on ACLR.

Offset	Adjacent power interference
5 MHz	45 dB
10 MHz	50 dB

station of another operator's base station operating on the adjacent band. In this case, the mobile is likely to be transmitting at its maximum power to reach its base station, while injecting the most interference into the other system's band.

Therefore, to minimize the inter-operator interference, the operator should coordinate in placing UMTS sites with other operators in the area who operate on adjacent bands. Site co-location with other such operators would help to avoid situations where effects from inter-operator interference due to using adjacent bands become a problem.

Moreover, in a multi-carrier UMTS deployment scenario, the operator should plan to place multiple carriers on the same physical sites or even antennas if possible. That would help to reduce multi-carrier interference problems within the network. If possible, an operator with multiple GSM bands deploying UMTS is advised to try to place its UMTS spectrum band for each market area between its GSM bands for a GSM co-location design. Then the GSM frequency plan should be adjusted to make sure that no BCCH frequencies are placed adjacent to the UMTS bands.

Further discussions on the mutual impact of two WCDMA operators on the network coverage and capacity are provided in Reference 22. There have been reports [24] that ACL was also experienced from nonadjacent bands in UMTS trial networks. This takes the vendors back to the standardization specification in which the masking and filtering requirements for Node B and mobiles were defined. It is widely believed that in the future, all vendors will need to develop filters with much sharper slopes. In fact, the specifications may be revised at some stage to include much more stringent modulation masking requirements [12].

References

[1] Mathar. R. and Niessen, T., 'Optimum Positioning of Base Stations for Cellular Radio Networks', *Wireless Networks*, **6**(6), 2000, 421–428.
[2] Tutschku, K. and Tran-Gia, P., 'Spatial Traffic Estimation and Characterization for Mobile Communication Network Design', *IEEE Journal on Selected Area in Communications*, **16**(5), June 1998, 804–611.
[3] Amaldi, E., Capone, A., and Malucelli, F., 'Planning UMTS Base Station Location: Optimization Models With Power Control and Algorithms', *IEEE Transactions on Wireless Communications*, **2**(5), September 2003, 939–952.
[4] Amaldi, E., Capone, A., Malucelli, F. and Signori, F., 'Models and Algorithms for Downlink UMTS Radio Planning', *Proc. IEEE Wireless Communications and Networking Conf.*, **2**, March 2003, pp. 827–831.
[5] Valkealahti, K., Hoglund, A., Parkkinen, J. and Hamalainen, A., 'WCDMA Common Pilot Power Control for Load and Coverage Balancing, *Proceedings IEEE PIMRC 2002*, **3**, pp. 1412–1416.
[6] 3GPP TS 25.133, V.350, 'Requirements for Support of Radio Resource Management'.
[7] 3GPP TS 25.101, 'UE Radio Transmission and Reception (FDD)'.
[8] Laiho, J., Wacker, A. and Novosad, T., *Radio Network Planning and Optimisation for UMTS*, John Wiley & Sons, Ltd, Chichester, 2002.

[9] TSGR 1 WG1 #8(99) FS8, 3GPP WG1 Contribution, 'Proposal to have Optional 20 ms RACH Message Length', source: Nokia.
[10] Holma, H. and Toskala, A., *WCDMA for UMTS*, John Wiley & Sons, Ltd, Chichester, 2000.
[11] Wacker, A., Laiho-Steffens, J., Sipila, K. and Heiska, K., 'The Impact of the Base Station Sectorisation on WCDMA Radio Network performance', *Proc. VTC 1999, Fall Conference*, Amsterdam, Netherlands, September 1999, pp. 2611–2615.
[12] Taaghol, P., 'Optimization of WCDMA', *Bechtel Telecommunication Technical Journal*, January 2004, pp 1–10.
[13] Sampath, A., Mandayam, N.B. and Holtzman, J.M., 'Erlang Capacity of a Power Controlled Integrated Voice and Data CDMA System', *IEEE 47th Vehicular Technology Conference Proceedings*, **3**, May 1997, Phoenix, Arizona, pp. 1557–1567.
[14] Proakis, J.G., *Digital Communications*, 3rd Edition. McGraw-Hill, New York, 1995.
[15] Viterbi, A.J., *CDMA Principles of Spread Spectrum Communication*, Addison-Wesley, Reading MA, 1995.
[16] Jakes, W.C., *Microwave Mobile Communications*, John Wiley & Sons, Inc., New York, 1974.
[17] 3GPP TS 25.848 Ver 4.0.0, March, 'Physical Layer Aspects of UTRA High Speed Downlink Packet Access'.
[18] Dehghan, S., Lister, D., Owen, R. and Jones, P., 'WCDMA Capacity and Planning issues', *IEE Electronics & Communication Engineering Journal*, June 2000, pp. 101–118.
[19] Wacker, A., Laiho-Steffens, J., Sibila, K. and Jasberg, M., 'Static Simulator for Studying WCDMA Radio Network Planning Issues, *Proceedings VTC 1999, Spring Conference*, Houston, Texas, May 1999, pp. 2436–2440.
[20] Labedz, G. and Love, R., 'A new time based outage criterion for the forward and reverse links of DS-CDMA cellular systems', *Proceedings VTC 1998*, Toronto, Canada, pp. 2182–2186.
[21] Laiho-Steffens, J., Sipila, K. and Wacker, A., 'Verification of 3G Radio Network Dimensioning Rules with Static Network Simulations', *Proceedings. VTC 2000, Spring Conference*, Tokyo, Japan, May 2000, pp. 478–482.
[22] Wacker, A. and Laiho, J., 'Mutual Impact of Two Operators WCDMA Radio Network on Coverage, Capacity and QOS in a Macro Cellular Environment', *Proc. VTC 2001, Fall Conference*, Atlantic City, NJ, October 2001.
[23] 3GPP TS 25.211, 'Physical Channels and Mapping into Transport Channels'.
[24] Joyce, R.M., Graves, B.D., Osborne, I.J., *et al* 'An Investigation of WCDMA Inter-Operator Adjacent Channel Interference', *IEEE, 3G 2003 Conference*, June 2003, London, UK.

6

The Layered and Multi-carrier Radio Access Design

A major challenge in WCDMA networks is to provide capacity, coverage, and quality for a heterogeneous mix of services with diverse bit rate and performance requirements. This chapter presents a quantitative analysis of service interaction caused by the multi-user interference effects. It then uses the results to conclude that layered radio architectures are the efficient approach to address the issue. The layered radio architectures implemented on single and/or multiple frequency carriers are a necessary mechanism to provide optimum capacity and service coverage in the multi-service scenarios of 3G networks. The practical guidelines and various scenarios in which multi-layered and multi-carrier radio access architectures are implemented will be discussed in this chapter.

6.1 Introduction

WCDMA technology provides a flexible radio transmission structure to provide services with many diverse bit rate and performance requirements. The 3GPP specifications release of 1999 allows for data rates of up to 384 kbps, with later releases (Release 5 and up) specify data rates that can range from 600 kbps to multiple Mbps up to 14.4 Mbps (after channel coding) using the high speed downlink packet access, HSDPA, feature [1]. WCDMA uses 5-MHz channel bandwidths, which are shared by users or services communicating over the channel. Each user's service transmission over the channel is separated from the rest by multiplying it and spreading it with mutually orthogonal pseudo random bit sequences over the 5 MHz channel band. However the asynchronous reception of the transmissions from mobiles results in significant loss of code orthogonality at Node B. This in turn results in some residual mutual interference between the users and services in the detection process. A similar process but to a lesser degree happens on the downlink side in the reception process within the mobile stations. The partial loss of code orthogonality in certain environments on the downlink, and hence the reduced multi-user interference, is due to the synchronous transmission from the same source (Node B) to the users in the cell. Any loss of code orthogonality in this case is due to multipath delay spread profile [2, 3]. Another source of service interaction that occurs in WCDMA, but

---

UMTS Network Planning, Optimization, and Inter-Operation with GSM   Moe Rahnema
© 2008 John Wiley & Sons, (Asia) Pte Ltd

on the downlink side only, is the sharing of the Node B power among all the users and services placed on the same 5 MHz chunk of bandwidth within the cell.

The above-mentioned mechanisms of service interactions can complicate the efficient provision of services of varying bandwidth (bit rate) and QoS requirements on the same carrier. This is particularly made clear by considering that the high bit rate data services will generally take much more power from the Node B due to the reduced processing gains on one hand, and the smaller symbol time on the other (for at least comparable E_b/N_0 requirements). This raises the level of interference created by the high bit rate services in the cell. The higher power required for the higher bit rate services will also affect the limited link budgeting on the downlink side and result in reduced coverage. As a consequence, the effect of mixing services with bandwidth and performance requirements that may be too far different can lead to high network costs in terms of the required number of base stations (Node Bs). These interactions between services of varying requirements are more quantitatively illustrated in the next section.

6.2 Service Interaction Analysis

The interaction between services of varying bit rate and QoS requirements is studied here by first evaluating the effect of adding a single connection of arbitrary requirement to a cell. The QoS requirement will simply be represented in terms of a required minimum E_b/N_0. The metrics of observation chosen are the load factors, the required power at input to the receiver, and the receiver noise rise. These quantities are often used to analyze the effects of traffic loading and interference in CDMA systems [3].

The following notation is used for this analysis:

- E_b = energy per bit, in mW
- H = an auxiliary variable to be defined later in the section
- I = multi-user interference power received
- P_N = the effective thermal noise power at the receiver, mW
- P_R = the power at input to the receiver, mW
- R = service bit rate, in kbps
- W = the chip bandwidth in KHz (= 3840 KHz)
- Δ = the change in a quantity
- α_k = the orthogonality factor on connection k (on the downlink side)
- υ = the service activity factor
- η = the load factor
- ρ = the service required E_b/N_0 at input to the receiver

For illustration, we will present the analysis in terms of the impact on the downlink of adding a single service (connection) to the cell.

Noise Rise Increase Expression
From the definition of the load factor

$$\eta = \frac{I}{I + P_N}$$

and solving for I gives

$$I = \frac{\eta \cdot P_N}{1 - \eta} \tag{1}$$

From the definition of the signal to noise ratio, ρ, we have

$$\rho = \frac{E_b}{(I + P_N)/W} \tag{2}$$

The incremental interference due to adding a connection k with the required signal to noise ratio of ρ, with a bit rate of R and an activity factor of υ, at a point in the cell with a downlink orthogonality factor of α_κ is

$$\Delta I = E_b \cdot R \cdot \upsilon \cdot (1 - \alpha_\kappa) \tag{3}$$

Combining Equations (1) and (2), solving for E_b, placing into Equation (3), and dividing by P_N, gives the following expression for the normalized incremental noise rise due to adding the single connection to the cell:

$$\frac{\Delta I}{P_N} = \frac{H}{1 - \eta} \tag{4}$$

in which H is defined by

$$H \equiv \frac{\upsilon(1 - \alpha_k)R \cdot \rho}{W} \tag{4a}$$

Updated Load Factor Expression

The cell load after adding the connection is

$$\eta_{updated} = \frac{I + \Delta I}{I + \Delta I + P_N} \tag{5}$$

Substituting for ΔI and I from Equations (4) and (1), simplifies Equation (5) to

$$\eta_{updated} = \frac{\eta + H}{1 + H} \tag{6}$$

Signal Power Expression

The signal power required at receiver input for a cell load of η and a service E_b/N_0 requirement of ρ, can be derived by rewriting Equation (2) in the form

$$\rho = \frac{P_R/R}{(I + P_N)/W}$$

Table 6.1 Service addition scenario.

Parameter	Case 1 AMR voice	Case 2 HSDPA
E_b/N_0	7 dB	7.7 dB
R	12.2 kb/s	768 kb/s
υ	0.6	1
α	0.5	0.5

Substituting for I from Equation (1), and solving for P_R, provides the following expression for the normalized receiver required input power:

$$\frac{P_R}{P_N} = \frac{\rho.R}{W} \frac{1}{1-\eta} \qquad (7)$$

Numerical Presentations

The effects of adding a connection for two different types of services were considered on the cell load factor and the noise rise on the downlink. The two services considered were AMR 12.2 kbps voice and HSDPA data at 768 kbps. The parameters assumed for these services are given in Table 6.1. The orthogonality factor at the location of adding the connection was assumed to be 0.5 for both cases.

The normalized required receiver power using Equation (7) is plotted in Figure 6.1 for the two cases of adding the voice (case 1) and the HSDPA data connection (case 2) versus the cell loading (the cell loading before adding the connection). The required receiver power is seen to increase rapidly with higher cell loadings at the time of adding the new connection. The data also show that the power required to add the much higher rate data service is about

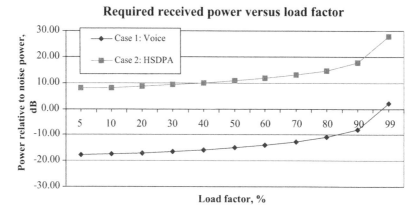

Figure 6.1 Comparison of receiver power requirements.

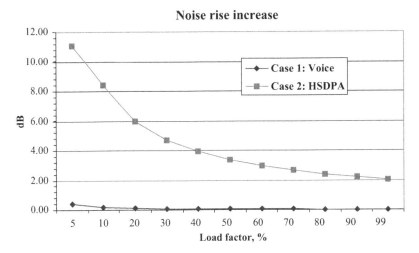

Figure 6.2 Comparison of noise rise increases.

27 dB more than the value require to add the voice connection. Figures 6.2 and 6.3 plot the noise rise increase and the resulting new cell load as each of the two connections is added. We notice that the addition of the high bit rate data connection causes a much greater increase in both the noise rise and the cell load. The cell load increases as a result of the increase in the noise rise. Starting at an initial cell load of say 30%, the figures show that the incremental increases in the noise rise and the load factor are about 0.1 dB and 0.3% in the case of voice (hardly any increase) and 4.7 dB and 26% in the case of the data connection (case 2). Now, with the addition of the data connection, the cell load increases from 30–56% (Figure 6.3). Then at the new value of the cell load (56%), the power that would be required

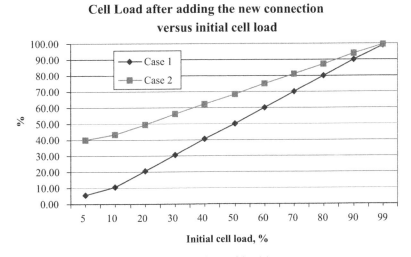

Figure 6.3 Comparison of load increases.

to support a new voice or the high-speed data connection would increase by about 3 dB (Figure 6.1). As the cell load increases, more and more power will be required to support new connections. This results in the rapid consumption of the limited base station power on the downlink, and that of the mobile on the uplink, thus shrinking the service coverage range.

What we have shown here is that the addition of the much higher data rates (compared to say AMR voice) raises the cell load and the noise rise much more rapidly, which in turn results in a much higher power required to support new connections. This positive feedback loop mechanism acts to bring the cell rapidly to its saturation point of the pole capacity. That is the point where the cell load has approached its limit of 100%, where an infinite amount of power would be required to support any new connection [3].

This analysis shows that the cell capacity can soon be consumed by just a few high rate data services in a mixed-service scenario, resulting in limited service coverage range through consuming the limited link budget path loss margins.

6.3 Layered Cell Architectures

The layered approach is based on using layered hierarchically structured radio cells to help eliminate the service interaction problems. Different layers are used to segregate services on the basis of service bit rates, service perform requirements, user mobility, and offered traffic load [4, 5]. In this way services that will overwhelm each other with respect to interference and the load placed on the network are separated into separate layers. There are alternative strategies that may be used to implement the layered hierarchical radio coverage architecture. The exact method will depend on several factors, which include the service mix scenario, service traffic distribution over the area, and the RF morphology of the service area. This chapter discusses two different strategies that are normally used to implement the layered radio coverage for service segregation. These are based on sharing the same carrier (same bandwidth) or using separate carriers (splitting the total available band) for different services. Each may suit a particular scenario and environment as discussed in the following sections.

6.3.1 Carrier Sharing

In the carrier sharing strategy, micro-cellular layers are deployed in hot spot areas and areas where the high bit rate data services may be localized. In this case, micro-cells may be deployed using the same carrier as that allocated to the contiguous macro-cell layer. Experience from IS 95 CDMA systems proves that localized capacity increases can be achieved if the path loss is steeper in the micro-cell areas than in the macro-cell [4]. The suitability of this technique will require that the local RF propagation conditions force the local traffic to the micro-cell layer. Any terminal operating to the non-preferred macro-cell layer in the area will cause interference problems. If the micro-cell area is well isolated from the nearby located macro-cells, with a CPICH signal level received by about 20–25 dB more than that of the surrounding macro-cells, then a single carrier may be shared between the two layers. In this case, the relative placement of the two layer sites and the local morphology should ensure sufficient isolation to prevent the uplink of micro-cells from blocking or de-sensitizing the macro-cell base station receivers. The selection of layers for carrier reuse should be done carefully and on a limited basis in order to limit the interference levels in the network. In various reuse scenarios [6], it is concluded that the best capacity improvements are achieved by reusing a macro layer carrier in the highly loaded micro-cells, which are far enough from macro-cells operating on the same carrier.

However, carrier sharing with micro-cells requires that the macro-cell layer be planned with consideration to future micro cell deployment from the outset. Thus it is unlikely to be suitable for providing hot spot coverage in an existing macro-cell layout, such as in an overlay scenario, depending on the relative positions of the existing sites relative to the hot spots where the demanding services may be localized.

6.3.2 Multi-carrier Design

In the single carrier approach, as the cell load increases the incremental transmit power required from each additional user grows as was illustrated earlier. This is because of the exponential increase in the receiver interference floor with cell loading. Therefore, it is inefficient to assign large quantities of power to a single carrier in the hope of being able to support more users or more services. For instance as argued in Reference 6, a cell configured with a 40 W power and a maximum allowed propagation loss of 155 dB, results in a capacity of 66 speech users with a single carrier. The cell capacity increases to 114 users if the same power is shared across two carriers.

The multi-carrier strategy is the more practical solution particularly when multiple services with a wide range of bit rates and QoS are to be provided over the same area in a CDMA system. It offers better flexibility in capacity and better system performance [7]. In the multi-carrier approach, service segregation is achieved by assigning each different service category to a different carrier. Low bit rate services, in particular speech are allocated to the macro-cellular layer. Such services are also typically associated to high mobility users. High bit rate services such as HSDPA on the contrary may be expected on limited coverage areas where the deployment of micro-cells are justified, and are associated with much lower mobility. Service segregation is then achieved by implementing the microcellular layer on different carriers. Then the base station equipment capability to force traffic to the micro-cell layers (operating on a different frequency) despite the radio conditions can be exploited to force the effectiveness of the micro-cells. In fact, certain vendors provide base station equipments that have features to direct radio resource connection setups for HSDPA to a dedicated HSDPA frequency layer.

With different carriers, the management of handover zones becomes challenging especially for the fast moving traffic [6, 8]. This is tackled by handing over the faster low data rate traffic to macro-cell and serving the slow moving high rate traffic with micro-cell.

The benefits of implementing the micro-cellular layer on a different carrier include releasing soft capacity in the macro-cellular layer (i.e., for voice services) through reducing both the in-cell noise rise and the out-of cell interference on the uplink.

However, this approach is not without drawbacks. If the high rate data services are not localized, ubiquitous coverage of the services will require a large deployment of micro-cells. Otherwise, hard handovers between carriers will be the option. Hard handovers may also be required in certain situations to balance load between the layers. But hard handovers will involve the use of compressed mode. This does cause a reduction in the overall capacity of the cell due to the noise rise increase, which results from increased power requirement during the compressed mode.

Using separate carriers for the micro-cell layer will also help to prevent the micro-cell users from blocking and de-sensitizing the macro-cell site receivers. This also helps to release soft capacity in the macro-cell layer by reducing uplink interference from the micro-cells.

References

[1] 3GPP Specifications 23.107, Release 5 and up.
[2] Parsons, D., *The Mobile Radio Propagation Channel*, John Wiley and Sons, Ltd, Chichester, 1996.
[3] Holma, H. and Toskala, A., *WCDMA for UMTS*, John Wiley & Sons, Ltd, Chichester, 2000.
[4] Hao, Q., Soong, B., Gunawan, E., Ong, J., Soh, C. and Li, Z., 'A Low Cost Cellular Mobile Communication System: A Hierarchical Optimization Network Resource Planning Approach', *IEEE Journal on Selected Area in Communications*, 1997, 1315–1326.
[5] Wu, J., Chung, J. and Yang, Y., 'Performance Study for a Microcell Hot Spot Embedded in CDMA Macrocell System', *IEEE Trans. Vehic. Tech.*, **48**(1), Jan 1999, 47–59.
[6] Laiho, J., Wacker, A. and Novosad, T., *Radio Network Planning and Optimisation for UMTS*, John Wiley & Sons, Ltd, Chichester, 2002.
[7] Roh, J. et al., 'Estimate of Uplink Capacity in Overlaid Macro-cell/Multi-micro-cell CDMA WLL System', *Proc. Asia Pacific Conference Commun.*, **1**, 1999, pp. 486–489.
[8] Ogawa, T.H.K. and Yoshida, H., 'Optimum Multi-layered Cell Architecture for Personal Communication with High Degree of Mobility', *Proc. VTC 1994*, pp. 644–648.

7

Utilization of GSM Measurements for UMTS Site Overlay

The incumbent GSM operators migrating to UMTS can utilize the GSM RF measurements collected by the live network to guide the optimal integration of 3G. The measurements performed on live GSM traffic reflect real user behavior and result in a much more accurate prediction than simply relying on link budget calculations or cell planning tools. This chapter will discuss and formulate the methodologies for deriving path loss and interference geometry models based on measurements collected by the GSM networks to guide the UMTS overlay process. Models are presented for deriving path loss data for driving Monte Carlo simulations to estimate capacity and coverage, and for estimating expected interference geometry and soft/softer handover overheads in UMTS site co-location with existing GSM sites. The measurement based models presented here are expected to provide realistic guidelines for the engineering of UMTS site overlays in existing GSM networks, and to help save time and effort in extensive drive test and propagation data collections.

7.1 Introductory Considerations

Link budgeting arguments provided in Chapter 5 showed that with moderate uplink load factors of 50%, the coverage range for 144 kbps in UMTS 1800 band is about the same as for GSM voice at 1800 band. This means the incumbent GSM operators can implement basically a one-to-one overlay of UMTS in their existing network to complement the GSM coverage with more voice capacity and data services. Then, measurements from the existing GSM networks can be used to guide and characterize the expected traffic, interference, and RF coverage that would result in a UMTS overlay over the existing GSM site-geometry.

The built-in mechanisms in some of the existing GSM BSS equipment provide an edge for the in-operation measurement of path loss and interference geometry in the network. Such mechanisms can be used to collect and process sufficient data from the live network and characterize the interference and path loss geometries in the network where the users are and from where they actually make their calls. It is well known that capacity in UMTS is more driven by power and impacted by interference than is the case in GSM [1, 2]. Therefore, both the interference and path loss geometries obtained from the existing GSM networks will provide very useful information as to how a UMTS overlay may be expected to perform. Furthermore, the results from such measurements after proper post processing, as discussed in this chapter, can be used to optimally select or adjust sites for UMTS overlay, and design for optimum link budgeting and capacity planning.

A well tuned propagation model to estimate path losses with sufficient accuracy– particularly in dense, heavily built-up areas such Manhattan, New York, and in hilly areas such as San Francisco – is critical to UMTS planning and optimization. Most of the existing propagation tools, such as the Okumura Hata, modified Hata, and COST 232 Hata, and their variations such as the COST 231 Walfisch-Ikegami models incorporating street canyon effects in micro-cellular propagation modeling, are based on statistical fitting of mathematical formulas to sample measurement data collected mostly through drive testing in selected environments [3–5]. These models also have certain restrictions such as the applicable RF parameter ranges, and after all provided limited resolution data. Since drive testing is time consuming and can provide only limited data, the resulting tuned models are expected to provide only rough estimates of the actual propagation environment for each specific cell site.

The importance of knowing the detailed propagation conditions within small area resolutions, to within 5 meters or so, for UMTS planning has led to the pursuit and further development of ray tracing models in recent years [6]. However, ray tracing requires detailed building and morphology data, which is not always easily available or is expensive to get. Furthermore, it is very processing demanding and can considerably slow down practical simulation assisted planning.

Given the limitations of the commonly used approaches to propagation modeling, this chapter proposes methodologies based on collecting and characterizing measurements from existing live GSM networks. Measurements obtained by mobiles operating in the live network can be used to collect sufficient data, and processed to derive models that can suit specific planning purposes for each specific site area. In order to fully and efficiently exploit such measurements on a large scale for UMTS site overlay planning, certain post processing routines and steps have to be scripted and implemented to routinely collect and characterize the measurement for easy and efficient use in network planning and model tuning. These routines and steps are proposed and outlined in the following sections.

7.2 Using GSM Measurements to Characterize Path Losses in UMTS

The imbedded mechanisms of certain vendors' base station equipment can be used to obtain measurements on path losses on both uplink and downlink. The path loss is defined as the loss between the radio base station and the mobile station reference point. The path loss measurements collected by the live GSM network can be properly scaled and used to determine analytically the average needed output power for the base station and the coverage degree for different services, particularly when the same traffic distribution is expected for WCDMA as for GSM. The calculated required base station power can also be used to check that the expected power does not exceed maximum allowed limits for a certain downlink load. A significant benefit of the path loss measurements obtained from the live GSM network compared to prediction tools is that the real traffic distribution is used. Even though the cell is covering large areas, the majority of the traffic might be generated close to the site, which is not seen in a cell-planning tool. The planning tools usually assume a uniform distribution of the traffic, and use propagation path loss models that are not much valid close to the site. Moreover, the GSM measurements will provide a more realistic characterization of RF path losses in critical areas (such as in heavily built-up areas, dense urban, undulating hilly areas, etc.) where the existing RF propagation models are not expected to provide adequate accuracy.

The path loss distribution obtained from the existing GSM networks are also very useful in telling if the sites are placed where the traffic is generated. If the path loss distribution is concentrated to relatively small values, the site is placed where the traffic is located. But if the majority of the traffic is caught far away from the cell, a new WCDMA site or other measures might be worth considering.

Adjustment to GSM Path Losses

In order to convert the measured GSM path losses, $L_{sa,\ GSM}$, to a calculated $L_{sa,\ WCDMA}$, the following must be taken into consideration:

1. The base station reference points in GSM and WCDMA (they may be different)
2. The antenna gain differences in GSM and WCDMA
3. The signal attenuation frequency dependence.

Therefore, the path losses obtained from the GSM frequency band should be corrected for the frequency difference (between the GSM and the UMTS band), GSM antenna gain effects, and feeder losses as there is likely to be a difference between antenna gains and feeder losses in GSM and UMTS before used to calculate environmental RF path loss data for UMTS planning and site engineering. The corrections are done simply by the following formula:

$$\text{Path loss used for UMTS} = \text{path loss obtained from GSM network} \\ + \text{correction due to frequency difference} \\ - \Delta G_{ant} - GSM_{feed} \quad (1)$$

in which ΔG_{ant} denotes the difference in antenna gain between GSM and WCDMA,

$$\Delta G_{ant} = G_{WCDMA} - G_{GSM}, \quad (2)$$

and GSM_{feed} is the feeder loss in the GSM antenna system. Equation (1) assumes that the radio base station reference point in WCDMA is at the antenna input point (bypassing the feeder losses).

The correction due to frequency difference is obtained from the free-space path loss coefficient. The frequency corrections thus obtained are +7 dB from GSM 900 to UMTS at 2100 MHz, and +1 dB from GSM 1800 MHz to UMTS at 2100 MHz. The correction factor from GSM 850 MHz to WCDMA 1900 MHz is also +7 dB.

The following sections discuss how the corrected path losses obtained in the GSM network can be used to obtain path loss related characterizations for UMTS planning.

7.2.1 Local Cumulative Path Loss Distribution

A multi-purpose local cumulative path loss distribution (LCPLD) can be obtained by post processing a large number of path losses obtained for a cell, or an area composed of a number of contiguous cells (with similar morphography) and presented numerically in database and graphical format. The LCPLD can then be used for the following two purposes:

- **Generating path loss inputs to Monte Carlo Simulation tools.** The cumulative path loss distribution for the cell/area can be used to draw sample path losses for use in Monte Carlo simulations, which will be performed for link budgeting/coverage-capacity design and verification on UMTS simulation modeling tools such as NetAct, Aircom, Planet, etc. Monte Carlo simulations are very critical to the link budgeting process particularly on the downlink side in UMTS and for working out designs that will provide convergence between coverage and capacity. The cumulative path loss histogram will generate Monte Carlo sampling inputs for path loss distributions for the simulation in the most natural way, as it uses data that are obtained in accordance to where the actual users make calls in the cell. This would provide much better accuracy than sampling path losses from a tuned propagation model in accordance to the random user locations generated within the simulation tool (as normally done).
- **Area coverage probability estimation and verification.** The cumulative path loss distribution for the area/cell can also be used along with the results for maximum path loss output from a link budgeting process to estimate the area coverage probabilities for different services. This is achieved by using the cumulative path loss distribution to check for which path loss (to the cell border) the desired area coverage probability in UL and DL is achieved. For instance, for a 95% area coverage probability, the 95-percentile of path loss is checked from the cumulative path loss distribution. The path loss distribution represents the area coverage probability since statistics are collected from UEs positioned in the whole cell. The path loss distributions can likewise be used to construct designs that will meet certain marketing recommended service area coverage probabilities.

7.2.2 Model Tuning

The measurements obtained on path losses for a cell or a set of contiguous cells with similar morphography from the existing GSM networks can be used along with any drive test RF measurements to more accurately tune up appropriate models (such as the generalized Hata or COST 231 models) to obtain a deterministic path loss-distance formula. The path loss-distance formula can then be used along with the maximum path loss allowed output from a link budgeting process to estimate the service coverage ranges.

7.3 Neighbor-Cell Overlap and Soft Handover Overhead Measurement

If a neighboring cell is indicated to have more than say 30% coverage overlapping, it is expected to cause too many unnecessary soft handovers per user on the average, and thus create excess soft handover overheads for a UMTS overlay in the area. Excess soft handover overheads are detrimental to downlink capacity in UMTS. This information can then help to take RF corrective measures on the UMTS overlay cells (such as lowering antennas, down-tilting, re-directing, and refining specs on beam widths) to reduce the amount of neighboring cells coverage overlap.

The inter-cell dependency measurement mechanism of vendors' GSM base station equipment can be used to obtain measurements to characterize the soft handover areas in a UMTS overlay scenario. The neighboring cells with potential for handover participation with measuring cells can be identified by thresholding the difference between the received signal strength (RSSI) from the measuring cell and the neighbor cell against a soft handover threshold value. The procedure is as follows.

Let Si be the measuring cell's BCCH signal as measured and averaged by a user in cell i, and Sj be the RSSI measured and averaged by that user on the BCCH signal from a potential neighbor cell j. Denote

$$D_{meas}(i,j,n) = Si - Sj, \quad \text{for measurement sample } n, n = 1, 2, \ldots .N \quad (3)$$

Then, define a handover overhead matrix HOM (i,j) as follows:

$$\begin{aligned} HOM(i,j,n) &= 1, \text{if } D_{meas}(i,j,n) \leq \text{'Soft handover threshold'(maybe set to 6 dB)} \\ &= 0, \quad \text{otherwise} \end{aligned} \quad (4)$$

The 'soft handover threshold' is logically determined by the cell specific ADD_Window parameter in the active set. Extensive simulations and field trials have shown that hardly any gain is achieved in the soft handover process when the signal difference between two handover legs drops to below about 5–6 dB [7, 8]. Therefore a reasonable value to use for the threshold in Equation (2) is 6 dB. This value should in principle be fine-tuned based on RF measurement observations in each cell location.

Perform the averaging over the number of data samples

$$HOM_{ave}(i,j) = \frac{\sum_{n=1}^{N} HOM(i,j,n)}{N} \quad (5)$$

The averaging should be performed for each cell pair i,j over a large number of sample measurements N, to give the averaging statistical significance. Two hours of data collected over the busy hours of the area is recommended. The averaged matrix obtained in this way (i.e., the HOM$_{ave}$ (i,j), is the same matrix that is referred to as the inter-cell dependency matrix, ICDM(i,j), by some base station equipment vendors.

The value given by HOM$_{ave}$(i,j) will then represent the percentage soft handover of cell i with cell j based on the handover threshold value used in Equation (4). If this overhead is more than say 30% (a reasonable criteria to use), cell i is said to have excess handover overhead with its neighbor cell j. Then, appropriate corrective measures such as down-tilting or lowering of the antenna, re-directing or changing antennas, should be performed on cell j to reduce this excess coverage overlap when using the site for UMTS overlay.

To obtain the total handover overhead for cell i, we sum the HOM$_{ave}$(i,j) matrix for cell i, over all its geographic cell neighbors j:

$$\text{Handover overhead for cell I} = (1/N_{neighbor}) \sum_{j=1}^{N_{neighbors}} HOM_{ave}(i,j) \qquad (6)$$

which will be less than 30% if action is taken to make sure that each of the matrix elements HOM$_{ave}$(i,j), j = 1, 2, ...N$_{neighbor}$ are less than 30%.

7.4 Interference and Pilot Pollution Detection

In UMTS, each cell transmits (on a continuous basis) a uniquely coded pilot signal, named CPICH, which carries cell specific information and determines the nominal RF coverage of the cell. Pilot pollution arises when the mobile can detect and hear equally powered CPICH signals or multiple strong CPICHs with their multipath components. Basically pilot pollution is caused in areas when none of the received CPICH signals is dominant enough. In that case, pilot detection is heavily influenced by interference from other cells and handover ping-pong is also a possible outcome. It is therefore important to plan sites in such a way to reduce inter-site interference, and establish cell dominance in each coverage area with respect to pilot signal power.

The measurements defined in the previous section can be repeated with a different threshold (interference threshold) for non-neighbor listed cells of a cell in a critical area to identify potential interfering cells.

Then, corrective action should be taken on the placement of the UMTS overlay sites to remedy the problem. In the meantime, GSM performance should also be checked in areas meeting the stated UMTS threshold criteria for any performance problems (by checking the KPIs), and action taken on the GSM sites as well if found to be necessary. The procedure for identifying cells that will act as a strong source of interference to a certain cell in a UMTS overlay scenario are outlined below.

Let cell i be a cell in a critical area where it is to be examined for the presence of strong interference from other cells in local or even remote area. Select cell i as the measuring cell for measurement collections. Let cell j, j = 1, 2,...k, be all the geographically non-neighboring cells that are suspected to be within the RF visibility range of cell i (all non-neighboring cells within, say, a radius of 5–10 kilometers). Let Si be the measuring cell's BCCH signal as measured and indicated by a user in cell i, and Sj be the RSSI measured by that user on the BCCH signal from cell j. Then as before denote

$$D_{meas}(i,j,n) = Si - Sj, \quad \text{for measurement sample } n, n = 1, 2, \ldots N \qquad (7)$$

Then, define an inter-cell interference matrix Intf (i,j,n) as follows:

$$\begin{aligned}\text{Intf}(i,j,n) &= 1, \text{if } D_{meas}(i,j,n) \leq \text{'interference threshold'} \\ &= 0, \text{ otherwise}\end{aligned} \qquad (8)$$

A reasonable value for the interference threshold setting would be 10 dB for UMTS. A co-channel interference at 10 dB would also hurt GSM, and therefore any cells injecting this level of interference should also be acted on for GSM optimization, particularly if performance problems are indicated by the KPIs for the area.

As in the previous section, the above matrix will be averaged to obtain the inter-cell dependency matrix between cell i and its potentially interfering cells j, j = 1,2,..... We will call the resulting matrix $\text{Intf}_{\text{ave}}(i,j)$, as given by

$$\text{Intf}_{\text{ave}}(i,j) = \frac{\sum_{n=1}^{N} \textit{Intf}(i,j,n)}{N} \tag{9}$$

The averaging should be performed for each cell pair i,j over a large number of sample measurements N, to give the averaging statistical significance. Two hours of data collected over the busy hours of the area is recommended.

Then the above matrix elements for all potential interfering cells j, j = 1,2,... are threshold against a 'significance threshold', which can be set to say 5%. All cells j meeting the criteria are then characterized as interfering cells to cell i in a UMTS overlay scenario. This means corrective actions should be taken for the UMTS overlay sites such as antenna down-tilting, re-directing, lowering, or if possible power reduction to prevent from generating excess interference to cell i.

References

[1] Gilhousen, K.L., Jacobe, I.M., Padovani, R., Viterbi, A.J., Weaver, L.A., and Weatley, C.E., 'On the Capacity of a Cellular CDMA System, *IEEE Trans. on Vehic. Tech.*, **40**(2), May 1991, 303–312.

[2] Dehghan, S., Lister, D., Owen, R. and Jones, P., 'W-CDMA Capacity and Planning Issues', *Electronics and Communication Engineering Journal*, June 2000, 101–118.

[3] Hata, M., 'Empirical Formula for Propagation Loss in Land Mobile Radio Services', *IEEE Trans. Vehic. Tech.*, **VT-29**(3), August 1980, 317–325.

[4] COST 231, 'Urban Propagation Loss Models for Mobile Radio in the 900 and 1800 MHz bands', TD(91)73, September 1991.

[5] Ikegami, F., Yoshida, S., Takeuchi, T. and Umehira, M., 'Propagation Factors Controlling Mean Field Strength on Urban Streets'. *IEEE Transations on Antenna and Propagation*, **AP-32**(8), August 1984, 822–829.

[6] Rizk, K., Wagen, J.F. and Gardiol, F., 'Two Dimensional Ray Tracing Modeling for Propagation Prediction in Microcellular Environments', *IEEE Trans. Vehic. Tech.*, **46**(2), May 1997, 508–517.

[7] Laiho, J., Wacker, A. and Novosad, T., *Radio Network Planning and Optimisation for UMTS*, John Wiley & Sons, Ltd, Chichester, 2002.

[8] Buot, T., Zhu, H., Schreuder, H., Moon, S. and Song, B., 'Soft Handover Optimization for WCDMA', *Fourth International Symposium on Wireless Personal Multimedia Communications*, 9–12 September 2001, Aalborg, Denmark.

8

Power Control and Handover Procedures and Optimization

Handover and power control are two critical functions in WCDMA systems. Handovers are needed not only for mobility handling in WCDMA, but are also essential for maintaining the connection to the best servers for the stable efficient operation of the power control. Handover helps to lower the transmit power requirements by maintaining connections to the best servers. Handovers of the soft kind in WCDMA provide macro-diversity gains against fast fading, which further helps to lower the transmit powers. Any reductions in the transmitted powers translate into capacity. The capacity of WCDMA cells is limited by the multi-user interference floor and power sharing on the downlink. Thus any reduction in transmit power translates one way or another into additional capacity. This is why power control plays an important role in making sure that both the mobile and the base station transmit just the power that is necessary to meet the target E_b/N_0 requirements at the other end of the link. Power control also resolves the near far problem by adjusting the transmit power so that the target C/I is evenly satisfied for users near the site and those at unfavorable radio conditions such as at the cell boundaries. This chapter discusses the details of the 3GPP Specifications for the procedures, mechanisms, and parameters that are used to control the handover and power control functions in WCDMA. In the meantime, some of the specific implementation aspects that are left to the vendors' discretion and competition are also pointed out. The optimization aspects of these two important functions in the WCDMA systems are also discussed in detail.

8.1 Power Control

Power control is a critical function in CDMA based systems for resolving the near-far problem on the uplink as discussed in Chapter 2. It is also a function that couples the system level performance with the radio link level performance. This is so as the closed loop power control tries to set the transmit powers to just achieve the minimum required E_b/N_0 ratios at the receiving ends. In consequence, the minimization of the transmit powers results in reduced interference, which in turn translates into increased system capacity. The basic function of power control, and the justification for the necessity to perform power control in WCDMA, was given in detail in Chapter 2. Here, we will discuss the detailed power control specifications, parameters, and issues for WCDMA/UMTS. Power control (PC) in UMTS consists of open loop PC [1], inner-loop PC (also called fast closed loop PC) [2, 3], and outer-loop PC [1] performed in both UL and DL and slow PC applied to downlink common channels. The open loop PC is responsible for setting the initial uplink and downlink transmission powers when user equipment (UE) is accessing the network. The inner-loop power control is used to adjust the transmit power on a 1500 Hz basis. The outer-loop PC estimates the received service quality and uses the information to adjust the target signal-to-noise ratio (SIR) for the fast closed loop PC to maintain the required quality. These are discussed in some detail in the following sections.

8.1.1 Open Loop Power Control

Open loop power control is not expected to be as effective in all situations for the FDD mode of WCDMA. This is the case because the uplink and downlink channels operate on different frequency bands (separated by at least 140 MHz) and the Rayleigh fading in the uplink and downlink is independent. That is why the open loop power control is only used in the initial power setting on channel access to compensate roughly for the path loss. However, open loop power control is more effective in the TDD mode of WCDMA because the uplink and the downlinks are more reciprocal (implemented on the same frequency band). The open loop power control is implemented in both the UE and the base station (Node B). The open loop sets the Tx power level based on the Rx power received on the other link and compensates for path loss and slow fading. The dynamic range for the open loop power control is about 74dB range (24dBm to –50dBm).

8.1.1.1 Uplink Open Loop Power Control

In the initial access to the network, the mobile terminal measures the signal received on P-CPICH and uses that to calculate the initial powers for the first PRACH preamble as well as for the DPCCH before starting the inner-loop PC. The power of the initial PRACH preamble is calculated from Reference 1:

$$PRACH\_initial\_power = CPICH\_TX\_Power - CPICH\_RSCP \\ + UL\_wideband\_power + UL\_required\_C/I, \text{ in dB} \quad (1)$$

where the transmit power of P-CPICH and the required C/I in the uplink are set by the network planner and read by the user terminal on the BCC channel. The received total wideband power, measured at the Node B, is also broadcast on the BCCH channel, and is used to estimate the 'I' in the denominator of C/I. CPICH_TX_Power – CPICH_RSCP provides the path loss in dB on the downlink, which is assumed to be the same for the uplink.

For setting up the first DPCCH channel, the UE starts the uplink inner-loop PC at a power level calculated by

$$DPCCH\_initial\_power = \text{uplink } DPCCH\_power\_offset - CPICH\_RSCP \quad (2)$$

where CPICH_RSCP (the received signal code power) is measured by the UE. DPCCH_power_offset is calculated by the admission control (AC) function in the RNC, provided to the UE at connection setup or reconfiguration, and calculated as

$$\text{DPCCH\_power\_offset} = \text{CPICH\_TX\_power} + \text{UL\_wideband\_power} \\ + (C/I)_{\text{DPCCH}} - 10 \cdot \log_{10}(\text{SF}_{\text{DPDCH}}) \quad (3)$$

where $(C/I)_{\text{DPCCH}}$ is the initial target C/I produced by the admission control for that particular connection and SF_{DPDCH} is the spreading factor of the corresponding DPDCH.

8.1.1.2 Downlink Open Loop Power Control

The downlink open loop PC is used to set the initial power of the downlink channels based on the downlink measurement reports from the UE. Basically, the measurements on the received E_c/I_0 reported by the UE are used to calculate the path losses on the DL, which is then used to guide the scaling of the required E_b/N_0 for the downlink DPDCH channel.

For calculating the initial radio link power of a soft handover branch, it is sufficient to scale the transmitted code power of the existing radio link(s) by the difference between the P-CPICH power of the cell with the existing link(s) and the P-CPICH power of the cell with the diversity branch. When a radio bearer is modified, the initial downlink transmit power is obtained by scaling the power of the old link by the new user bit rate and the new required downlink E_b/N_0.

8.1.2 Fast Closed Loop Power Control (Inner-loop PC)

The inner-loop power control is a fast closed loop PC that operates at 1500 Hz and relies on the feedback information from the opposite end of the link to maintain the received signal to noise ratio (SIR) at a target value set by the outer-loop power control. The fast closed loop power control is implemented both for the UL and the DL in UMTS. The SIR measurements performed on the receive end of the link are compared to the target SIR, and a power control command for increasing or decreasing the transmit power is transmitted to the transmitting end to either increase the power or reduce the power by a certain amount (PC step size) depending on whether the received SIR is below or above the target. Simulation results [4, 5] have shown that the optimum power control step size varies with the UE speed. The simulation results show that a power control step size of 1 dB can effectively track a typical Rayleigh fading channel up to a Doppler frequency of about 55 Hz (30 km/h). For higher speeds of up to 80 km/h, a PC step size of 2 dB has shown better results. Simulation results [4, 6] have also indicated that for UE speeds greater than 80 km/h, the closed loop fast power control can no longer follow the fades (due to the delays in the power control loop and increased signaling errors) and just introduces noise into the transmission (on the uplink). This adverse effect on the uplink performance could be reduced if a PC step size smaller than 1 dB is used. Also, the results have shown that for UE speeds lower than 3 km/h, where the fading rate of the channel is very small, a smaller PC step size is more beneficial.

Closed loop power control takes care of the near-far problem on the uplink, where it is expected to always result in some gain [7]. However, closed loop fast power control results in higher average transmission power as a result of the more variations introduced in the transmitted power to track the channel fast fades. When the MS speed is low, the fast power control is able to compensate the fast fading. This increases the average transmitted power, and has led to the definition of the power rise concept. The power rise is defined as the ratio of the average transmission power in a fading channel to that in a non-fading channel when the average received power level is the same in both the fading and the non-fading channels with fast power control. This power rise implies the need for a margin referred to as the power control headroom in the uplink link budget, in order for the mobile station transmission power to maintain adequate closed loop fast power control [7]. The power control headroom is a margin for the fast power control to track the fast fading and is specific to WCDMA radio link budget. Without this margin, the fast power control exceeds the ceiling established by the sum of the lognormal fade margin, and the handover gains at the cell borders and cannot control the mobile station power to maintain the service quality. In the cell border, the fast power control starts to hit the maximum Tx power (21 dBm/24 dBm) in the mobile to maintain quality. The power control headroom also helps to maintain link quality at the cell border.

Simulation results [5] with a two-path (2 tap) ITU Pedestrian channel model, with controllable second path strength, have shown that the fast closed loop power control gain is highly dependent on the amount of diversity present on the link. The results show that more gain is achieved from the fast power control loop when there is less diversity. The results also show higher variations in the transmission power and hence higher average transmitted power when less diversity is present [5]. The higher variations in the transmission power for the case with less diversity indicate that the power control loop is tracking the fast channel fades. When diversity is present, and in particular with high mobile speeds, the diversity itself helps to stabilize the effective received signal power against the fast channel fades, rendering the power control less critical. This means that the gain achieved from well-designed soft handover in cell border areas reduces the PC headroom requirement and hence improves the coverage. Furthermore, the diversity effects from soft handover or any other means help to reduce the variations in the transmit power that would otherwise result from the power control process. The reduced transmit powers then result in increased capacity.

The power rise resulting from the closed loop fast power control adds to the interference on the uplink of the users in adjacent cells. This is a side effect of the closed loop fast power control on the uplink. Diversity can help to mitigate this effect as explained above.

8.1.2.1 Closed Loop Fast Power Control Specifics on Uplink

The uplink closed loop power control is used to set the powers of DPCCH and its corresponding DPDCHs (if present). The Node B receives the target SIR from the uplink outer-loop power control, which is implemented in the RNC, and compares it with the estimated SIR on the pilot symbols of the uplink DPCCH for every slot. Depending on whether the received SIR is larger or less than the target, the base station transmits a TPC command 'down' or 'up' to the UE on the downlink DPCCH.

Two alternative algorithms named algorithm 1 and 2 have been specified for the UE by 3GPP [1] for processing the TPC commands sent by the Node B. The selection of which algorithm to use can be set in the RAN. The power control step size for algorithm 2 is fixed at 1 dB, whereas for algorithm 1 it can be set to 1 or 2 dB. In algorithm 1, the UE processes and acts on each TPC command received in each slot and is designed for tracking channel fades that are fast but slow enough so they can be tracked. Algorithm 2 collects TPC commands over 5 slots and makes a majority hard decision. It is more suitable for tracking channel fades that are too slow (say UE moving at less than 3 km/h) or channel fades that are too fast (mobiles moving faster than say 80 km/h). During transmission of DPCCH PC preamble frames, which is a parameter that can be set and is sent before the DPDCH is started on the uplink, the TPC commands sent by the base station are followed by algorithm 1, enabling the uplink transit power to converge more rapidly and stabilize faster before the normal PC is set in.

With macro-diversity, when the UE is connected to multiple base stations in soft handover, the power control information received from different base stations may contain different information. To increase the reliability of the received power control commands, each algorithm has a specific way of processing the collective information received and reaching a decision on whether to increase, decrease, or maintain its power level. In algorithm 1, the UE performs a soft symbol decision W_i on each of the power control commands TPC_i (where $i = 1, 2, \ldots, N$, where N is greater than 1 and is the number of TPC commands from radio links of different base stations), derives a combined TPC command, TPC_cmd, as a maximum likelihood function of all the N soft symbol decisions W_i, and changes its transmit power accordingly by the defined PC step size. In algorithm 2, the UE makes a hard decision on the value of each TPC command from each base station, for 5 consecutive and aligned time slots. This results in N hard decisions for each of the 5 slots. The UE first determines one temporary TPC command, TPC_temp_i for each of the N sets of 5 TPC commands as follows:

- If all 5 hard decisions within a set are '1', $TPC\_temp_i = 1$.
- If all 5 hard decisions within a set are '0', $TPC\_temp_i = -1$.
- Otherwise, $TPC\_temp_i = 0$.

Finally, the UE derives a combined TPC command for the fifth slot, TPC_cmd, as follows:

- TPC_cmd is set to -1 if any of TPC_temp_1 to TPC_temp_N are equal to -1.
- Otherwise, TPC_cmd is set to 1 if $\frac{1}{N}\sum_{i=1}^{N} TPC\_temp_i > 0.5$.
- Otherwise, TPC_cmd is set to 0.

where TPC_cmd of 1, -1, and 0 signals increases, reduces, or leaves unchanged the transmit power by the defined PC step size, respectively.

The 3GPP Standard [2] requires that the UE be able to reduce its transmit power on DPCH to at least -50 dBm. This results in a dynamic range of about 70 dB assuming a maximum transmit power of 21 dBm for the UE.

8.1.2.2 Closed Loop Fast Power Control Specifics on Downlink

The closed loop fast power control is used to set the power of the downlink DPCH. The downlink received SIR is estimated using the pilot channels of the downlink DPCH. The estimated SIR is compared to a target SIR set by the outer-loop power control running in the RNC. If the estimate is greater than the target, the UE transmits the TPC command 'down' to the base station. Otherwise the UE transmits the TPC 'up'. If the parameter DPC_MODE = 0, the UE sends a unique TPC command in each slot, otherwise the UE repeats the same TPC command over three slots. UTRAN estimates the transmitted TPC command TPC_{est} over three slots to be 0 or 1, and updates the power every three slots.

The DPC_MODE is a UE parameter that is controlled by the RAN. The uplink DPCCH channel is used to transmit the TPC commands, which simultaneously control the power of the DPCCH and its corresponding DPCCH on the downlink. The relative power differences between the DPDCH and the TFCI, TPC, and pilot fields of the downlink DPCCH are determined by the power offset parameters PO1, PO2, and PO3, respectively. The downlink power control step size is an RNC parameter that can be set to 0.5, 1, 1.5, or 2 dB. The minimum mandatory step size that the base station must be capable of is 1 dB. Other step sizes are optional [1]. To help avoid power drifting, all base stations in SHO with a terminal must use the same power control step size.

Downlink Power Drifting
The single TPC command sent by the mobile is received by all the base stations when the user is in SHO. Each base station may then detect the single command differently due to errors on the air interface. However, the base stations are designed to process and act on the TPC command received independently since combining at the RNC would incur too much delay as well signaling in the network. This can result in the possibility that one base station may lower its transmission power to the mobile while another increases it. This behavior leads to a situation where the downlink powers start drifting apart, referred to as power drifting. Power drifting, which is a result of the closed loop fast power control, degrades the downlink soft handover performance and is not desirable. An effective way to control power drifting is presented in Reference 8. The method is based on having the RNC periodically send a reference power to the base stations in SHO and having each base station make small corrections to its transmission power towards the reference power. The RNC derives the reference power from the transmission code power level of each soft handover connection that is averaged at the corresponding base station over periodic time intervals of say 500 ms (corresponding to 750 TPC commands) and transmitted to the RNC. In this way, a small correction is periodically performed by each base station towards the reference power received from the RNC. This method helps to reduce the amount of power drifting.

8.1.3 Outer-Loop Power Control

The purpose of the outer-loop power control is to generate the target SIR (signal to noise ratio) for the inner-loop power control. The target SIR is the value that helps maintain the quality requirement in terms of BLER or BER for the bearer service under power control. The outer-loop operation is needed for each DCH belonging to the same RRC connection. To maintain steady quality, the outer-loop power control adjusts the target SIR when the mobile speed or the radio environment changes. The outer-loop power control helps to prevent any unnecessary power rise by setting the target SIR to just what is necessary to maintain the desired service quality for the connection. The frequency of the outer-loop power control is in the range 10–100 Hz. Typical values for the adjustment of the step size in adjusting the target SIR is in the range 0.1–1.0 dB.

For the uplink outer-loop power control, the uplink quality is observed after macro-diversity selection combining is done in the RNC. Depending on the type of the radio bearer, the PC entity uses either a BLER estimate, computed in the macro-diversity combining unit (MDC) according to CRC bits of the selected frames, and/or BER estimates calculated at the base stations. Then the target SIR for the inner-loop power control is adjusted by the difference between the BLER/BER estimate and the target BLER/BER multiplied by a step size.

The downlink outer-loop power control is implemented in the UE. The UE uses proprietary algorithms to adjust the target SIR value so that the quality target (BER) signaled by the RNC for the bearer is obtained. The UE sets the E_b/N_0 within the range allocated by the RNC when the physical channel has been set up or reconfigured. It should not increase the E_b/N_0 target value before the closed loop power control has converged on its initial E_b/N_0 setting. The UE makes this determination by comparing the averaged measured E_b/N_0 to the E_b/N_0 target value. The UTRAN can always signal a new maximum allowed range for the outer-loop power control function beyond which the UE should not increase the target E_b/N_0.

For the CPCH channels, the quality target signaled by the RNC is the downlink DPCCH BER, otherwise a BLER target is provided to the UE by the admission control function within the RNC.

8.1.3.1 Estimating the Received Quality

A reliable measure of the received data quality is critical to the proper operation of the closed loop power control function. The received data quality as explained earlier is used by the outer-loop power control to set and adjust the necessary target E_b/N_0 for the connection. Vendors may implement a few different approaches for deriving the necessary quality from the received signal. These include:

- The use of the cyclic redundancy check (CRC) to check whether there is an error in the data frame received.
- Estimated raw bit error rate (BER) before the channel decoder.
- Soft information from Viterbi decoder when convolutional encoding is used by transmitter.
- Soft information from Turbo decoder, for instance BER or FER after an immediate decoding iteration, etc.

The CRC based approach is suited for services where errors are allowed to occur frequently, at least once every few seconds, such as in non-real time packet data services where frame errors of up to 10–20% can occur before the retransmissions and in speech service where a FER of 1% provides the required quality. In AMR speech where the frame length is 20 ms, a FER of 1% means that on the average one error can occur every 2 seconds.

The use of the average raw BER can be a misleading measure of quality. This is because the final FER after decoder will depend on the bit error rate distribution, which in turn depends on the multipath fade profile, the mobile speed, and also the receiver algorithms in the detection process.

In high quality services where the required FER can be as low as or less than 10^{-3}, errors will occur very infrequently. Therefore, if the received quality is estimated based on CRC detected frame errors, the adjustment of the target E_b/N_0 and convergence to the optimum value will be slow and take a long time. Therefore, the soft frame reliability information is expected to provide a better measure of quality for adjusting the target E_b/N_0 for the high quality services [5].

8.1.3.2 Settings of the Maximum and Average Target E_b/N_0

In certain situations, such as at the cell edge, increases in the target E_b/N_0 may not help the service quality. Since at the cell edge the mobile may be transmitting at its maximum transmit capability, it cannot respond to increases in the target E_b/N_0 requested by the outer-loop power control to improve quality. The service quality (BLER) goes down at the cell edge as a result of signal weakening and increased interference from neighboring cells. Therefore, the outer-loop PC keeps blindly raising the target E_b/N_0. But the increased target E_b/N_0 can cause unnecessary increases in the transmission power and hence interference to other users in situations when the mobile moves back closer to the site. This will continue until the target E_b/N_0 re-converges to what it should be at the new locations.

This situation implies that technique is needed to intelligently limit the dynamic range for the target E_b/N_0 output from the outer-loop power control. One simple way is just to have the network set a maximum limit for the target E_b/N_0 when the bearer for the service is set up.

The average received E_b/N_0 to achieve the required service quality is what is used in link budget planning. The closed loop power control helps to significantly reduce the required received average E_b/N_0 compared to cases with no fast power control. Link level simulations performed with and without power control in the ITU Pedestrian channel model A show that the required average received E_b/N_0 reduces by about 8 dB, 5.8 dB, 3.7 dB, 1.9 dB, and 0.2 dB at Doppler frequencies of 5 Hz, 20 Hz, 40 Hz, 100 Hz, and 250 Hz, respectively [9]. The Doppler frequencies simulated corresponded to mobile speeds of 3, 11, 22, 54, and 135 km/h, respectively. In these simulations, no SHO links were used (SHO diversity gains would significantly lower these gains, of course). The gains are seen to be quite significant at the slow mobile speeds of 3, 11, 22, and 54 km/h due to the fact that the closed loop power control is able to track the fades. The gains diminish at the very high mobile speed of 135 km/h. In GSM systems, where no fast power control is used, we have a loss instead of a gain due to the deep Rayleigh feeds, and a margin called the 'fast fading margin' is used in the link budget to compensate for them.

The diminishing of the fast power control gain in the required average E_b/N_0 at high mobile speeds points to the need for a 'fast fading margin'. At higher speeds, such as 120 km/h, the fast power control becomes imperfect, and little correlation is left between the channel and the power control commands. To still achieve the required received quality, either a slight fast fading margin should be added to the link budgeting or a slightly higher average received E_b/N_0 is required. Simulation results [9] show some of the trends in the dependency of the required average E_b/N_0 on the mobile speed for 64 kbps CS data with a 1% BLER quality requirement. The results are shown in Table 8.1. Therefore, in setting the average required E_b/N_0 for a service the possible speed ranges of the mobiles in the areas should be taken into account.

8.1.3.3 Power Control in Compressed Mode

The compressed mode is used in single receiver handsets to execute measurements for an inter-system (handover to GSM, for instance) or inter-frequency handover. UE that are capable of compressed mode enter the compressed mode under the control of RNC when necessary. In compressed mode, the information transferred normally in 10 ms is compressed in time to open up a transmission gap for inter-frequency measurements. In compressed mode, the transmit power is instantaneously increased to compensate for the

Table 8.1 E_b/N_0 targets at various terminal speeds (for BLER 1% and 64 kbps CS data traffic) [9]. [Laiho, J., Wacker, A. and Novosad, T., *Radio Network Planning and Optimisation for UMTS*, 2002. © John Wiley & Sons Limited. Reproduced with permission]

Terminal speed (km/h)	Target average E_b/N_0 UL, dB	Standard deviation, dB	Target average E_b/N_0 DL, dB	Standard deviation, dB
3	3.75	1.00	5.72	0.78
20	3.75	1.10	5.09	0.93
50	3.70	1.46	5.87	1.47
120	4.10	1.86	5.99	1.90

reduced processing gain due to the higher data transmission rate. This means that the required target E_b/N_0 is raised during the compressed mode. During compressed mode, the fast power control loop is not active during the gap period and the effect of interleaving is decreased. This also results in the requirement for a higher target E_b/N_0 to maintain the service quality. During the recovery period after a transmission gap, the power control step size is normally made higher by the system than that for the normal mode to speed up the power control convergence. However, this can lead to an increased interference level and hence an increased outage probability for other users in the system.

8.1.4 Power Control Optimization

The WCDMA power control algorithms and the update rates are specified by 3GPP. It basically remains with the manufacturers to ensure that the inner and outer power control loops are designed to perform efficiently and the latter is based on accurate measurements of QoS. There are however a few things that can be controlled or set by the network planner. These are the choice of the power control step size, within the range implemented by each vendor, and the choice of algorithm 1 or 2. As was explained earlier, for high mobile speeds approaching 80–100 km/h, it is best to set the power control step size as small as possible. The choice of the algorithms, between 1 and 2, would depend on the environment and the speed range of the terminals in the cell. Algorithm 2 operates at a much lower rate and is more suitable for tracking channel fades that are too slow (say UE moving at less than 3 km/h) or channel fades that are too fast (mobiles moving faster than say 80 km/h). In the case of selecting algorithm 2, the power control step size is fixed at 1 dB.

The power settings of the primary pilot and the synchronization channels, P-CPICH and P-SCH, have a strong influence on the radio access network performance. There is no power control on these channels, and the operator sets the powers manually. Some guidelines on power settings on these channels were given in Chapter 5 on radio site planning.

8.2 Handover Procedures and Control

The handovers in WCDMA can be divided into two categories: soft and hard handovers. In the case of a soft handover, the mobile is simultaneously connected to several base stations in the transition region between two cells. In the case of a hard handover, the mobile is connected to one base station at a time.

Soft handover is possible only between intra-frequency cells, i.e., cells using the same frequency. Hard handover, however, can be performed between two intra-frequency cells (belonging two different RNCs that may not allow soft handover between them, due for instance to congestion on the Iur interface, or other reasons), two cells operating on different WCDMA carriers (inter-frequency handovers), or between WCDMA and GSM. Handovers between WCDMA and GSM networks are called inter-system handovers. Inter-system handovers are generally referred to as handovers between networks that use different radio access technologies. For example, handovers can also occur between a WLAN and UMTS in which case it is still called an inter-system handover. The subject of handovers between WCDMA and GSM will be discussed in detail in Chapter 11.

8.2.1 Neighbor Cell Search and Measurement Reporting

The handover control function requires continuous neighbor cell search and measurement reporting by the UE. The neighbor cell search should be carried out continuously in both the idle and the connected modes. If in the process the UE detects a candidate cell that has not been defined as a neighboring cell, it has to decode the cell's BCCH to identify the cell before it can report the measured E_c/I_0 of the cell detected as a neighboring cell to the RNC. This information should include, for a WCDMA cell, the downlink scrambling code, location area code (LAC), and cell identity (CI). For a GSM neighboring cell, the information will contain the GSM frequency channel number of the BCCH and the base station identity code (BSIC).

The measurement reporting by the UE will contain the results of neighboring cells' signal strength and E_c/I_0 measurements needed by the network for a handover decision. The RNC may request the UE to execute and report the following different types of basic HO measurements, depending on the handover type:

- Intra-frequency measurements
- Inter-frequency measurements
- Inter-system measurements
- UE internal measurements

The measurement types are controlled independently as defined on a cell-by-cell basis. Multiple measurement types can be active simultaneously. The real time (RT) and non-real time (NRT) bearers may use different parameters sets for collecting the measurements.

8.2.1.1 Intra-frequency HO Measurements

The UE reports the results of the intra-frequency measurements to the RNC when the criteria for the measurements, broadcast on the BCCH, are fulfilled. The HO decision is made by the RNC. If the handover cannot be executed, the UE continues to measure the neighboring cells but changes to a periodic reporting of the results, using separate measurement criteria as transmitted to it.

8.2.1.2 Inter-frequency and Inter-system HO Measurements

These measurements are made only when ordered by the RNC, using separate measurement criteria as transmitted to the UE. When they are initiated, the UE periodically reports the results to the RNC. The parameters controlling the measurements are the reporting duration and the reporting interval. The events triggering the inter-frequency and inter-system measurements are not part of the Standards. The RNC may initiate these measurements under various circumstances [9] such as the following:

- Average DL transmission power of a radio link approaching its maximum limit
- Quality deterioration report from uplink outer-loop power control
- Quality deterioration report from MS
- Unsuccessful SHO
- Unsuccessful radio access bearer set up
- Restricted SHO capability in the area
- Requested bearer bit rate is higher than allowed in the area
- Frequent SHOs (cell size and MS speed do not match)
- Radio recovery management initiates forced HO
- Hierarchical cell structure (multiple WCDMA carriers) in the area
- Hierarchical cell structure composed of WCDMA and GSM systems
- Restricted intra-system HO capability in the area

8.2.1.3 UE Internal Measurements

The UE internal measurements reported to the RNC consist of two categories. These are the measurements on the UE transmit power and the UE RX-Tx time differences. The latter is used to adjust the timing of the frame transmission on the downlink DPCH. The transmit power measurements are used by the RNC to trigger off inter-frequency and inter-system measurements. The events that can trigger the UE internal measurements are controlled by parameters, which are partly common to all cells in the RNC and partly specific to individual cells. These parameters are defined as follows [10]:

- Event 6A: UE Tx power becomes larger than an absolute threshold
- Event 6B: UE Tx power becomes less than an absolute threshold
- Event 6C: UE Tx power reaches its minimum value
- Event 6D: UETx power reaches its maximum value
- Event 6E: UE RSSI reaches the UE dynamic receiver range
- Even 6F: UE Rx-Tx time difference for a radio link included in the active set becomes larger than an absolute threshold
- Event 6G: UE Rx-Tx time difference for a radio link included in the active set becomes less than an absolute threshold

8.2.1.4 BTS Measurements

The BTS measurement reports may also be used to trigger inter-frequency or inter-system (GSM) measurements. The BTS measurements are also used to balance the power control (uplink and downlink) of the diversity branches during SHO. The BTSs send the measurement report to the RNC for each radio link at regular intervals of, for instance, 500–1000 ms. These measurement reports include the following:

- Total received wideband energy within the band
- Total power transmitted into the cell

- Average DL transmission power of the dedicated physical channels
- Average measured uplink SIR of the dedicated physical channels
- Uplink SIR target currently used on the dedicated physical channel

The first two measurements are used by the RNC in performing admission and load control as discussed in Chapter 9.

Filtering of Measurement Results

In reaching the optimum conclusions and handover decision, it is important to have stable accurate measurements of the P-CPICH and E_c/I_0 that reflect the true path loss geometry in the area. This will also result in optimum active sets in the case of soft handovers. This requires that the rapid fluctuations due to multipath fading be taken out from the measurements. However, the frequency of the signal fast fading component will depend on the channel multipath profile as well as the mobile speed. For low mobile speeds of up to just a few km/h and one dominant path, the fast fading fluctuations may contain deep fades with long lasting duration. This makes them practically impossible to eliminate by just filtering as shown by simulation results [9]. Long filtering lengths of the order of 1 second can always help in such situations, but that would introduce unacceptable delay into the handover process. Large handover delays can increase the interference in the network by keeping the connection with the wrong base station. This is particularly critical in smaller micro-cells where the best server can change quickly. In such situations, time hopped antenna transmit diversity can help to stabilize the received signal by introducing more multipath signal components [11].

For the P-CPICH Ec/I0 measurements used in handover decisions, the technical specification [12] defines a measurement period of 200 ms. This means that the reported measurement is the average of the measured quantity over the last 200 ms period. With this much filtering, the fast fading for mobile speeds in the range of 50 km/h can be easily eliminated as has been shown by simulation [9].

8.2.2 Hard Handover

In hard handovers, all the existing (old) radio links of a mobile station are released before the new radio links are established. This means a short disconnection of the bearer. This can slightly affect the quality of real time connections depending on the length of the service gap due to link swapping, whereas for non-real time data users, it is lossless. Typical service gaps, for instance in GSM handovers (which are hard HOs), are 60–80 ms as indicated by measurements. The GSM-WCDMA handovers have typically shown service disconnections in the range 80–150 ms. These are so short that they do not degrade the speech quality as long as they do not occur too frequently.

Hard handovers can also occur between intra-frequency cells belonging to the same WCDMA networks. This can happen for instance when the two cells belong to different RNCs which do not allow soft handovers for some reason. Intra-frequency hard handovers are mobile evaluated handovers (MEHO). Simple criteria for such hard handovers could be based on the averaged P-CPICH E_c/I_0 values of the serving cell and the neighboring cells and a HO margin (hysteresis) used as a threshold to prevent repetitive (Ping pong) HOs between cells. Intra-frequency hard handovers help to maintain service quality and prevent excessive interference when soft handover cannot be carried out.

Inter-frequency handovers (IF-HO) is another category of hard handovers that can take place between different carriers of a WCDMA network. This happens for instance when the neighboring cells use different carriers, or in layered networks where handover may occur from a macro-cell to a co-located micro-cell that happens to be on a different carrier frequency, or vice versa. The RAN handover control functions should be able to support IF-HOs that may involve a single Node B, a single RNC, or two different RNCs. The IF-HO is a network evaluated handover (NEHO) because the evaluation and decision take place in the RNC. The RNC uses the known configuration of the network in the area (such as the frequencies assigned, neighbor cell definitions, layers, etc.) to command the mobile station to start inter-frequency measurements and report the measurement results periodically. The handover decision is then made by the RNC based on processing the measurement results and any other control parameters involved. The inter-frequency measurements require that the mobile station either be equipped with a dual receiver or support the compressed mode.

8.2.3 Soft (and Softer) Handovers

Soft handover (SHO) types have a special importance in CDMA based systems, due to their close tie to power control. For the power control to work effectively, the system must ensure that each mobile station is connected to the base stations with the strongest signals at all times, otherwise a positive power feedback problem can destabilize the entire system. SHO means that UE may be connected to two or more base stations when moving from one cell to another. In general, application of SHO makes it possible to have seamless (transparent) handover and improved coverage. In soft and softer handover, the mobile station always keeps at least one radio connection to the network. During soft handover, the UE is simultaneously connected to two or more cells belonging to different Node B (BTSs) of the same or different RNCs. In softer handover the UE is connected to at least two cells under one Node B. The soft and softer handovers are only possible within one carrier frequency.

8.2.3.1 WCDMA SHO Algorithm and Procedures

The WCDMA soft handover algorithm [10] uses the E_c/I_0 measurements for use in the handover decisions. The algorithm organizes the cells into two groups, which are defined as follows:

- Active set: The cells in the active set form a soft handover connection to the mobile station.
- The neighbor set, also called the monitored set: This is the list of cells that the mobile station continuously measures but whose pilot E_c/I_0 are not strong enough to be added to the active set.

WCDMA uses a dynamic threshold for the received pilot E_c/I_0 in the handover decisions. The advantage of the dynamic threshold over the fixed threshold, which is used by IS95A, is that it provides for signal location dependent scaling. The problem with IS95A is that the same threshold is used in weak as well as strong signal strength locations, which means that for optimum settings, lower thresholds should be planned for areas with weaker signal strength. This is automatically taken care of in WCDMA by using dynamic thresholds that are defined relative to the strongest pilot or alternatively relative to the weighted averaging of the strongest and the other pilots in the active set. Since the actual comparisons are made with the signal to noise + interference ratio (E_c/I_0), the relative based thresholds also take the interference levels in different locations into account. The improvements achieved in the handover performance by using a relative threshold have been verified [13] by comparing the handover performances of IS-95A with that of IS-95B, which uses relative threshold.

In connected mode, the UE continuously monitors the P-CPICH E_c/I_0 of the cells defined in the neighbor and active lists. The MS compares the measurement results (after averaging over a number of samples) with the HO measurement control thresholds provided by the RNC for the DCH connection, and sends a measurement report back to the RNC when the reporting criteria is fulfilled. Thus the SHO is mobile assisted, with the criteria and the decision algorithm located in the RNC. UE in connected mode that has not received dedicated measurement control criteria uses the measurement control messages sent in SYSTEM INFORMATION on the BCCH. The RAN may divide the measurement control criteria in SYSTEM INFORMATION into idle and connected modes. If connected mode information is missing, the connected mode UE uses the same measurement control message as for idle mode. In idle mode, the RAN may also request the UE to append radio link related measurements to the RRC establishment signaling message in order to support immediate SHO (macro-diversity links) links at the start of a call.

Based on the measurement reports received from the UE, the RAN commands the UE to add or remove cells from its active set. Basically, a mobile station enters the SHO state with a neighboring cell when the signal strength of the cell exceeds a certain threshold range but is still below the signal strength of the current cell.

The UTRAN handover control supports soft and softer handovers for both real time (RT), and non-real time (NRT) radio access bearers. Soft and softer handovers can be supported at the same time.

The SHO process involves all the RRM functions. The reference transmit power calculated in the RNC and used in the power drifting prevention procedure (discussed in Section 8.1 on power control) is also a function under handover control. When a new SHO branch is to be allocated, the Admission Control (AC) function discussed in Chapter 9 is needed for the downlink power allocation on the link. If the AC rejects the HO branch, it may initiate a forced call release, forced IF-HO, or a forced IS-HO. The Resource Manager (see Chapter 9) allocates the downlink spreading code for the HO branch, and releases it likewise when the branch is removed from the cell. The load control updates the downlink load information of the cell when the new SHO branch for a real time bearer is added or removed.

8.2.3.2 Measurement Reporting in Support of SHO

In connected mode, the UE continuously monitors the P-CPICH E_c/I_0 of the cells defined in the intra-frequency neighbor and active lists, and evaluates the measurement reporting criteria. A dedicated measurement reporting criteria is normally communicated to the UE in connected mode, or otherwise the UE uses the connected mode measurement control messages transmitted on the BCCH. Before the measurements are reported to the RAN, the UE performs an averaging on the latest values taken. The number of values taken into the averaging process is a UE specification parameter. Examples of intra-frequency measurement reporting events that would be useful for intra-frequency handover evaluation are given in the next few sections [10]. The UEs do not need to report all these events. The listed events are the toolbox from which the UTRAN can choose the reporting events that are needed for the implemented handover evaluation function or other radio network functions. The UTRAN communicates to the UE through the DCCH, using the measurement control messages, which events should be reported.

Reporting any of the following events may typically result in the active set update (a branch is added, removed, or replaced), as decided in the RNC. In case the active set update cannot take place because of, for instance, lack of capacity or HW resources, the mobile changes to periodic reporting. In this case, the UE sends a measurement report every reporting interval to the network until the active set update has taken place, or the measurement criteria are no longer met. However, the specifics of the decisions for the active set update based on the measurements received are not defined in the UTRAN technical specifications. Thus, it is up to each manufacturer to decide how the network reacts on the measurement reports that the mobile stations are transmitting.

Reporting Event 1A: A P-CPICH Enters the Reporting Range

When event 1A is ordered by UTRAN in a measurement control message, the UE sends a measurement report when a primary CPICH enters the reporting range as defined by the following formula:

$$10 \cdot Log M_{New} \geq S \cdot 10 \cdot Log \left(\sum_{i=1}^{N_A} M_i \right) + (1-S) \cdot 10 \cdot Log M_{Best} - (R + H_{1a}), \quad (4)$$

The variables in the formula are defined as follows:

- M_{New} is the measurement result of the cell entering the reporting range.
- M_i is a measurement result of a cell in the active set.
- N_A is the number of cells in the current active set.
- M_{Best} is the measurement result of the strongest cell in the active set.
- S is a parameter sent from UTRAN to UE, and will be called here the weighting coefficient.
- R is the reporting range.
- H_{1a} is the hysteresis parameter for the event 1A.

The additional window of cells in event 1A is configured with the reporting range parameter (R), which is common to many reporting events and an optional hysteresis parameter (H_{1a}). Event 1A is used by UTRAN to order the UE to add the new radio link to its active set. A 'report deactivation parameter', which specifies the maximum size of the active set, is used to control the occurrence of event 1A.

Reporting Event 1B: A P-CPICH Leaves the Reporting Range

When this event is ordered by UTRAN in a measurement control message, the UE sends a measurement report when a P-CPICH leaves the reporting range as defined by the following formula:

$$10 \cdot Log M_{Old} S \cdot 10 \cdot Log \left(\sum_{i=1}^{N_A} M_i \right) + (1-S) \cdot 10 \cdot Log M_{Best} - (R + H_{1b}), \quad (5)$$

where

- M_{Old} is the measurement result of the cell leaving the reporting range.
- R is the reporting range.
- H_{1b} is the hysteresis parameter for the event 1B.

The reporting range, R, together with the hysteresis parameter, H_{1b} is called the drop window.

Event 1B may be enhanced with a drop timer, which is configured with the time-to-trigger parameter. If the timer is used, the weakening cell must continuously stay below the reporting range for the given time period before the UE may send a measurement report.

In case of events 1A and 1B, when the weighting parameter S is non-zero, the measurement criteria involve comparing against the total energy in the active set as contributed by each cell. Under certain circumstances when a particular cell's received E_c/N_0 is very unstable within the reporting range, it may be desirable to exclude it from affecting the measurement criteria. For that case, a neighboring cell parameter can be specified for each cell to indicate whether the cell is allowed or not in the calculations to affect the reporting range when it is in the active set. Furthermore, for each cell that is monitored, a positive or negative offset can be assigned. The offset is added to the measurement quantity before the mobile station evaluates if an event has occurred. In general, this cell-offset mechanism provides a means to bias the reporting of an individual primary CPICH; for example, it can be seen as a tool to move the cell border.

Reporting Event 1C: A Non-active P-CPICH Becomes Better than an Active One
This event is triggered, if it has been ordered by UTRAN, when a P-CPICH that is not included in the active set becomes better than a P-CPICH that is in the active set. The event is used to replace cells in the active set. It is activated if the number of active cells is equal to or greater than a replacement activation threshold parameter that UTRAN signals to the UE in the MEASUREMENT CONTROL message. This parameter indicates the minimum number of cells required in the active set for measurement reports triggered by event 1C to be transmitted. The event may be equipped with a hysteresis parameter, called a replacement window, requiring the new cell to be better than the worse cell by this value.

Reporting Event 1D: Change of Best Cell
This event is triggered, if ordered by UTRAN, when any P-CPICHs within the reporting range becomes better than the current best P-CPICH plus an optional hysteresis parameter.

Reporting Event 1E: A P-CPICH Becomes Better than an Absolute Threshold plus an Optional Hysteresis Value
This event is triggered, if ordered by UTRAN, when a P-CPICH becomes better than an absolute threshold plus an optional hysteresis value.

Reporting Event 1F: A P-CPICH Becomes Worse than an Absolute Threshold minus an Optional Hysteresis Value
This event is triggered, if ordered by UTRAN, when a P-CPICH becomes worse than an absolute threshold minus an optional hysteresis value.

Time-to-Trigger Mechanism

To prevent frequent event reporting, which can be triggered due to lots of possible neighboring cells, each of the reporting events can be connected with a time-to-trigger parameter. If a time-to-trigger value is used, the measurement criteria have to be fulfilled during the entire time until the timer expires before the event can be reported to the network.

8.2.3.3 SHO Gains

Soft handover contributes to the link budget gains through two mechanisms. These are the gains against lognormal, or the slow shadow fading, and the gains against the fast multipath fading. The gains against shadow fading as estimated in Reference 14 (page 211) are achieved because the slow fading (caused by signal shadowing from terrain and environmental effects) on the radio links to or from two neighboring base stations are not completely correlated. Besides, the handover process also makes it possible to switch to the best server in the area. The gains against fast multipath fading are achieved as a result of the macro-diversity combining of multiple links [15]. SHO provides Rx diversity on the uplink by separate decoding at each Node B and then performing selection combining at the RNC. On the downlink, and with softer handover also on the uplink, it results in increased received signal detection through maximal ratio combining (using mobiles' Rake receivers) and improved fading statistics. Thus, SHO handover helps to mitigate the effects of fast fading by reducing the required E_b/N_0 relative to a single radio link, due to the effect of macro-diversity combining. The macro-diversity gains against multipath fading are expected to be (and confirmed by simulations [9]) larger in poor multipath diversity receiving conditions. In the uplink, SHO is expected to always result in some gain. This is the case since the mobile transmit power over the air is simply collected by multiple base stations and combined to enhance the received E_b/N_0, for instance. In the downlink, the overall benefits achieved in the system performance will depend on the tradeoffs between the macro-diversity gains achieved in the downlink and the reduced system capacity due to increased interference [16]. The increased interference is due to the additional powers contributed by multiple BTSs involved in the soft/softer HO branches. These will be discussed in more detail in Section 8.2.3.4 on SHO performance optimization.

The results of uplink simulations carried out with ITU Pedestrian A and Vehicular A channel models and presented in Reference 9 show some of the trends in the SHO gains achieved against multipath fading. These simulations were carried out at different Doppler speeds corresponding to mobile speeds of 3, 122, 12, 22, 54, and 135 km/h. The simulator model included SHO with two BSs, with the power differences between the two soft handover branches set at 0, 3, 6, and 10 dB. In all simulations the BER = 10^{-3} performance was searched, and selection combining was used at the receiver. The results, which show the gains achieved in the average received E_b/N_0 with the fast power control on, are shown in Tables 8.2 and 8.3. These results show that the SHO gains are highest when the power level difference between the two soft handover branches is the least, namely 0 in the stated simulations. The SHO gains vanish when the power difference between the two handover branches is at 10 dB for the Pedestrian channel model, whereas this loss of any gain happens when the level difference is at 6 dB for the Vehicular channel model. The SHO gains are seen to be larger for the Pedestrian A Channel than for the Vehicular A channel. This is

Table 8.2 SHO gains in averaged received power for the ITU Pedestrian A Channel with power control [9].
Kari, S., Mika, J., Jaana, L.S. and Achim, W., "Soft Handover Gains in A Fast Power Controlled WCDMA Uplink," *IEEE VTC'99*, 1999, pp. 1594–1598. (© 1999 IEEE)

Mobile speeds, km/h	Level difference between SHO links				Single link received E_b/N_0 (dB)
	0 dB	3 dB	6 dB	10 dB	
	SHO gain in received power (dB)				
3	1.6	0.7	0.3	0.1	4.9
11	1.6	1	0.5	0	5.7
22	1.7	0.8	0.3	0	6
54	1.4	0.5	0.2	0	6
135	1.3	0.1	0.1	0	6.3

expected as the Pedestrian A channel has less multipath diversity. Notice that the highest gains were obtained for the Pedestrian channel.

Similar simulations show the trend in the SHO gains when no power control is used. The results were calculated from data presented in Reference 9 and are given in Table 8.4 for the Pedestrian channel A model. These results show much higher gains without power control used. This is natural as the power control helps to introduce diversity by compensating for the multipath fast fading. This in essence implies that soft handover helps to reduce the power control headroom, which then helps to improve the coverage. Furthermore, the results in Table 8.4 show that the SHO gains also diminish in this case with the increases in mobile speed and the power level difference between the two soft handover branches. Higher mobile speeds introduce more multipath and hence more multipath gain through Rake receivers, reducing the effectiveness (or to put it better, the need) of diversity through soft handover.

8.2.3.4 SHO Performance Optimization

Soft handover optimization is a critical aspect of WCDMA network optimization as it has significant influence on system quality, capacity, and coverage. SHO optimization should consider the impacts on the uplink and downlink. On the uplink, as mentioned in the

Table 8.3 SHO gains in averaged received power for the ITU Vehicular A Channel with power control [9].
Kari, S., Mika, J., Jaana, L.S. and Achim, W., "Soft Handover Gains in A Fast Power Controlled WCDMA Uplink," *IEEE VTC'99*, 1999, pp. 1594–1598. (© 1999 IEEE)

Mobile speeds, km/h	Level difference between SHO links				Single link received E_b/N_0 (dB)
	0 dB	3 dB	6 dB	10 dB	
	SHO gain in received power (dB)				
3	1.1	0.3	0.1	0	6
11	1.2	0.4	0.2	0	6.3
22	0.7	0.2	0.1	0	6.1
54	0.8	0.1	0	0	6.2
135	1.1	0.1	1	0.1	6.6

Table 8.4 SHO gains in averaged received power for the ITU Pedestrian A Channel without power control [9].
Kari, S., Mika, J., Jaana, L.S. and Achim, W., "Soft Handover Gains in A Fast Power Controlled WCDMA Uplink," *IEEE VTC'99*, 1999, pp. 1594–1598. (© 1999 IEEE)

Mobile speeds, km/h	Level difference between SHO links				Single link received E_b/N_0 (dB)
	0 dB	3 dB	6 dB	10 dB	
	SHO gain in received power (dB)				
3	4.6	2.9	1.7	0.7	13.1
11	4.5	3	1.1	0.4	11.5
22	3.3	1.7	0.6	0	9.7
54	2.5	0.9	0.1	0	7.9
135	1.3	0.5	0.1	0	6.5

previous section, it always results in increased link performance, and thereby reduced interference. The only fall offs on the uplink are the increased signaling load and use of additional processing power in the network base station equipment. For the downlink side, the situation is not as simple. There, the macro-diversity gain depends on the radio conditions and the number of available RAKE fingers in the UE. In the downlink, each new SHO connection also contributes to increasing the interference floor in the cell and the neighborhood. When the increased interference exceeds the diversity gain, the SHO net system performance gain diminishes. This happens particularly when there is an increase in the number of links with poor soft handover gains (such as SHO links with large power level differences). In addition, the number of reserved orthogonal codes increases due to the soft handover, which can lead to a shortage of the limited available DL codes. It is therefore very important to make sure that all the BTSs contribute positively to the received signal at the UE, to reduce the HO overhead on downlink. These considerations, particularly the ones on the downlink, not to mention the additional transmission backhaul overheads, indicate the need to limit the SHO load or probability in the network to an optimum range.

Typically, the SHO parameters should be set so that the soft handover probability is in the 30–40% range [5]. This means that on average 30% of the connections in the network at any time are in soft handover. The typical ranges for the SHO probabilities are in the range 5–15%. The SHO probability can be controlled either by reducing the SHO windows or by setting a lower limit for the total base station transmitted power in the RNC admission control (see Chapter 9). By lowering the total transmitted power limit for the base stations, the downlink admission control function will reduce the number of SHO branches accepted.

Cell Overlap Optimization
Cell overlap here is defined here as the overlap between adjacent cells with regards to the primary pilot, P-CPICH, coverage. Since the E_c/I_0 and RSCP measurements on the cells pilot signals are used in the handover decision, controlling the degree of cell overlap is critical to keeping the handover probabilities within the acceptable range. This can be done by proper assignment of pilot signal power with regards to the local radio environment and nominal coverage design, and providing adequate down-tilting to keep the pilot signal within the expected range. Too much cell overlap causes too many users in the cells to be in soft handover simultaneously, thus increasing the SHO probabilities and overhead beyond the

optimal limits. Typically, the amount of cell overlap between any two adjacent cells should not exceed about 30%. And the number of neighboring cells that overlap each other with this degree should not exceed 2–3 to keep the SHO probabilities within the 30–40% range.

The Add and Drop windows are also two very important parameters that need to be tuned up to optimize the SHO gains and minimize the overheads. The effect of various parameters on the SHO probabilities is studied in Reference 17.

Add Window
The Add window determines the relative difference between the pilot signal levels (E_c/I_0 and RSCP) of cells, as received by the UE, which can be included in the active set. The Add window size is determined by the 'reporting range parameter, R, and the hysteresis value, H. It is important to size up the add window so that only the cells that can contribute most to the SHO gains are included. Those are cells that have their E_c/N_0 signal levels within less than about 6 dB from each other as indicated by the simulation results discussed in the previous section. The Add window should at the same time be made small enough so that the SHO overhead is kept within the acceptable range.

Drop Window
The Drop window is typically set relative to the Add window and is slightly larger by the difference between the hysteresis values (H parameters) that are set for each. If the Drop window is made too large, the cells with E_c/I_0 that are growing way apart stay in the active set, resulting in reduced uplink and downlink capacity. On the other hand, if both the Add and the Drop windows are made smaller than the required margins, frequent HO due to ping-pong effects takes place, which results in increased signaling overhead and loss of capacity.

Replacement Window
The Replacement window determines the relative threshold that is used to trigger the reporting event 1C, which can result in branch replacement. A cell can replace the weakest SHO branch in a full active set with a new cell whose P-CPICH E_c/I_0 is equal or greater than the replacement window size. If the window is too large, the branch replacement happens too slowly, which prevents the best server from entering the active set. This will result in a non-optimal active set, leading to excess transmit power and interference. The increased interference results in reduced capacity and quality on the radio links, particularly on the downlink. A too small replacement window causes rapid replacements and ping pong effects, which deteriorate the network capacity through increased SHO signaling overheads.

Active Set Size and the Weighting Coefficient
The size of the active set determines the maximum possible number of SHO branches. If the adjacent cell overlaps and Add and Drop window sizes are kept within the optimal ranges, the size of the active set on the high side will not be too significant. However, the combination of too large active set size and wrong values for other HO parameters can result in non-optimal SHO branches, increased overhead, and interference reducing network capacity and service quality. Typical active set sizes in the range 2–3 cells are considered practically optimal.

The active set weight coefficient S used in Equation (4) results in Add or Drop thresholds that are calculated relative to a weighted measure of the best server and the sum energy of the signals in the active set. If S is set to 0, the Add and Drop thresholds are determined relative to only the best server in the active set. The optimum value of this parameter would depend on the interference geometry and the traffic distribution in the area. A non-zero value of this parameter helps to obtain a more stable handover performance in situations with changing interference and traffic pattern. For instance, simulations [18] have confirmed that when the handover thresholds are determined by the total energy of the signals in the active set (a non-zero parameter for S, basically), it results in more stable handover boundaries and probabilities in changing interference situations. The simulations have shown that only modest performances are obtained when only the relative thresholds (to the best server) are used.

References

[1] 3GPP TS 25.214, 'Physical Layer Procedures (FDD)'.
[2] 3GPP TS 25.101, 'User Equipment (UE) Radio Transmission and Reception (FDD)'.
[3] 3GPP TS 25.104, 'Base Station (BS) Radio Transmission and Reception (FDD)'.
[4] Baker, M.P.J. and Moulsley, T.J., 'Power Control in UMTS Release 99', *3G Mobile Communication Technologies*, Conference Publication No. 471, Philips Research Laboratories, UK, 1999, pp. 36–40.
[5] Holma, H. and Toskala, A., *WCDMA for UMTS*, John Wiley & Sons, Ltd, Chichester, 2000.
[6] Holma, H., Soldani, D. and Sipila, K., 'Simulated and Measured WCDMA Uplink Performance, *Proceedings of VTC Fall Conference*, Atlantic City, Seattle, October, 2001.
[7] Sipila, K., Laiho-Steffens, J., Wacker, A. and Jasberg, M., 'Modeling the Impact of the Fast Power Control on the WCDMA Uplink', *Proceedings of VTC 1999 Spring Conference*, Houston, TX, May 1999, pp. 1266–1270.
[8] 3GPP TS 25. 433, 'UTRAN Iub Interface Node B Application Part (NBAP) Signaling'.
[9] Laiho, J., Wacker, A. and Novosad, T., *Radio Network Planning and Optimisation for UMTS*, John Wiley & Sons, Ltd, Chichester, 2002.
[10] 3GPP TS 25.331, 'RRC Protocol Specifications'.
[11] Kuchi, Kiran (Irving TX, USA), 'Hopped Delay Diversity for Multiple Antenna Transmission', United States Patent, 7065156, http://www.freepatentsonline.com/7065156.html.
[12] 3GPP TS 25.133, 'Requirements for Support of Radio Resource Management (FDD)'
[13] Homnan, B., Kunsriruksakul, V. and Benjapolakul, W., 'A Comparative Performance Evaluation of Soft Handoff between IS-95A and IS-95B/cdma2000', *IEEE APCCAS'2000*, 2000, pp. 34–37.
[14] Viterbi, A.J., *CDMA – Principles of Spread Spectrum Communication*, Reading, MA, Addison Wesley 1995.
[15] Kari, S., Mika, J., Jaana, L. S. and Achim, W., 'Soft Handover Gains in A Fast Power Controlled WCDMA Uplink', *IEEE VTC'99*, 1999, pp. 1594–1598.
[16] Lee. CC. and Steele, R. 'Effects of Soft and Softer Handoffs on CDMA System Capacity', *IEEE Trans. Vehic. Tech.*, **47**(3), August 1998, pp. 830–840.
[17] Liu, Z.P., Wang, Y. F. and Yang, D.Ch., 'Effect of Soft Handoff Parameters and Traffic Loads on Soft Handoff Ratio in CDMA Systems', *Communication Tech. Proceedings. ICCT 2003*, 2003, **2**, pp. 782–785.
[18] Laiho-Steffens, J., Jasberg, M., Sipila, K., Wacker, A. and Kangas, A., 'Comparison of Three Diversity handover Algorithms by Using Measured Propagation Data', *Proceedings of VTC'99 Spring Conference*, Houston, TX, 16–19 May 1999, pp. 1370–1374.

9

Radio Resource and Performance Management

The WCDMA system is planned to provide a large set of services with different quality and bandwidth characteristics. Traffic for non-real time services (NRT) coming from the PS domain and IP world are not well predictable. Moreover, their usage may vary a lot according to the transmission delay as experienced by the end user. The real time streaming and conversational services place different requirements on the network resources. On the other hand, the changes in the traffic distribution, user mobility, and varying radio propagation environments will impact the radio capacity and coverage available from the network. It is the task of the radio resource management to dynamically track, measure, and control the network radio resource allocations to various services so that the network can continue to operate under the planned tradeoffs between coverage, capacity, and quality performance for different service categories. In 3G networks, where services of varying bit rates and bandwidth requirements are involved, hardware resources such as the available number of channel elements do not yield a good measure of the load unlike in IS-95 where the service is mainly voice. This means that most often the air interface load results in blocking of the traffic to ensure service quality before hardware capacity limits (available channel elements) are reached. Therefore, load measurement and control must be based on measurements of the various radio resource utilizations such as the total transmit power, total noise floor, code usage, etc. Radio resource management functions consist of admission control, congestion control, channel switching and bearer reconfiguration, code resource management, and packet scheduling. Most of the specific procedures and algorithms for these functions are not subject to 3GPP standardization, and are considered a major differentiating factor between equipment manufacturers. The standards have specified certain measurements and the required accuracy, the usage of which are critical in the design and implementation of RRM functions. Power control and handover are commonly classified under RRM, but the power control functions in particular have procedures that are mostly specified by the 3GPP Standards. Power control and handover are critical

---

UMTS Network Planning, Optimization, and Inter-Operation with GSM   Moe Rahnema
© 2008 John Wiley & Sons, (Asia) Pte Ltd

background-running functions used to keep the WCDMA radio access operating effectively, optimally, and efficiently. For this reason, an entire chapter (Chapter 8) was allocated to covering those two topics in detail. In this chapter, the focus is on the measurements and the typical procedures and algorithms that are used to manage the allocation of the UTRAN transmission power and code resources to different services, and the management of the resulting wideband interference so that each service is handled in accordance with its urgency, priority, and required performance.

9.1 Admission Control

The objective of the admission control is to check whether a new connection request should be admitted based on the available resources and its impact on existing connections. Thus admission control provides a preventive measure for overload. The admission checking is fundamentally performed at several levels as discussed in the following sections.

9.1.1 Processing Admission Control

At the processing level, the admission control checks to see if the required processing resources are available for the connection. The exact procedures and the parameters that are checked will depend on the vendor's specific hardware/software architecture.

9.1.2 Radio Admission Control

The radio admission control is responsible for controlling the utilization of radio resources by accepting or rejecting requests for their usage. Those requests are initiated when setting up new connections, reconfiguring existing connections, or performing (soft) handovers.

In principle, a new call should only be accepted if the closed loop power control mechanism is able to reach a new equilibrium in which the required signal-to-interference ratio at the receiving ends are met for all connections including the new one. However it is not practical to admit a connection and then see if the conditions are met. Therefore almost all admission control algorithms are based on using measurements collected by the base stations to estimate if there is adequate margin left in the resources before accepting a new connection. In some algorithms, such as the ones presented in References 1 and 2, the additional load of the new connection is not considered, and the measured existing load is compared against a threshold. The acceptance thresholds are tuned to limit the dropping probability. The setting of the threshold must consider the variations in the added load due to varying user locations and radio propagation conditions. The threshold should be set low enough to account for the worse possible scenarios and thereby help to minimize the outage probabilities. The safety margin of the admission thresholds can be reduced by estimating the additional loads due to the new connection [3, 4]. This approach would result in non-uniform traffic admission in accordance with terminal locations. Terminals closer to the site taking smaller resources (for instance, less downlink power) are then more likely to get accepted. A survey of alternative admission control algorithms is provided in Reference 5. In the following two sections, the specifics of admission control on the uplink and downlink are discussed. In the approach given, the added load of the new connection is considered in the admission processes.

9.1.2.1 Uplink Radio Admission

The uplink radio admission control checks the availability of the radio resources on the uplink. This is determined by the uplink interference level, which is the total wideband received power, $P_{rxTotal}$, within a single carrier band. Specifically, the impact of the new connection to the total wideband received power at Node B is evaluated. The received wideband power at Node B gives the total interference plus noise floor at the base station receiver, and its measurement and reporting is a requirement of 3GPP standardization. The Standards [6] require the received wideband power at Node B be measured at measurement intervals of 100 ms, with an accuracy of +/− 4 dB within the range of −103 to −74 dBm, and reported to the controlling RNC. The controlling RNC then estimates the increase of the new connection to the total received uplink wideband power, δI, and checks the result against certain thresholds. The uplink check is then passed if the potential resulting received total wideband power is below the threshold chosen. Mathematically, the connection is allowed if

$$P_{rxTotal} + \delta I < P_{rxTarget} - \text{Offset}_{Service\_category} \tag{1}$$

where $P_{rxTarget}$ sets the interference margin (cell radius) and $\text{Offset}_{Service\_category}$ is a service category dependent positive offset parameter. The operator can set the $\text{Offset}_{Service\_category}$ parameters in such a way as to achieve different admission criteria policies for real time guaranteed services, such as conversational voice, and non-guaranteed services, such as background and Internet web services (best effort services). The offset parameters help the operator to configure different thresholds for admitting different services and hence achieve differentiated access based on service category. This effectively results in the reservation of a different degree of resources for each service class. The operator can favor soft/softer handover connection requests by setting the $\text{Offset}_{Service\_category}$ to zero or relatively smaller values (compared to say the threshold for new connections) and thus provide lower admission margins for the users in the communication process. In this way, a less strict admission policy is configured for handover connections than the new calls. This provides a guarantee for the calls in mobility and helps the overall network performance because soft handovers always help to improve the uplink performance. The $\text{Offset}_{Service\_category}$ parameters allow the setting of different admission thresholds for different service categories and the adjusting of specific ratios between the services at certain load conditions. Generally, the $\text{Offset}_{Service\_category}$ parameter provides the mechanism to achieve different tradeoffs between coverage and capacity. For instance, one can set the admission threshold below the interference margin ($P_{rxTarget}$) for the lower bit rate bearers to improve the coverage of higher bit rate service in exchange for lower capacity.

The target interference margin $P_{rxTarget}$ parameter, measured relative to the effective thermal noise power at the base station receiver, is related to the target uplink load factor by the following relation:

$$P_{rxTarget} \eta_{UL} / (1 - \eta_{UL}) \tag{2}$$

in which the target load factor should be set carefully to provide a reasonable tradeoff between call quality and call blocking. High values result in fully loaded cells with degraded call quality and increased call drop rates. On the other hand, lower values improve call quality and drops but increase the call blocking rates.

The increase in the wideband received power, ΔI, caused by the new connection can be estimated by considering that the carrier to noise power ratio for the connection can be expressed as

$$\frac{C}{I_T} = \frac{\Delta I}{I_T} \qquad (3)$$

where I_T is the total noise + interference power within the band and C is the carrier power required for the new connection (which is just ΔI).

But from the definition of C/I, we can write

$$\frac{C}{I_T} = \nu \frac{E_b R}{N_T W}$$
$$= \nu \rho \frac{R}{W} \qquad (4)$$

where N_T is the total wideband noise + interference power spectral density in the band, W is the chip rate, R is the new connection bit rate, ν is the connection activity rate, and ρ is the service signal to noise ratio requirement.

Combining Equation (3) with Equation (4), and solving for ΔI, gives

$$\Delta I = I_T \nu \rho \frac{R}{W} \qquad (5)$$

as the contribution of the new connection to the wideband power received at Node B.

The Load Factor Based Criteria

Some vendors may use the uplink load factor instead of the uplink total received wideband power as the metric to check for radio admission on uplink. The total received wideband power has the advantage that its measurement includes the interference caused by other cells but is time averaged over a specified period. The uplink load approach includes the interference from other cells through an assumed value for the f factor (see Chapter 5 for the derivation of the uplink load factor), but avoids the smoothing operation in the time averaging process done in the received wide band power approach.

When the load factor is used as the criteria for admission process, the resulting uplink load factor is checked against load factor thresholds to determine the admissibility of the new connection, that is if

$$\eta_{UL} + \Delta L < \eta_{UL\_threshold} \qquad (6)$$

in which the load factor increase due to the new connection, ΔL, is approximately given by

$$\Delta L \approx \frac{1}{W} \rho \nu R / I_{total} \qquad (7)$$

(see the uplink load derivation in Chapter 5), where W is the chip rate, R is the new connection bit rate, ν is the connection activity rate, and ρ is the service signal to noise ratio requirement.

Likewise, different load thresholds are used for different service categories as stated in the case with the total received wideband power.

9.1.2.2 Downlink Radio Admission

The available base station downlink transmitted carrier power is a limited resource due, for instance, to the Multi Carrier Power Amplifier (MCPA) capability. Therefore, it is important to monitor the amount of downlink transmitted carrier power used in a cell. The measurement and reporting of the downlink transmitted carrier power is a requirement of 3GPP standardization. The Standards require that the measurements be performed every 100 ms intervals with an accuracy of + or −5% in the range of 5–95% of the Node B total transmit power capability per carrier [6].

Once the uplink radio admission criteria are met for establishing or reconfiguring a new radio connection, the RNC checks the downlink radio admission requirements. Based on the downlink power measurement reported from each Node B, the RNC determines whether the addition/reconfiguration of the radio link is acceptable or not. For this purpose, it first estimates the additional power required for the addition/reconfiguration of the radio link. This can be estimated by the following equation:

$$\text{Tx}_{\text{Powadd}} \approx \frac{C}{I_T} + (E_c)_{transmitted} - \left(\frac{E_c}{I_T}\right)_{received} \tag{8}$$

where $\text{Tx}_{\text{Powadd}}$ is the additional transmit power required from Node B to support the new connection, C/I_T is the required carrier to noise + interference ratio for the connection at user terminal ($= \nu \rho_W^R$), $(E_c)_{transmitted}$, is the transmitted power on the pilot signal (which the Node B knows), and $(E_c/I_T)_{received}$ is the pilot signal to noise + interference power ratio at the user terminal that is measured by the UE and reported to Node B.

The RNC uses the additional power required to check the saturation state of the Node B transmit power if the new connection is to be allowed. Since the downlink transmit power needed by a user can vary significantly during the call, depending on the user mobility within the cell area, margins should be provided in setting the saturation levels. Also, just as in the case of uplink checking, different thresholds are normally used to provide differentiated access for different service categories. The varying threshold mechanism according to the service categories may be implemented through the use of offset parameters tuned for each service. Generally, the admission criterion for the downlink can be expressed by the following relation:

$$\text{Connection admitted if}: \text{Tx}_{\text{Powadd}} + \text{Tx}_{\text{PowBef}} < \text{Tx}_{\text{PowAdm}} + \text{Tx}_{\text{PowAdmOffset}} \tag{9}$$

where $\text{Tx}_{\text{PowBef}}$ is the measured total Node B transmit power before the new connection is allowed, $\text{Tx}_{\text{PowAdm}}$ is a nominal setting for the total Node B allowed transmit power, and $\text{Tx}_{\text{PowAdmOffset}}$ is a service category dependent offset from that nominal value, which can be positive or negative.

In properly setting the threshold parameters in Equation (9), services are viewed from the perspective of controllability or degree of controllability with respect to the transmit power required. One classification would be real time traffic, which requires a certain guarantee of bit rate and quality, and non-real time best effort packet traffic (such as Internet web traffic,

FTP, email, etc.). The offset $Tx_{PowAdmOffset}$ may be set to zero for the guaranteed non-handover connections, to make the Tx_{PowAdm} the threshold set for the non-controllable power and ensure sufficient reserved capacity for it. The best effort traffic has either no guarantee on bit rate or only a minimum bit rate guaranteed. This means that the transmit power allocated for this class of traffic can always be reduced in congestion cases. Therefore, this class of traffic can be given the highest admission threshold by using a positive offset value for the parameter $TxPow_{AdmOffset}$. In this way, the traffic is also admitted in temporarily high loaded situations and uses the excess power when the peak loads subside. For certain classes of traffic with very low priority in comparison to others, a negative offset may be used to set a much stricter admission requirement. Then the very low priority category of services is only allowed when the network has much excess capacity.

The optimum settings of the threshold parameters would depend on the traffic mix and the radio propagation behavior in the area. The optimum values should be based on extensive trials and measurement data collections. The equipment vendors usually have guidelines on the optimal settings of these parameters based on their specific algorithms, field trials, and measurements.

In addition to the automated functions of admission control, the operator may also take certain measures to limit the number of simultaneous users for a certain service such as HSDPA. The HS-DSCH is a shared transport channel. If a very large number of users are simultaneously assigned to this channel, the throughput experienced by each user can become very poor, which then brings a very low quality to each user. Therefore, it may be beneficial to the operator's business to limit the number of users that can be allocated to the HS-DSCH in a cell. Downlink admission strategies have been investigated in References 7 and 8.

9.2 Congestion/Load Control

Even though the admission control function tries to control the traffic admitted into the cell to keep things under check, overload and congested situations can still happen. This can then negatively feedback to reduced capacity, increased interference, and reduced service quality. The admission control is basically designed to achieve a balance between overload and efficient utilization of the radio resources. For this balance, it considers the average statistical service requirements, radio conditions at the time of connection request, and admission thresholds that are not too strict in all cases, for instance for non-real time non-guaranteed services. In fact, the unpredictable bit rate characteristics and behavior of the non-real time best effort traffic can always result in congestion when the dimensioning is performed based on statistical averaged requirements for efficiency purposes. Moreover, the changing radio conditions will have an impact on the required transmitted power on the downlink as triggered by power control loops. This means that mechanisms are also needed to monitor the load from the admitted connections and take actions when overload situations occur.

The radio congestion control function detects and resolves overload situations on the monitored radio resources. As mentioned earlier, the downlink transmitted carrier power and the uplink received wideband powers are the resources that are monitored by admission control. These same resources are also monitored against certain congestion monitoring thresholds to detect the onset of congestion. Congestion control works closely with the admission control. When a radio overload situation occurs, congestion control takes action to try to solve the congestion, which may include ordering the admission to restrict admitting new connections. In the case of downlink cell congestion, all requests for new connections may be blocked. In the case of uplink cell congestion, only the non-handover requests for resources are blocked. Handover requests can still be allowed since the handovers on the uplink side can help to decrease the uplink interference problem due to the macro-diversity gain. The macro-diversity gain results in less transmit power required from the mobile stations, which then helps the uplink interference problem caused by congestion. A good discussion of load control strategies is provided in References 9 and 10.

9.2.1 Congestion Detection Mechanisms

Radio congestion is detected on both uplink and downlink by generally checking the received wideband power and the carrier transmitted power against certain thresholds, respectively. The threshold for congestion detection on the downlink is set to the highest admission threshold offset plus a new congestion offset parameter, Cong_Offset. A hysteresis time is also normally used to prevent actions on very short momentary peaks in the load that would lead to unnecessary blocking of traffic.

Therefore, for the downlink side, congestion occurs if

$$Tx_{Total} > Max(Admission_{Threshold}) + Cong\_Offset \qquad (10)$$

for a duration larger than a congestion hysteresis time, Cong_Hyst_Time, where Tx_{Total} is the total carrier transmitted power.

On the uplink, the received wideband power, $P_{rxTotal}$, is checked against a congestion threshold, Interference_Threshold, and congestion occurs when the received wideband power is found to be larger than the threshold for a duration larger than a hysteresis time, Interf_Hyst_Time. That is, uplink congestion occurs, if

$$P_{rxTotal} > Interference\_Threshold \qquad (11)$$

for a duration > Interf_Hyst_Time.

As in the case of the downlink, the purpose of the hysteresis parameter Interf_Hyst_Time is to prevent the detection of congestion due to short duration peaks in the received wideband power. The unnecessary detection of congestion results in unnecessary blockage of calls and should be avoided.

9.2.2 Congestion Resolving Actions

Congestion resolving actions may include a number of different possible actions including:

- Ordering the admission control to restrict admitting new connections.
- Down switching the bit rate of the non-real time best effort non-guaranteed services to a lower rate.
- Down rating a high-speed circuit switched data connection to a lower rate, for instance from 384 kbps to 128 or 64 kbps.
- Refusing soft handover branch requests on the downlink when congestion occurs on the downlink side.

- Dropping best effort connections.
- Switching to a lower AMR mode for voice connections.
- Restricting the number of compressed mode radio links.

Downlink or uplink congestion is considered resolved when the carrier transmitted power, or the received wideband powers, are found to be less than the respective thresholds for the duration of the hysteresis times involved.

9.3 Channel Switching and Bearer Reconfiguration

Chapter 1 provides a good discussion of the various channel states and configurations in the UMTS radio access network. Basically, UTRAN provides either dedicated physical channels (DCH) or common physical channels on the user radio access side. The dedicated channels, DCHs, are provided on both uplink and downlink in the Cell_DCH state. The dedicated channels are used to support circuit switched or packet switched services. The common channels such as RACH on the uplink and FACH on the downlink are provided in the Cell_FACH state. These are shared channel resources that do not require long set up times before data transmission and are suited for the infrequent transmission of small packets of data. There is also the high-speed downlink shared channel, HS-DSCH, which is used to support very high rate packet data services in shared modes (among multiple simultaneous users). The HS-DSCH is basically a DCH channel on which adaptive modulation to 16 QAM and adaptive coding is used to achieve high throughputs in the range of multiple Mbps.

For the transfer of real time information, the UE is automatically placed in the Cell_DCH state and is assigned a real time radio access bearer (RAB) through a DCH channel. When the UE is assigned a RAB for transmitting non-real time packet switched data, the need of a dedicated channel depends on the effective bit rate used. In that case, for exchanging small amounts of data, the UTRAN can avoid setting up a DCH. Instead, the UE is placed in the Cell_FACH state, in which case uplink data are transmitted through a RACH and downlink data through a FACH, for instance. The use of Cell_FACH state frees up dedicated resources used on a DCH channel, which can then be used by other users requiring higher data rates.

During the course of a non-real time data transmission session, the UE can also transition between the Cell_FACH and Cell_DCH states through dynamic channel switching as specified in 3GPP TS 25.331 and 25.931. Traffic volume measurements performed by both the RNC (on the downlink) and the UE (on the uplink) are used to trigger the proper allocation of radio resource bearers for the transmission of non-real time data. Traffic volume measurements in the UE are controlled by the measurement control information transmitted by the UTRAN. The UE measures uplink channel traffic volumes and sends the measurements to the serving RAN. UE measurement reporting can be periodic or event triggered. In the latter case, a measurement report is sent when the uplink measured effective traffic exceeds the threshold levels sent by the RNC. The traffic volume is measured, for example, by checking the average buffer size or the number of RLC PDUs transmitted in regular time intervals. The RAN uses the measurement reports from the UE to dynamically transition the UE between the Cell_FACH, and Cell_DCH state depending on the indicated traffic volumes. Likewise, the RAN measures the downlink data volumes by checking the averaged buffer sizes against certain thresholds, or by measuring the number of RLC PDUs transmitted over certain intervals of time.

When the RLC buffer payload becomes smaller than an absolute threshold, with associated reporting thresholds, time-to–trigger, and pending time-after-trigger parameters, the RNC transitions the UE between the Cell_FACH and Cell_DCH states depending on the measured traffic intensities. A transition from CELL_DCH to Cell_FACH is triggered when traffic volume measurements indicate a low traffic and/or when a measurement report indicating low traffic volumes is received from the UE and vice versa. The UE stays in Cell_FACH as long as the traffic demand remains low, for example, when the user has finished downloading a web page and is reading it. When a high traffic demand is detected, the UE transitions back to Cell_FACH, for example, when the user starts to download the next web page. The operator may define different thresholds for the interactive and background RAB, in order to keep the UEs on the interactive RABs longer in Cell_DCH state. This can result in better service quality and reduced delays for the interactive data transmission sessions by reducing the time to obtain access to DCH resources.

Transitions from Cell_FACH state can also be made to the Cell_PCH state in which the UE stays in the 'always on' state. When no data are detected for an inactivity time period, (determined by the expiration of an inactivity timer) the UE transitions to URA_PCH state, which helps to save on power consumption. This is usually the case when the user stops the active data session but remains 'always on'. When data transmission is detected in the URA_PCH, the UE transitions to the Cell_FACH state to allow transmission of short data packets, which is usually the case when the user reactivates a previously inactive session. If the user then starts to send more data, the UE transitions to CELL_DCH when higher traffic demand is detected. For faster activation response times, direct transition between URA_PCH, and Cell_DCH states is optional, and may be configured by the OAM. When this transition is enabled, the CELL_FACH state is bypassed. Note that to release a connection, the UE has to transition to either a Cell_FACH or Cell_DCH state. The channel state transition diagram is illustrated in Figure 9.1.

In order to adapt the DCH radio bearer to the varying traffic conditions, the data rate on the dedicated channels is dynamically reconfigured [11, 12]. During an active CELL_DCH state transmitting data using a dedicated channel, certain packet switching services do not need the allocated resources for a constant high data rate. This is particularly the case in the downlink. A high data rate transmission requires channelization codes with lower spreading factors, which consumes a significant part of the channelization code tree. The channelization codes on the downlink are known to be one of the limited resources in UMTS. Therefore, dynamic RAB configuration has been specified [11]. Also, depending on the vendor's specific implementation, dynamic radio bearer reconfiguration can be made to respond to changes on the radio channel condition. For instance, signal strength measurements reported by the UE on the downlink channel can be used to switch the bearer to a lower rate and higher spreading code in cases of poor radio condition. Similar measures can be done on the uplink side through the signal quality measurements performed by the UTRAN. In this way the outage probabilities can be reduced.

In the dynamic RAB configuration, the effective data rates utilized on the DCH channels are measured over regular intervals of time. When the measured effective data rates fall below configurable thresholds, the RAB used is downgraded to a lower rate, which then allows the use of a higher spreading code. The higher spreading code not only results in improved gain in the link budget and hence less required transmission power but also helps more efficient usage of the channelization code tree.

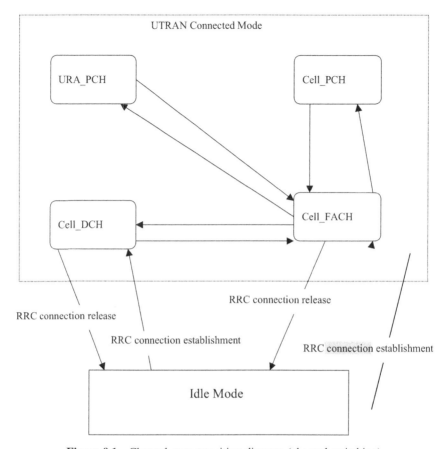

Figure 9.1 Channel state transition diagram (channel switching).

9.4 Code Resource Allocation

Two steps of channel coding are involved in WCDMA. The user signal is first spread by the channelization code and subsequently scrambled by a scrambling code. The spreading operation transforms every data symbol into a number of chips by multiplying each bit by an orthogonal variable spreading factor code (OVSF). This process increases the signal bandwidth. The resulting number of chips per symbol is called the spreading factor. The OVSF codes have different lengths, which result in different spreading factors. The different spreading factors result in different symbol rates. The channelization codes are used by CDMA systems to separate different user connections within the same cell, as first proposed in Reference 13. The codes are rows from a Hadamard matrix, based on the work by Hadamard from the end of the 19th century. The codes are perfectly orthogonal if they are perfectly aligned in time, and the orthogonality is preserved across different symbol rates. However, due to the multipath phenomenon, the codes lose their orthogonality on the uplink, and to some degree on the downlink. The perfect loss of orthogonality on the uplink is due to asynchronous transmission from users in different locations within the cell.

The scrambling or pseudo-noise (PN) codes are mainly generated by a linear feedback shift register. They have very low cross-correlation, which makes them suitable for cell and

Radio Resource and Performance Management

call separation in a non-synchronized system. The scrambling codes are used on the downlink to identify a whole cell. On the uplink a scrambling code is assigned to each call or transaction by the system.

9.4.1 Code Allocation on the Uplink

In the uplink, both the channelization and scrambling codes are assigned to the calls by the system. The uplink scrambling codes are assigned by the RNC on a per call basis in the connection establishment phase. There are 2^{24} long and 2^{24} short uplink scrambling codes (length 256 chips), which are divided between RNCs in the planning process. Each RNC thus has its own planned range. The I- and Q- branches are independently multiplied with an orthogonal spreading code. The resulting signals are then scrambled by multiplying them by a complex-valued scrambling code.

9.4.2 Code Allocation on the Downlink

The allocation of the downlink scrambling codes to cells is part of network planning and was discussed in Chapter 5. The downlink channelization code assignment is a function within the RNC code resource management. In the channel allocation process, the code resource manager within the RNC assigns a certain spreading code (the channelization code) to the downlink channel. The length of the spreading code chosen will depend on the data rate requested and the available codes at the time: the higher the data rate, the shorter the code that is selected. The tree of orthogonal downlink codes is shown in Figure 9.2. To maintain mutual orthogonality among the codes allocated, codes should be selected hierarchically from the tree. This means that when a code is used from the tree, the branch

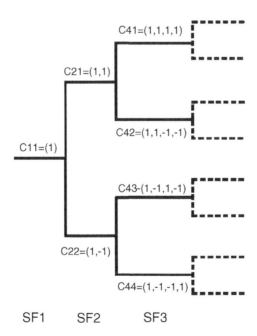

Figure 9.2 The orthogonal code tree.

codes of the used code as well as the codes with a smaller spreading factor on the path to the root of the tree will not be available. Therefore, the number of available codes at any given time is dependent on the spreading factors used and the position of the allocated codes in the tree. This results in the requirement to reshuffle and reallocate codes from time to time as codes are released to make more of the code tree available for new connections. When lower spreading factors are used to obtain a high-speed bearer connection, more codes are made unuseable. This problem can be resolved by multi-code assignment from codes with higher spreading factors to obtain a high rate bearer connection. The specifics of how to achieve high rate bearer connections is left to the vendor's discretion and is one differentiating factor in the manufacturer's efficient code utilization.

Multiple parallel codes may be assigned to the same connection on the downlink to achieve high data rates. The allocation and release of the channelization, as well as the scrambling codes to individual calls, is under the complete control of RNC. The code resource management within the RNC is also responsible for the re-allocation and reshuffling of the code tree to prevent fragmentation of the downlink code tree.

9.5 Packet Scheduling

Packet scheduling is basically concerned with selecting the channel type and the users for transmitting packet data, which belong predominantly to non-real time or best effort services. The full packet scheduler functionality is generally implemented in the RNC. Scheduling on certain channels such as the high-speed DSCH is most often implemented in the Node B, as will be discussed in Section 9.5.3. Packet data can be transmitted on dedicated channels, common channels, or shared channels such as the HS-DSCH and HS-USCH channels. The packet scheduler in the RNC selects channels for the transfer of packet data based on bearer parameters and service type (such as throughput and delay requirements), performance of different channels, amount of data to be transmitted, status of different channels at the time, etc. Common channels are usually selected for the transmission of one or a few small packets in unacknowledged mode. The common channels do not have a feedback channel associated with them. Many users may share the common channels such as the RACH, FACH, and CPCH in a time division manner. When a dedicated DCH channel is selected, the associated channel in the reverse direction is used for the transmission of packet acknowledgements and signaling information. This is also the case for the downlink high-speed DSCH. The uplink control channel for the HS-DSCH is called HS-DPCCH, which is also used to indicate the channel quality to Node B for adaptive modulation and coding. The possibility of using the HS-DSCH channel depends on the user equipment radio access capability. This channel is optional. However, the implementation of DCH and common channels, RACH, and FACH are mandatory and is supported by all UMTS user equipments.

Sharing of DCH or the HS-DSCH among multiple users can be done through either time division or code division scheduling. A combination of code and time scheduling is mostly implemented on the HS-DSCH.

9.5.1 Time Scheduling

In time scheduling, one user or a few are allowed to use the transport channel (DCH or the HS-DSCH) at a time for a specified interval of time. This results in the efficient usage of the limited downlink channelization codes since one or few codes can be used by several users in due time. However, although time division scheduling can result in high data rate and therefore shorter transmission time and delay for each user, it can also result in increased scheduling delay in case of congestion. Congestion situations when many users are sharing the channel can increase the scheduling delays depending on the load of each user and the time allocated to each. Priority mechanisms can be implemented to reduce the delays for the more urgent applications or higher priority users. Time division scheduling can also result in more bursty traffic due to the short transmission times and the higher data rates. This can then cause rapid fluctuations in the interference level on the network depending on the number of users and their arrival patterns. The rapid variations in the interference level can affect the performance of real time users. In time division scheduling, the overhead of setting up connections (which can take up to a few frames) for each user is also enhanced when compared to the short transmission times that are allocated per connection. In time division scheduling, the higher rates of data transmissions result in higher power requirements, which, can in turn lower the coverage range of the services.

9.5.2 Code Division Scheduling

In code division scheduling, multiple users can share and use the channel simultaneously, by differentiating each user's transmission through a different channelization code. This results in low bit rates for several simultaneous users of the channel, but reduces the overall data transmission delays particularly when the number of simultaneous code users is not large. The lower transmission bit rate in code division scheduling results in long uplink coverage range, and also lowers the requirement on the user terminal capabilities. Code division scheduling reduces the overhead of connection setup time due to the longer transmission times allocated to the users and the more parallel call set ups.

9.5.3 Scheduling on the HS-DSCH Channel

Packet scheduling on the high-speed downlink shared channel, HS-DSCH, is very critical and an important differentiating factor in vendor equipments. Packet scheduling on HS-DSCH must achieve high utilization of the channel, while at the same time providing fairness among users, and the interactive and background traffic classes [14]. The HS-DSCH is a high capacity shared channel (see Chapter 1), which can support aggregate data rates up to 14 Mbps using 16 QAM modulation in favorable radio conditions. The channel codes and the transmission power allocated to the HS-DSCH in a cell are dynamically shared between users in the time and code domain. The short transmission time intervals (TTI) used by the channel reduces round trip delays, reduces packet scheduling granularity, and results in better tracking of time varying channel conditions. This feature is exploited for fast link adaptation. The Node B receives the Channel Quality indicator (CQI) report and power measurements on the associated HS-DSCH channel. Based on this information, it determines the transmission data rate. Thus the DSCH may use channelization codes whose spreading factor may vary from frame to frame depending on the indicated radio conditions on the link. In this case, the channelization codes taking care of the spreading are allocated from the same branch of the code tree to simplify the terminal implementation. In the frame by frame operation, the TFCI field within the associated DPCCH (control channel) informs the receiver of the spreading code used as well as any other transport format parameters for the DSCH. In HSDPA, users close to the Node B are generally assigned higher modulation with higher code rates (smaller spreading factor) as a result of better radio channel conditions.

The short transmission intervals used on the DSCH channel also provide the opportunity for implementing intelligent fast channel dependent scheduling among users [15]. This is implemented through the HSDPA MAC layer close to the air interface in the Node B to make the channel scheduling decisions almost instantaneous. The channel dependent scheduler can decide for each TTI which users to assign the channel to, and in close cooperation with the link adaptation function, which modulation (QAM or 16 QAM) and how many codes should be used (in case of multi-code transmission, the spreading factor is the same for all the codes). This helps to increase the overall throughput achieved from the channel in contrast to channel blind round robin scheduling among the users sharing the channel. The objective of the channel dependent scheduler is to use the resource to transmit to users in favorable instantaneous radio conditions. In this way, user throughput fairness is sacrificed to maximize the cell throughput. This strategy results in reduced transmitted power, which in turn translates into less interference and more capacity and quality on the downlink side. Therefore, the emphasis for HSDPA is placed on channel scheduling rather than power control.

However, a practical scheduling strategy should also be geared to maintaining some degree of fairness among users while exploiting the short-term variations in the radio channel conditions. For example, user fair throughput is achieved at the other extreme by providing them with equal throughput wherever they are located. This is achieved by providing the low C/I (Carrier to Interference ratio) users with more channel resources (for example, codes). Fair throughput leads to low total cell throughput. This scheduling is useful for guaranteed bit rate services such as streaming. As a compromise between the two extreme scheduling policies, proportional fair (PF) scheduling can be used based on prioritizing users in accordance to the ratio of the instantaneously achievable bit rate based on users channel condition, and the moving average data rate received by the user over a time period. In this way, user diversity is exploited by giving each throughput proportional to the 'goodness' of the channel experienced, while maintaining some degree of long-term fairness among users. This reduces the possibility of persistently neglecting users who experience a bad channel condition. Scheduling algorithms on this basis [16] tries to achieve a fair balance between the delays experienced by users and the overall system throughput.

The PF scheduling algorithm also has the advantage of automatically giving priority to users with short flows. Short flow connections are prevalent over the Internet in web browsing applications. Short flow users experience a lower time averaged data rate and therefore are prioritized in the PF scheduling process, which takes into account the average rate achieved. However, in the presence of large number of users with long flows, short flows from users with a mixture of short and long flows do not get the priority. To improve on this scenario, modifications of the PF scheduling algorithms are proposed in Reference 17, which try to exploit diversity across all users with long and short flows while retaining differentiating for short flows. Other algorithms based on considering the QoS requirements among users are investigated in Reference 18.

The proportional fair algorithm requires very fast updating, because its viability relies on the tracking of fast fading. This requires the user scheduling functionality to reside in Node B instead of RNC because sending the channel measurement information to the RNC for scheduling purposes will be too slow. Also, as scheduling is based on a time division basis, the shorter the time slots used, the more accurate the scheduling would be. For this reason, the smaller TTI of 2ms is used in HSDPA.

In the end, what makes the differentiating factors among different vendors' scheduling strategies is what becomes most obvious under heavy system loads and diverse application platforms.

9.5.4 Integration with Load Control

Advanced packet scheduling implementations should be closely integrated with the load control functions. Since the delays for non-real time best effort data are not to be guaranteed, their load is controllable. Therefore, congestion control can feed information to the packet scheduler to reduce the load for the non-real time traffic when the cell gets congested. The packet scheduler can feed information on the controllable load from non-real time traffic to the admission control. The admission control can then use that information in the decisions to admit new real time connections that have higher priorities.

References

[1] Knutsson, J., Butovitsch, T., Persson, M. and Yates, R., 'Evaluation of Admission Control Algorithms for CDMA Systems in a Manhattan Environment', *Proceedings of 2nd CDMA International Conference*, CIC'97, Seoul, South Korea, October 1997, pp. 414–418.
[2] Huang, C. and Yates, R., 'Call Admission in Power Controlled CDMA Systems', *Proceedings of VTC'96*, Atlanta, GA, May 1996, pp. 1665–1669.
[3] Holma, H. and Laakso, J., 'Uplink Admission Control and Soft Capacity with MUD in CDMA', presented at *Vehicular Technology Conference*, VTC'99, Fall 1999, 50th IEEE.
[4] Outes, J., Nielsen, L., Pedersen, K. and Mogensen, P., 'Multi-cell Admission Control for UMTS', presented at *Vehicular Technology Conference*, VTC 2001, Spring 2001, 53rd IEEE.
[5] De Alwis, P.M., 'Call Admission Control and Resource Utilization in WCDMA Networks', submitted for Master's Thesis, University of Canterbury Christchurch, New Zealand, February 2005.
[6] 3GPP TS 25.133 V4.0.0, 'Requirements for Support of Radio Resource Management'.
[7] Knutsson, J., Butovitsch, P., Persson, M. and Yates, R.D., 'Downlink Admission Control Strategies for CDMA Systems in a Manhattan Environment', presented at *Vehicular Technology Conference*, VTC'98, 1998, 48th IEEE.
[8] Dahlman, E., Knuttsson, J., Ovesjo, F., Persson, M. and Roobol, C., 'WCDMA – The Radio Interface for Future Mobile Multimedia Communications', *IEEE Trans. Vehic. Tech.*, **47**(4), November 1998, 1105–1118.
[9] Mueckenheim, J. and Bernhard, U., 'A Framework for Load Control in 3rd Generation CDMA Networks', *Proc. IEEE GLOBECOM 2001*, San Antonio, November 2001.
[10] Bernhard, U., Jugl, E., Mueckenheim, J., Pampel, H. and Soellner, M., 'Intelligent Management of Radio Resources in UMTS Mobile Communications Systems', *Bell Labs Technical Journal*, March 2003, **7**(3), 109–126.
[11] 3GPP TS 25.331, 'Radio Resource Control (RRC) Protocol Specification.'
[12] Jugl, E., Link, M., Mueckenheim, J. and Pampel. H., 'Performance Evaluation of Dynamic Data Rate Adaptation on the UMTS Dedicated Channel', *Proc IEE 3G 2003*, London, June 2003.
[13] Adachi, F., Sawahashi, M. and Okawa, K., 'Tree Structured Generation of Orthogonal Spreading Codes with Different Lengths for Forward Link of DS-CDMA Mobile', *Electronic Letters*, 1997, **33**(1), 27–28.
[14] Malik, S.A. and Zeghlache, D., 'Improving Throughput and Fairness on the Downlink Shared Channel in UMTS WCDMA Networks', *European Wireless*, http://www.ing.unipi.it/ew2002/schedule/tuesday.htm#Resource.
[15] Fragouli C., Sivaraman V. and Srivastava M.B., 'Controlled Multimedia Wireless Link Sharing via Enhanced Class-based Queuing with Channel State Dependent Packet Scheduling', *Proc. INFOCOM 98*, April 1998, San Francisco, California.

[16] Jeon, W.S., Jeong, D.G. and Kim, B., 'Design of Packet Transmission Scheduler for High Speed Downlink Packet Access Systems', *IEEE-VTC 2002*, **3**, 1125–1129.
[17] Chan, M.C. and Ramjee, R., 'Improving TCP/IP Performance over Third Generation Wireless Networks', *Proceedings IEEE INFOCOM'04*, 2004.
[18] Huang, V. and Zhuang, W., 'QoS-Oriented Packet Scheduling for Wireless Multimedia CDMA Communications', *IEEE Transactions on Mobile Computing*, **3**(1), January–March 2004, 73–85.

10

Means to Enhance Radio Coverage and Capacity

There are a few site re-engineering techniques that can be performed to obtain incremental improvements in radio coverage and capacity for handling the growth in the traffic. The site re-engineering solutions assume that the initial site densities have been planned carefully to allow capacity upgrades without requiring the interleaving of new sites. This can be accomplished if the initial site densities in different locations are planned to be consistent with the known and projected future trends in the traffic densities and distributions from planned services. Two major important mechanisms for providing improved capacity and coverage are the layered radio access architecture and sectorisation. The layered radio access approach, which includes the multi-carrier expansion strategy, helps to increase capacity by providing micro-cells at traffic hot spots through re-use of macro-cell carrier frequencies or through the multi-carrier strategy. The multi-carrier strategy reduces the interference floors in each carrier by splitting the traffic in a given area over two or more carriers. This helps to extend coverage and improve capacity in the meantime. When the interference floor reaches a near saturation point, adding more base station power to the same carrier cannot overcome the infinitely growing power required to support additional traffic. The multi-carrier strategy solves this capacity limit problem. The layered radio access and the multi-carrier design, along with the resulting benefits, were discussed in detail in Chapter 6.

Sectorisation helps to increase the site capacity and coverage by confining the multi-user interferences to smaller regions, and provides a measure of isolation between site sectors. The higher transmit antenna gains of sectorised cells further help to extend the service coverage ranges. Sectorisation is also often driven by the nature of the coverage geographies required in the practical world. Therefore, the sectorisation design and benefits were covered in detail in Chapter 5 on radio site planning and optimization.

The objective of this chapter is to discuss other remaining site re-engineering mechanisms that can help to improve capacity and coverage as it becomes necessary. These mechanisms include the implementation of the HS-DSCH for improved packet service throughputs on the downlink, the use of secondary scrambling code, the use of MHAs (mast head amplifiers) and RRHs (remote radio heads), higher order receiver and base station transmit diversities, multi-beam and adaptive antennas, and possible use of repeaters. In order to find the right and optimum solution in each case, an analysis should be performed to find out the limiting links with respect to coverage and capacity. This can be done by simple considerations of the uplink link budget equations and parameters, and the downlink base station power consumed as was formulated in detail in Chapter 5. It is commonly thought that the service coverage is uplink limited, whereas in fact the system capacity may be either uplink or downlink limited. The uplink or downlink limiting factors in the capacity depend on the site configuration, the traffic mix, and the terminal performance. The uplink capacity limited cases can occur in less populated areas where the network is planned with a rather low uplink cell load for the benefit of extended cell coverage range and therefore reduced site count. Then additional traffic results in exceeding the uplink margin provided in the initial plan to handle the increased interference at the base station. Thus the maximum uplink load is reached before the base station runs out of transmit power.

A downlink coverage limited situation is expected in micro-cells with highly sensitized base stations (through the use of MHAs, for instance), which implement asymmetric data rate services (such as web downloads) with limited base station power. A downlink capacity limited situation can occur in micro-cells of the urban area where the network is planned with a higher uplink load to increase capacity. In this case, the base station runs out of power before the maximum planned uplink load is reached and before it is heavily loaded on the downlink. Thus a downlink capacity situation requires consideration of the uplink load and downlink power budgets. Understanding what the limiting factor is (coverage or capacity) and on which link(s) is essential to defining a proper solution. A solution may improve coverage, capacity, or both. It is important to also understand the impact of each in the process of selecting an optimum solution.

10.1 Coverage Improvement and the Impact

Service coverage improvement is achieved by improving parameters in the link budget. These include increasing the base station sensitivity through MHAs, lowering E_b/N_0 requirements on the uplink through the use of receiver antenna diversities, sectorisation, and multi-beam antennas to confine interference, and the use of repeaters for extending coverage to special places.

Improving the service coverage allows users from a farther distance to communicate with the cell on the uplink. This will result in a higher average base station transmit power requirement per downlink connection. This simply moves the system closer to becoming downlink capacity limited if the system capacity is uplink limited to begin with. But if the system capacity is downlink limited, improving of the service coverage leads to a loss in system capacity, because in the latter case, more base station power is consumed to support users that can come in from the farther distances.

10.2 Capacity Improvement and the Impact

The mechanisms to improve the cell capacity in an existing radio plan depend on whether the cell capacity is uplink or downlink limited. If the cell capacity is uplink limited, then the uplink load equation needs to be improved. This can be done, for instance, by using base station antenna diversity to reduce the service E_b/N_0 requirements.

If the cell is downlink capacity limited, the mechanisms to improve the capacity include the use of base station power boosters (higher power amplifiers), lowering the service E_b/N_0 requirements on the downlink (through better performing terminals, terminal receiver antenna diversities, or base station transmit antenna diversity), implementing the high-speed downlink shared channel (HS-DPSCH) for packet data services, and use of a secondary scrambling code when the system capacity is hard limited by the number of available downlink channelization codes. The use of multi-carrier implementation, sector-isation, and multi-beaming to confine interference, will help to improve capacity in both uplink and downlink capacity limited scenarios. The use of HS-DPSCH may impact the cell coverage on the downlink, if enough base station power is not provided particularly in a scenario where MHAs are also used to increase base station sensitivity and range.

Sectorisation, multi-beaming, and adaptive antennas will also improve the coverage, and this can also impact a downlink capacity limited case negatively. The net improvement in capacity is dictated by the tradeoffs in the capacity improvement as indicated through the downlink load (or base station transmit power equation), and the downlink capacity loss as a result of the gains achieved on the uplink link budget (see Chapter 5).

10.3 HSDPA Deployment

The High-Speed Downlink Packet Access (HSDPA) implemented on the shared HS-PDSCH channel implements adaptive modulation between QAM and 16-QAM depending on channel state and multi-code transmission to increase the downlink data transmissions to much higher rates than is possible with the DCH channels of Release 99. Downlink data rates up to a maximum of 14 Mbps are achievable with HSDPA. This allows high-speed downloads of packet switched data such as web searching and FTP applications. The implementation of the HSDPA is a good solution for increasing the downlink capacity, depending on the traffic mix and the HSDPA capable terminal penetration in the area. The use of the HSDPA is not likely to impact the uplink coverage or the uplink traffic much.

The HSDPA requires a small amount of base station power for the associated downlink control channel, HS-SCCH. This overhead can be minimized by using the power control feature on HS_SCCH. The power control feature adjusts the transmission power on the HS-SCCH according to the required power level at the UE. This helps to reduce the average power overhead required for the HS-SCCH (compared to using a fixed transmit power level). Vendor studies estimate the power control feature on HS-SCCH to provide HSDPA cell throughput gains of 5–20% (compared to using fixed HS-SCCH power) depending on the environment and the total power allocated to HSDPA.

10.4 Transmitter Diversity

Transmit diversity implemented at the base station (Node B) with two transmit antennas helps to improve the downlink performance. The transmit diversity requires two transmit antennas at the base station and transmit diversity receiving algorithms (discussed in Section 10.4.2) in the mobile terminal. The two transmit antennas at the base station can be formed from a single cross-pole antenna or two vertically polarized antennas. The 3GPP WCDMA air interface specifications [1] have made the transmit diversity capability mandatory for the user terminals, but optional for the base station. This means that the operators have the choice of whether or not to provide transmit diversity.

10.4.1 Transmit Diversity Benefits and Gains

Transmitter diversity at the base station provides a means to achieve performance gains similar to those obtained with mobile station receiver diversity, but without the complexity of a mobile-station receiver antenna array. Simulations [2] have shown negligible performance gains of 0.5–1.5 dB under ITU vehicular channel model A with speeds of 3–50 km/h, but with significant gains of up to 3 dB in the case of the ITU pedestrian channel model A at speeds of 3 km/h. The gains have been measured in terms of the reductions in the required downlink E_b/N_0 for 12.2 kbps AMR speech and 144 kbps data services with a 1% target BLER. Generally, just as with receive diversity, the gains achieved with transmit diversity are the largest in low mobility environments when the performance of time and multipath diversity is poor. Transmit diversity then provides a rather simple capacity upgrade solution through application of two transmit antennas or a single cross-pole antenna at the base station.

The improvements achieved in the downlink E_b/N_0 requirements with transmit diversity impact both the downlink system capacity and service coverage. The service coverage improvement is critical in downlink coverage limited scenarios such as in micro-cells with relatively base station transmit power. The capacity gains achieved with transmit diversity in downlink capacity limited cases are also significantly higher in micro-cells than in macro-cells as confirmed by simulation results [2]. The micro-cells have shown capacity gains of around 60% on average compared to a 30% average gain achieved in macro-cells. Therefore, transmit diversity schemes are seen to be particularly appropriate for micro-cells where other techniques such as adaptive antennas or multi-beaming are less suitable due to the larger angular spread of received signals at the mobile station [3].

10.4.2 Mobile Terminal Requirements

The optimum operation of transmit diversity requires base station transmitter's knowledge of the downlink channel formed by the two transmit antennas. The uplink channel estimations cannot be used effectively due to the channel frequency separations on the uplink and downlink in the FDD mode. Therefore the open loop based on channel reciprocity is not implemented.

However, 3GPP has specified two techniques one of which uses an open loop approach based on space-time coding for implementation in the mobile station. This basically results in generating the symbols at the receiver that are proportional to the sum of the channel powers received from each of the two transmit antennas. This is robust and is easy to implement in the mobile station. The space-time coding exploits diversity in both the spatial and temporal domains in an open loop fashion [2, 4, 5]. The space-time coding makes the two transmitted signals orthogonal to each other allowing rather simple detection at the mobile terminal. The principle is illustrated through the following mathematics. Let r(t) and r(t + T) be the signal received at the mobile terminal in consecutive time intervals t and t + T. Then,

$$r(t) = r_1 = S_1.h_1 + S_2.h_2 + n_2 \qquad (1a)$$

$$r(t+T) = r_2 = -S_2^*.h_1 + S_1^*.h_2 + n_2 \qquad (1b)$$

where h_1, and h_2 are the channel impulse response of the two transmit antennas and n_1 and n_2 are the combined noise plus interference received during the two time intervals. The mobile then estimates the symbols by linear combination according to

$$\hat{S}_1 = h_1^* . r_1 + h_2 . r_2^* \qquad (2a)$$

$$\hat{S}_2 = h_2^* . r_1 - h_1 . r_2^* \qquad (2b)$$

Using Equations (1a) and (1b) yields approximately

$$\hat{S}_1 \approx (|h_1|^2 + |h_2|^2) S_1$$

$$\hat{S}_2 \approx (|h_2|^2 + |h_2|^2) S_2$$

which results in the generation of the symbols at the receiver proportional to the sum of the channel powers from each of the antennas.

The channel impulse responses h_1 and h_2 are each estimated independently at the mobile using two known symbol sequences that are transmitted on the primary pilot channel, P-CPICH, one on each of the two transmit antennas with the same power, and the same scrambling and channelization codes. The two symbol sequences transmitted are orthogonal to each other and of the form {A,A,A,A,A,...} and {−A,A,A, −A, −A,A...}. The first symbol sequence is the same as that used by a standard non-transmit diversity cell.

The 3GPP WCDMA air interface specifications have defined two closed loop modes of transmit diversity [1]. Both are based on using the feedback information (FBI) field provided within the uplink frame slot to communicate the relative phase shifts between the measured channel impulse responses from each of the two transmit antennas, at the slot rate of 1500 Hz. The base station transmitter then uses the feedback information received to generate complex weights for each transmit antenna so that the two signals received at the mobile terminal are as coherent as possible. In closed loop mode 1, only the phases are adjusted with a 1bit accuracy. In closed loop mode 2, the signal amplitude is also adjusted with an antenna specific weight factor as determined by the UE.

Finally, it is noted that the performance achieved with transmit diversity will be partly influenced by the accuracy with each vendor's terminal implementations estimate, the channel impulse responses h_1, h_2, and the received signal to noise ratio. This is one area where vendors can differentiate their terminal products from each other.

10.5 Mast Head Amplifiers

One of the factors that influences the uplink coverage is the base station reference sensitivity. The uplink coverage is improved by increasing the reference sensitivity of the base station subsystem as seen from the antenna ports. The base station subsystem reference sensitivity (for a service at a given quality) is determined by a number of factors, including the required E_b/N_0 for the service, the BTS receiver noise figure, and the losses associated with the feeder cables and jumpers that connect the BTS transceivers to the antenna ports. The latter can particularly become significant in the case of long feeder lengths where the BTS equipments are remote from the antennas. The (MHAs) are high gain low noise amplifiers that are placed

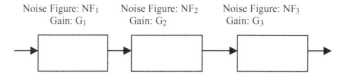

Figure 10.1 The cascaded receiver subsystem components.

either externally at the antenna input ports or integrated within the antenna base to reduce the effect of the feeder/jumper losses. If the MHAs are internally integrated within the antenna base, the antenna is then referred to as an active antenna. In either case, the presence of an MHA, which is an active device, requires a power supply that can be fed through antenna feeders using a bias T connection at the BTS end of the feeder cable. When a cross pole antenna is used, two MHAs are required, one at each of the antenna ports.

10.5.1 MHA Benefit on System Coverage

The MHAs help to increase the BTS sensitivity as 'seen' from the antenna ports by reducing the effective noise figure of the BTS system. The increased sensitivity of the BTS subsystem translates into uplink service coverage. The BTS subsystem here refers to the BTS plus the feeder and jumper cables and anything else in between such as diplexers. To quantify this, the Friis equation [6] can be used to evaluate the composite noise figure of the BTS system. The Friis equation calculates the effective noise figure of a system cascaded of active and passive components as shown in Figure 10.1, through the following relation expressed in linear scale (not dBs):

$$\text{system effective noise figure} = NF_1 + \frac{(NF_2-1)}{G_1} + \frac{(NF_3-1)}{G_1.G_2} + \frac{(NF_4-1)}{G_1 G_2 G_3} + \cdots \quad (3)$$

The number of stages to consider in the Friis equation is dependent on the configuration of the base station subsystem. Figure 10.1 shows three components that can consist of the MHA, the feeder and jumper lumped together (as there is no active component between them), and the BTS itself. If diplexers are present, which are normally considered passive components, their effects in terms of losses can be added to that of the feeder cable. The jumper losses are normally around 0.5 dB. The noise figure of a passive component such as the feeder cable plus jumper is simply equal to its loss. As the equation shows, the noise figure of the system can never be less than the noise figure of the first component (in this case the noise figure of the MHA), which is not scaled down.

Moreover, the higher the gain of the first component (the MHA), the closer the system noise figure is to that of the first component. So it is important for the manufacturers to minimize the noise figure of the MHAs, while maximizing its gain.

Using Equation (3), the gains achieved with an MHA, with a gain of 12 dB (typical) and a noise figure of 2 dB under various feeder + jumper loss scenarios, are given in Table 10.1. The diplexer and T bias connector are assumed to have losses of 0.5 dB and 0.1 dB, respectively. The BTS is assumed to have a noise figure of 3 dB.

Notice from the data given in Table 10.1, that the gains achieved in using an MHA increase with feeder losses as would be expected. Moreover, the composite noise figure of

Table 10.1 Benefits achieved in using MHAs as a function of the feeder + jumper loss.

Feeder + jumper (dB)	NF without MHA (dB)	NF with MHA (dB)	Benefit (dB)
1.0	4.5	2.3	2.2
2.0	5.5	2.4	3.1
3.0	6.5	2.6	3.9
4.0	7.5	2.8	4.7

Based upon an MHA with a gain of 12 dB and noise figure of 2 dB, a BTS noise figure of 3 dB, a diplexer loss of 0.5 dB, and a T bias loss of 0.1 dB

the BTS subsystem as seen from the antenna port is reduced to below the noise figure of the BTS itself, particularly for low feeder losses. Therefore, the MHA helps to reduce the effects of both the feeder losses and the BTS noise figure.

10.5.2 MHA Impact on System Capacity

The MHA impact on the cell capacity will depend on whether the cell is uplink or downlink capacity limited. If it is downlink capacity limited, the use of an MHA can result in reduced capacity. This occurs from the resulting increase in the uplink allowed path loss, which makes it possible for users farther from the site to get on, and hence use more of the downlink power, relatively. If the cell is uplink capacity limited, the use of the MHA should not have any significant effect on the capacity other than bringing the cell closer to being downlink capacity limited. This happens as the cell uses more of its power to support users farther from the site. It is noted that the MHA inserts a loss of around 0.5 dB in the downlink side, which affects the cell capacity accordingly in either case.

10.6 Remote Radio Heads (RRH)

The remote radio heads (RRH) are a solution for improving both coverage and capacity in space limited hot spots. The RRH basically are the RF amplifier and transceiver modules, which are separated from the base station baseband processing subsystem. The latter remain within the base station cabinet, which can be placed remotely from the radio site. The connection between the two is realized through optical fibers. This makes it possible to place cells at site locations that would normally require long runs of feeder cables between the antenna and the BTS equipment, incurring large losses on both the uplink and downlink. For example, an RRH can be placed next to a rooftop antenna and connected through optical fiber to the BTS cabinet housing the baseband modules located in the basement of the building. Similarly, RRHs can be placed in remote space limited hot spot locations on tower tops and fed optically through a remotely located BTS cabinet. The optical connection between the BTS cabinet and the RRH can be as long as 2 km in length.

10.6.1 RRH Benefits

In situations where long feeder cables are involved between the BTS equipment and the antenna, the use of RRHs results in significant gains on both the uplink and downlink. The gains achieved are simply the difference between the feeder losses of a conventional site configuration and one based on RRH. This difference can be as much as several dBs when the feeder losses associated with the conventional site configuration are large. The resulting gain translates into equal improvements in coverage (uplink) and capacity. This is unlike the unidirectional improvements achieved with MHAs where only the uplink link budget is improved. With RRH, the elimination of the large feeder losses also translates into increased EIRP (power) and hence capacity and coverage on the downlink side. Whereas the MHAs can be used to increase coverage at the expense of some loss in capacity in downlink capacity limited situations, the RRHs can be used to either increase coverage without the capacity loss or maintain the same coverage but increase the capacity. The increased EIRP can be used to support more users if the coverage increase is not needed.

On the practical side, the weight size and the increased wind loading of the RRHs should also be considered, and the requirement for the optical links between the base station cabinet and the RRH should be factored in.

10.7 Higher Order Receiver Diversity

Receiver antenna diversity generally reduces the effect of fading and enhances the signal to noise ratio obtained at the input to the detector. The improvement against fading is achieved due to the diversity effect in that the fading is not expected to be fully correlated between the antennas. The enhanced received signal to noise ratio is a result of combining the signals from antennas in a coherent fashion. The enhanced received signal to noise ratio results in lower E_b/N_0 requirements, which in turn result in lower average uplink transmit power requirement per connection. Therefore the receiver antenna diversity mechanism is a solution for increasing uplink coverage and performance as will be clarified further later.

The standard configuration for a WCDMA base station consists of a 2-branch Rx diversity, which is implemented with either a single cross-polarized antenna or two vertically polarized antennas. The higher order receiver diversity pertains to more than two receiver branches. A rather practical configuration for a higher order base station receiver diversity system would consist of a 4-branch Rx antenna configuration. This is implemented as either a single dual cross-pole antenna (two x-pole antennas within one radome) or two separate cross-polarized antennas located within a spatial distance of about a meter. The former offers the advantage of being much more compact, but the disadvantage of reducing the achievable gains by about 0.5–2.5 dB due to the correlation between the two x-pole antennas as a result of them being closely located [7].

10.7.1 Operation and Observed Benefits

The signals from a 4-branch receiver diversity system are detected and combined through a Rake receiver similar to multipath signals from a single antenna. The Rake receiver resolves and combines the signal energy from all the received resolvable multipath signals from all antennas. The coherent combining is achieved by pre-compensating for the channel induced phases in each signal component. Let r be the envelope of the coherently combined signals from M fingers over each of the four antennas, where M is assumed to be the maximum number of resolvable multipath signals from each antenna. Let $r_{i,j}$ be the complex signal received from the jth path over antenna i, and $h_{i,j}$ be the channel impulse response of the jth path from antenna i. Then r is obtained by

$$r = \sum_{i=1}^{4} \sum_{j=1}^{M} \hat{h}_{i,j}^{*} \cdot r_{i,j}$$

where $\hat{h}_{i,j}^{*}$ is the complex conjugate of the estimate of the channel impulse response $h_{i,j}$. Thus the Rake receiver is assumed to have 4.M fingers.

Simulation as well as field measurements presented in Reference 7 have shown that the 4-branch receiver diversity gives an average coverage gain of 3 dB compared to the traditional 2-branch reception in the ITU vehicular A channel model with two x-pole antennas spaced to provide low mutual correlations. With less multipath diversity present, such as with the ITU pedestrian channel model A, higher gains up to 4 dB have been reported [2, 7]. When the two x-pole antennas are spaced close to each other (within the same radome), the gains for the ITU vehicular channel model (particularly at the slower mobile speeds) are reported to be reduced by about 0.5–2.5 dB. The higher gains achieved in the case of the pedestrian channel model are attributed to the poor multipath and slow mobile speeds, which result in poor time diversity as provided by channel interleaving and error correction coding. Multipath and time diversity is poor for low mobility users who experience low levels of delay spreads. Such users can benefit more from the additional reception diversity provided by the 4-branch receiver diversity. Consequently the gain from the additional diversity with 4-branch reception is higher than in the ITU vehicular A.

The performance gain of the 4-branch over 2-branch reception diversity is attributed to the two-fold increase in the signal energy collection achieved in the coherent combining process. With the additional coverage gains of 3–4 dB achieved with the 4-branch diversity, the number of sites necessary for the same coverage probability is reduced from 33–41%, assuming a path loss distance exponent of 3.5. This reduction is site count is significant for improving coverage in the initial investments for UMTS deployment.

The gains achieved in the uplink link budget with the 4-branch receiver diversity translate into lower E_b/N_0 requirements for the same coverage or increased margin against fast fading with the same E_b/N_0. Therefore, additional gains of the 4-branch reception diversity also result in a lowering of either the average transmit power requirement per uplink connection or of the power rise (see Chapter 8) in the fast power control process. This translates into increased capacity on the uplink. In fact the simulations [7] have shown capacity increased from 85% in the ITU vehicular channel model A at speeds of 3km/h to 130% in the ITU pedestrian channel model A at 50 km/h over the 2-branch reception diversity. In these simulations, uncorrelated antennas were used. If the two antennas are correlated, the capacity gains would become smaller.

10.7.2 Impact to Downlink Capacity

If the system is uplink capacity limited, there is virtually no impact to downlink capacity. The additional gain achieved in the uplink capacity as just argued in the previous section helps to boost up the cell capacity. If the cell capacity is downlink limited, then the 4-branch receiver diversity results in a loss of capacity due to the improved coverage range achieved. The loss, however, will be less than in the case of using the MHA solution since the 4-branch receiver diversity will not incur the 0.5 dB insertion loss in the downlink.

10.7.3 Diversity Reception at Mobile Terminal

The benefits in implementing diversity reception through multiple antennas at the mobile terminal is similar to using transmit diversity at the base station. However, the transmit diversity at the base station is a more practical solution than using multiple antennas at the mobile terminal for diversity reception. Nevertheless, the local environment of the mobile terminal that is rich in scatters results in low signal correlation distances of the order of half a wavelength [8] as explained in Chapter 3. Therefore two antennas can be placed at a mobile terminal within a distance of about 8 cm and provide reasonable diversity. This cannot be much of a problem when data terminals are used in 3G to download data services. For the traditional handsets for voice communication, 2-branch receiver diversity may be feasible using a dual polarized patch antenna or a combination of a monopole and patch antenna as mentioned in Reference 2.

10.8 Fixed Beam and Adaptive Beam Forming

Beam forming is a solution that improves both the uplink and downlink performance. This results from the spatial confining of multi-user interference on both links. In addition, beam forming results in much higher gain antennas, which in turn result in improved performance on both links.

10.8.1 Implementation Considerations and Issues

There are two varieties of beam forming techniques, referred to as *fixed beam forming* (or 'switched beams') and *adaptive beam forming* (or adaptive antennas). In the fixed beam forming technique, a number of antenna elements are used in the base station to provide a number of fixed overlapping beams covering the required area. The system then detects the strength of the signal from a mobile and according to suitable algorithms switches the mobile from beam to beam as the mobile moves through the area to maintain gain in the direction of the mobile. In the adaptive beam forming (or user-specific beam forming), an antenna array panel is used at the base station, each element being properly phased with respect to each other, to create narrow beams aimed at each specific user. In this approach, the amplitude and phase associated with each antenna element is adjusted in real time to provide beam steering at each user. The fixed beam approach can be implemented in a relatively simpler manner by integrating analog phase shift component into the antenna panel. The adaptive beam forming (user-specific) approach is more complex and requires the real time tracking of user location and movement with fast control of the signal amplitude and phases for each antenna element for beam steering. The fixed beam technology is also more mature at this time than the user-specific adaptive beam approach.

WCDMA downlink physical layer performance requires the mobile terminal's accurate estimation of the channel impulse response and measurement of the received SIR. The 3GPP specifications define a reliable phase reference through the P-CPICH for the case when a single transmit antenna is used at the base station. When the fixed beam forming is deployed, the S-CPICHs are used to provide a separate and reliable phase reference for each beam. The downlink beam forming gains are dependent on the mobile terminals reception of the CPICHs [9].

The fixed beam approach is favored by the WCDMA specifications as it simplified matters from the practical side. The fixed beam approach has well specified functions for the mobile terminal and allows beam specific S-CPICH to be exploited. In contrast, the user-specific beam forming necessitates the use of the pilot sequence within the dedicated control channels, DCCH, which ends up reducing the link performance by 2–3 dB relative to when the P-CPICH can be used. The fixed beam approach allows primary and secondary scrambling codes to be assigned across the beams belonging to a cell, which helps to alleviate the limitations in the channelization code trees. The fixed beam technique also has the advantage of placing the minimum impact upon the radio resource management functionality.

The largest drawback with the adaptive user-specific beam forming is the increases in system complexity and the large impact upon the radio resource functions. In addition, user-specific beam forming does not provide significant performance over the fixed beam approach, further supporting the WCDMA's preference for the fixed beam approach [2].

10.8.2 Gains of Beam Forming

The spatial filtering required in beam forming results in higher gain antennas that in turn reduce the transmitted power. Furthermore, spatial filtering results in spatial confinement of multi-user interference, providing isolation between the different beams (in the fixed beam approach) and different users in the user-specific beam forming. These same benefits are also applicable on the uplink side. The reduced transmitted powers translate into reduced interference, which then translates into lower E_b/N_0 requirements. The reduced E_b/N_0 requirements result in improved coverage and capacity. Simulation results [2] for 4-beam configurations (using the fixed beam approach) for 12.2 kbps speech service with a BLER of 1% have shown improvements in the uplink E_b/N_0 requirements of between 4.3–5.9 dB under the ITU modified vehicular channel model A (at speeds of 3km/h, 50 km/h, and 120 km/h), and between 6–7 dB under the ITU pedestrian channel model A at 3 km/h relative to a 2-branch receiver diversity at the base station. The largest gains are achieved when the mobile terminal locations are in the direction of the maximum beam gain.

Simulation results [2] also indicate that beam forming provides an effective way for improving the downlink performance particularly in environments with a low angular spread of the radio propagation environment around the base station antenna array. For example, the reduction achieved in the downlink E_b/N_0 requirement relative to a single transmit antenna base station is reported to increase from 3.7 dB with a 20 degree of angular spread to 5.1 dB with a 2 degree angular spread in a 4-beam cell configuration.

10.9 Repeaters

Repeaters provide a cost effective easy solution for extending the coverage of an existing base station into remote low-density traffic areas, and special places such as tunnels, valleys, and occasionally buildings. Repeaters are normally placed between the base station donor antenna and the service coverage area. The repeater consists of the necessary amplification equipment and two antennas, one pointing to the base station antenna (donor) and the other to the coverage area (coverage antenna). The communication between the base station antenna and the donor antenna takes place normally through a directional radio link, which results in minimum interference and multipath effects.

10.9.1 Operating Characteristics

Repeaters are transparent to the parent base station, in the sense that the base station would not 'know' of the repeater's existence. The repeater simply receives the signals, amplifies them, and retransmits them in the uplink and downlink directions. Therefore all the RRM and power control functions in the RNC will operate transparent to the repeater. The link budget performance of the parent cell also remains the same. But a separate link budget must be evaluated for the repeater coverage area using the repeater's transmitter and receiver characteristics.

The WCDMA repeaters perform in the analog domain and therefore cannot regenerate the bit stream to clean up the noise before retransmission. This is as a result of the fact that contrary to GSM, in which at a given time slot and for a given frequency only one user's data is on the air, in CDMA systems, multiple users' data are simultaneously in transmission over the same frequency band but with different channelization codes (and scrambling code on the uplink). In order for the repeater to be able to regenerate the digital bit streams, it would need to know the scrambling and the channelization codes for each user, which is not the case. The amplification of noisy signals through the repeater and their passage through multiple receiver subsystems results in some degradation of the users' signals on both the uplink and downlink. This raises the E_b/N_0 requirements at both the mobile station and the parent base station to maintain the same service quality. The increases in the required E_b/N_0 on the uplink side can range from about 1–2 dB for speech services depending on whether a 2-branch receiver diversity antenna is provided on the repeater coverage side or not [2].

The delay through the repeaters is about 5 µs on uplink and downlink. This is too small compared to the WCDMA slot time of 666 µs to cause any problem in the functioning of the closed loop fast power control loops. However, it is quite large enough compared to the WCDMA chip time of 0.26 µs to significantly raise the level of multipath in repeater coverage areas where signals can also be received directly from the base station. Then the resulting large increases in the multipath and the unresolvable paths will cause loss of downlink code orthogonality and therefore a loss of downlink capacity. This issue should be carefully considered particularly when repeaters are used for the non-standard applications, such as to boost signals in the weak coverage areas visible to the parent base station antenna. In the non-standard applications, repeaters are used to help decrease interference in the system, since mobiles communicating via the repeater need a lower output power than without the repeater. If handled carefully, this can help the interference-limited systems such as WCDMA where reduced interference results in increased capacity.

To reduce the potential interference to nearby cells, On-Off repeaters have been considered for CDMA systems [10] in cases when they are just used to provide coverage to users who occasionally roam to dead spots. These repeaters switch on and work normally if an active user is within the coverage area, otherwise the repeater turns off automatically. Using a repeater like this reduces additional interference to the system compared to when the repeater is always on, which is good for an interference limited systems.

10.9.2 Repeater Isolation Requirements

In the installation and configuration of repeaters, it is important to ensure a sufficient amount of isolation between the repeater transmitter and receiver [11]. Since a repeater receives signals from a base station and amplifies them, under certain circumstances the repeater may act as an oscillator, with the coverage and donor antennas as the feedback path in the amplifier system [12]. To prevent oscillations in the system, the feedback must be lower than the amplifier gain by about 15 dB on the downlink and uplink side. Therefore the isolation between the repeater antennas must be 15 dB larger than the repeater gain. Since the repeater coverage and base station side antennas are usually mounted in opposite directions, it helps to choose both donor and coverage antennas that have a high front-to-back ratio [11].

The repeater isolation can also be highly affected by the nature of the surrounding environment. If the location is not properly selected, radio waves reflected and scattered by nearby obstacles, buildings, etc. can feed back into the repeater causing oscillation and collapse of the system. The waves transmitted by antennas are reflected by surfaces, depending on the materials. If there is a reflection from a building towards the pole with the mounted antennas, this can decrease the antenna isolation by more than 10 dB. The best isolations are usually provided when the repeater coverage antenna is isolated from the repeater donor antenna through a concrete building or hill.

10.9.3 Repeater Coverage and Capacity Evaluation

To determine the repeater coverage area, separate link budgeting must be performed for the repeater coverage area. The repeater link budget may different from the donor cell link budget in a number of parameters. These include the repeater receiver noise figure, E_b/N_0 requirements (which may differ), receiver antenna gain, feeder losses, and target loading. Depending on how these parameters may differ from that of the donor cell, the repeater coverage area may result in a lower maximum allowed path loss and hence coverage range. The repeater capacity will depend on the transmit power capability of the repeater on the coverage side that is influenced by its gain. The gain must be set to be 15 dB less than the isolation provided between the repeater antennas to prevent self-oscillations.

10.9.4 Impact on System Capacity

The repeater impacts the system capacity through the increased E_b/N_0 requirements as argued earlier, the possible loss of the downlink code orthogonality in boundary areas between the repeater coverage area and the parent cell, and the reduced downlink base station transmit power requirements for users connected to the cell through the repeater. The latter are a result of the better link budget on the direct radio link between the repeater and the donor cell antennas. The impact of these parameter changes to the system capacity will depend on whether the donor cell is uplink or downlink capacity limited.

If the cell is uplink capacity limited, the higher E_b/N_0 requirements for users linking to the repeater result in some loss of capacity. The extent of the increased E_b/N_0 requirement depends on whether the repeater is provided with antenna receiver diversity or not. If the system is downlink capacity limited, the increased E_b/N_0 requirement on the downlink for users linking to the repeater results in an increase in the downlink interference floor (loading) for both the repeater and the donor cell coverage areas. This increase in downlink loading is offset to the extent that less transmit powers will be required from the donor base station for users linked to the repeater due to the better link budget on the direct radio link between the repeater and the base station. Depending on the relative contributions of these two opposite effects, the system capacity will decrease or increase. The capacity in the repeater coverage area is also a function of the repeater transmit power capability, which is determined by its gain. However, users can be admitted when there is insufficient power due to lack of knowledge from the RNC side on the repeater instantaneous transmit power status, or even of its existence.

10.10 Additional Scrambling Codes

When the system capacity is downlink limited, the capacity is determined by the maximum available base station power and the service mix. The service mix impacts the downlink channelization code tree and the base station power consumption. Data services with higher bit rates (more than 64 kbps) consume more power (more power is needed to result in the same energy per bit), and also more code resources from the downlink orthogonal code tree as was explained in Chapter 9. The system can implement a high data rate connection through either a code with low spreading factor or multiple codes of higher spreading factor. In either case, the code resources can rapidly be used up particularly if certain mechanisms such as transmit diversity are implemented to reduce the downlink power requirements and help support more high speed data connections, for instance at 144 kbps or higher. To cope with the downlink orthogonal code availability in such scenarios, a second scrambling code can be used. The second scrambling code results in a second orthogonal code tree from which codes can be selected to support new connections in the cell until the capacity limits set by the base station power are reached.

The orthogonal codes from the two code trees will not be orthogonal to each other, as they are chosen independently from each other. The only isolation provided is through the different scrambling code containing each set. However, since the scrambling codes also lose some orthogonality on the downlink due to multipath effects, perfect isolation is not maintained between the two code trees. As a consequence, users assigned channelization codes from one code tree will experience the downlink interference (load) effect of users on their own code tree, as well as the more severe interference from users on the second code tree. The mutual severe inter-code tree interference is experience by more users if the users are distributed evenly between the two code trees (the two scrambling codes). Therefore to minimize the average interference experienced by each user, the connections should be assigned codes from one code tree until exhaustion. Then users for whom codes are not available from the first code tree should be assigned to the second.

10.11 Self-Organizing Networks

Self-organizing networks are based on the concept of re-configuring the network to follow the changing traffic distribution and user mobility pattern. This technology is most appropriate for CDMA-based systems such as UMTS where capacity is primarily interference limited. In many cases, just adding new equipment may not help to enhance the network capacity and coverage to where it is needed. But the dynamic adjustment of antenna tilt and base-station transmit power (traffic, signaling, and pilot channels), along with an adaptive set of resource management algorithms, are mechanisms that can help to release hidden system capacity. In self-organizing networks, as the user distribution changes over time, the network finds a new set of optimum parameters that best serve the users from both capacity and service quality points of views.

Several European Commission research and development initiatives on this topic have started to look into the complexities involved with any such system [13].

References

[1] 3GPP TS 25.101 and 25.211–25.214, Release 99, March 2000.
[2] Laiho, J., Wacker, A. and Novosad, T., *Radio Network Planning and Optimisation for UMTS*, John Wiley & Sons, Ltd, Chichester, 2002.
[3] Andersson, S., Hagerman, B., Dam, H., Forssén,U., et al., 'Adaptive Antennas for GSM and TDMA Systems', *IEEE Personal Communications*, June 1999, 74–86.
[4] Alamouti, S., 'A Simple Transmit Diversity Technique for Wireless Communications', *IEEE Journal on Selected Areas in Communications*, **16**(8), October 1998, 1451–1458.
[5] Tarokh, V., Seshadri, N. and Calderbank, A., 'Space-time Codes for High Speed Data Rate Wireless Communication: Performance Criteria and Code Construction'. *IEEE Transactions on Information Theory*, **44**(2), March 1998, p. 306.
[6] Friis, H.T., Noise Figures of Radio Receivers, *Proc. IRE*, July 1944, 419–422.
[7] Holma, H. and Tolli, A., 'Simulated and Measured Performance of 4-Branch Uplink Reception in WCDMA', *Proceedings of the IEEE Vehicular Technology Conference*, 8VTC'2001, May 6–9, 2001, pp. 2640–2644.
[8] Lee, W.C.Y., *Mobile Communications Engineering*, McGraw-Hill, New York, 1982.
[9] Tiirola, E. and Ylitalo, J., 'Performance Evaluation of Fixed-beam Beamforming in WCDMA Downlink', *Proc. VTC 2000 Spring*, Tokyo, Japan, May 2000, pp. 700–704.
[10] Choi, W., Cho, B.Y. and Ban, T.W., 'Automatic On-Off Switching Repeater for DS/CDMA Reverse Link Capacity Improvement', *IEEE Communication Letters*, **5**(4), April 2001, 138–141.
[11] 3GPP project, 'Technical Specifications Group, Radio Access Networks; UTRA Repeater: Planning Guidelines and System Analysis 8 (Release 4)', 3GPP TR 25.956-4.0.0 (2001–03).
[12] Pennock, S.R. and Shepherd, P.R., *Microwave Engineering with Wireless Applications*, Macmillan Press, 1998.
[13] Papaoulakis, N., Nikitopoulos, D. and Kyriazakos, S., 'Practical Radio Resource Management Techniques for Increased Mobile Network Performance', *IST Mobile & Wireless Telecommunications*, June 2003, Aveiro, Portugal.

11

Co-planning and Inter-operation with GSM

Operators with existing GSM networks face two major challenges and responsibilities in the deployment of WCDMA: first, maximum re-use of their existing GSM sites for co-locating the UMTS sites; and second, optimal integration of their GSM with the 3G network to realize resource pooling and achieve maximum resource utilization and service coverage. The first challenge requires consideration of RF interference geometries resulting from re-using clusters of the existing GSM sites and the necessary site re-engineering in co-locating UMTS. Site co-location also requires taking various measures such as proper frequency planning and proper antenna spacing or antenna isolation mechanisms to prevent or offset inter-modulation distortion effects between the two co-located systems. These issues are discussed in this chapter. A methodology for using the live network operational data collected by GSM equipment for guiding the selection and re-engineering of sites in UMTS co-location was detailed in Chapter 7.

The current chapter ends by discussing alternative inter-system handover strategies to ensure seamless inter-operation between the GSM and UMTS access networks in order to effect resource pooling to achieve maximum resource utilization and service coverage through the two systems.

11.1 GSM Co-location Guidelines

When co-locating the UMTS sites with the existing GSM sites, several measures should be taken to ensure the necessary RF isolation requirements between the two systems. The RF isolation requirements and the mechanisms to provide the mutual protection between the two systems are discussed in the following two sections.

11.1.1 The Isolation Requirements

Both GSM and UMTS transmitters generate spurious transmissions outside their allocated channel band, which can cause harmful interference to the receiver of the other system.

UMTS Network Planning, Optimization, and Inter-Operation with GSM   Moe Rahnema
© 2008 John Wiley & Sons, (Asia) Pte Ltd

Spurious emissions are emissions that are caused by unwanted transmitter effects such as harmonics emission, parasitic emission, inter-modulation products, and frequency conversion products, but excluding out of band emissions. This is measured at the base station RF output port.

The GSM Specification [1] sets an upper limit of −30 dBm of spurious emissions per 3 MHz of band in the 1–12.5 GHz frequency range (which includes the UMTS band) from a GSM BTS. This means that an interference level of up to −95 dBm/Hz or about −29 dBm in a 4 MHz effective UMTS channel bandwidth can occur. To reduce this interference to at least the maximum tolerable interference level of −118 dBm [2] in the 5 MHz Rx band of UMTS (equivalently −185 dBm/Hz), the isolation required between the two antennas is −95 − (−185) = 90 dB. In the case of GSM 1800, the BTS can have up to −96 dBm/0.1 MHz = −146 dBm/Hz spurious emissions at the antenna connector. This means that the isolation required in this case to bring the interference down to at least the maximum tolerable limit of −185 dBm/Hz in the UMTS receive band is −146 − (−185) = 39 dB. In practice, it is advised to use a well-calibrated spectral analyzer at each particular vendor's GSM equipment to measure the actual level of spurious emissions in the receive band of a co-located UMTS receiver. Then, the measured results can be used to find out the actual level of isolation that needs to be provided between the two systems at the site.

Likewise, the UMTS transmitters can generate spurious emissions outside their allocated bands of up to −96 dBm per 0.1 MHz, or −93 dBm per 200 KHz of the GSM channel bandwidth. This would result in a 28 dB isolation requirement between the transmit antenna of the UMTS and a co-located GSM receiving antenna, to bring the interference level down to at least the GSM channel thermal noise level of −121 dBm.

However, the isolation requirement to protect the GSM system from UMTS is more dominated by the blocking requirements for GSM receivers. The blocking characteristic is a measure of the receiver ability to receive a wanted signal at its assigned channel frequency in the presence of an unwanted interferer on frequencies other than those of the adjacent channels. To protect the operator GSM 1900 from UMTS interference, an isolation of 43 dBm is recommended between the antenna connections of the two systems. This stems from the consideration that a UMTS transmitter can transmit a maximum power level of about 43 dBm. Thus the necessary isolation to reduce this to a specified tolerance level of 0 dBm within the GSM channel bandwidth is simply 43 dB.

To prevent the GSM system from blocking a co-located UMTS receiver, an isolation of 58 dB is required between the two systems. This is based on the consideration that the specified tolerance level for a CW (continuous wave) interfering signal in UMTS band is 15 dBm, whereas the GSM transmitter can generate about 43 dBm over a 200 KHz (almost a CW to UMTS) channel bandwidth.

The full scenarios for possible interference between the UMTS and GSM (1900) systems and the necessary isolation requirements are clarified in Table 11.1.

11.1.2 Isolation Mechanisms

To obtain the isolation of 43 dB to protect the GSM system from UMTS, physical isolation of the antennas can be used. With single band antennas used for each, a horizontal separation of 0.9–1 m or a vertical separation of 1.8–2.0 m is recommended. In the antenna-sharing scenario, 43 dB isolation should be provided by the diplexers used to de-multiplex the signals to and from the GSM and UMTS BTSs.

Table 11.1 Isolation requirements between GSM 1900 and UMTS antennas [based on data taken from References 1 and 2].

Transmitter	Effect on frequency, MHz	Level, X dBm/X MHz	Parameter affected	Allowed interference limit, X dBm/X MHz	Required isolation (dB)
UMTS, spurious	1920–1980 (FDD UL)	−96/0.1 equivalently −80/4	**UMTS BTS sensitivity**	−121/4	41
UMTS spurious	1850–1910 (GSM UL USA)	−96/0.1 equivalently −80/4	**GSM BTS, sensitivity**	<−110/0.2, Typical	17
UMTS, main	2110–2170, 1930–1990 (FDD DL)	43/4, Typical	**GSM BTS, Blocking**	0/CW	43
GSM (all bands), spurious	1920–1980, 1850–1910 (FDD UL)	−95 dBm/Hz	**UMTS BTS sensitivity**	−121/4, typically	90
GSM, main	1930–1990 (GSM DL USA)	43/0.2, Typical	**UMTS BTS Blocking**	<−15/CW	58
UMTS, main	2110–2170, 1930–1990 (FDD DL)	43/4	**UMTS BTS Blocking**	<−40/4	80

Obtaining the isolation requirements of 90 dB, to protect UMTS from the GSM 900 system, through antenna spacing alone will be difficult. Specific filters at both the GSM 900 and the UMTS sides are recommended to protect each system's BTS from the other system's emissions. Filters at the UMTS BTS side will be imperative when:

- Dual band antennas are used for which the isolation between the GSM and the UMTS bands is only about 30 dB.
- Wideband antennas are used for which there is basically no isolation between the GSM and the UMTS bands.

For the dual band antenna (two antennas within one panel), the diplexers should be selected carefully to ensure that the combination of the BTS side and the antenna side diplexers will provide the necessary isolation requirement of about 90 dB as indicated in the previous section.

The dual band solution with two diplexers offers the advantage of allowing the independent setting of the gains and tilts for the GSM and UMTS antennas.

11.1.3 Inter-modulation Problems and Counter-measures

Inter-modulation is a process, which generates an output signal containing frequency components not present in the input signal. Inter-modulation distortions (IMD) can be generated both inside an offender or a victim system.

In the case of two input frequencies, $v_{in} = \cos w_1 t + \cos w_2 t$, output will consist of harmonics $m.w_1 + n.w_2$, where n and m are positive or negative integers.

Most harmful are third order $(|m| + |n| = 3)$ products. Inter-modulation distortions can result from the following:

- Nonlinearities of active components such as amplifiers under normal operation
- Nonlinearities of passive components
- Antennas
- Feeders
- Connectors
- Antenna mismatching
- Reflected wave entering the power amplifier
- Damaged feeders resulting in mismatching
- Loose connectors resulting in mismatching, reflections, and rectification
- Aging of the components increases IMD

The IM products resulting from component aging may be avoided by installation, site administration, and maintenance recommendations.

The power of the third order inter-modulation product, P_{IMD3} from two equal power tones, each with a power of P_{indBm} can be calculated from the results given in Reference 3 with the formula

$$P_{IMD3} = 3G_{dB} + 3P_{indBm} - 2IP_{3dBm} \qquad (1)$$

where G_{dB} is the gain of the active component (in which inter-modulation takes place, such as a power amplifier) and IP_{3dBm} is the third order intercept point for the component, defined as the output power at which the fundamental power intersects the third order inter-modulation power. This point is usually a hypothetical point since the power amplifier has already saturated at this output power. For a passive component, $G_{dB} = 0$.

The inter-modulation distortions resulting in the active components are generally much stronger than those resulting in passive components. Typical IM3 suppression values for power amplifiers are –30 to –50 dBc depending on frequency spacing and offset, whereas the typical values for passive elements are –100 to –160 dBc

Harmonic distortion can be a problem in the case of co-siting of GSM900 and WCDMA. IMD products or harmonics generated inside the offender system can be coupled through the air and fall into the victim system's band. Moreover, signals from the offending system can be coupled through the air and go into the victim receiver causing IMD inside the victim system.

The GSM 900 DL frequencies are in the range 935–960 MHz where the second harmonics may fall into the WCDMA TDD band (1900–1920 MHz) and into the lower end of the FDD band. The second harmonics can be filtered out at the output of GSM900 BTS. The GSM1800 IM3 products ($f_{IM3} = 2f_2 - f_1$) can hit into the WCDMA FDD UL RX band if $1862.6 \leq f_2 \leq 1879.8$ MHz and $1805.2 \leq f_1 \leq 1839.6$ MHz. These frequencies correspond

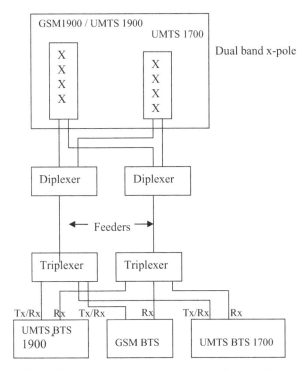

Figure 11.1 Cross pole dual band antennas configured with triplexer.

to the GSM 1800 channels of 512 to 684 (f_2) and 799 to 885 (f_1). The inter-modulation products resulting in the BTS active components can be filtered out in the GSM1800 BTS TX. IM products generated inside a WCDMA receiver cannot be filtered out.

The passive IM products cannot be filtered out in BTS TX if they are generated in feeder lines and connecting elements such as the diplexers/triplexers and connectors after the filtering unit of BTS.

Then, if they fall within one of the co-located system's receiver band, they can create new sources of interference. This can happen for instance if the GSM 1900 system or the UMTS 1900 system is co-located with UMTS 1700/2100 system. In this case, depending on where in the bands the operator obtains its licenses for UMTS, the third order inter-modulation between the GSM or UMTS and the UMTS 2100 transmit bands can create emissions in the 1700 band and fall within the receiver band of the UMTS 1700/2100 system. Such a scenario is possible particularly if the feeders are shared between the GSM or UMTS 1900 system and the UMTS transmitter operating on the 2100 band, and at best should be avoided. Otherwise, MHAs should be provided on the UMTS 1700 receive band to boost the receiver signal and help it to overcome the interference that can add to it downstream the feeders from inter-modulation effects (such as possible in the antenna configuration given in Figure 11.1).

11.1.4 Antenna Configuration Scenarios

There are various scenarios that can be used to provide the UMTS antenna system in a co-located GSM site. Each scenario presents advantages and disadvantages. These include

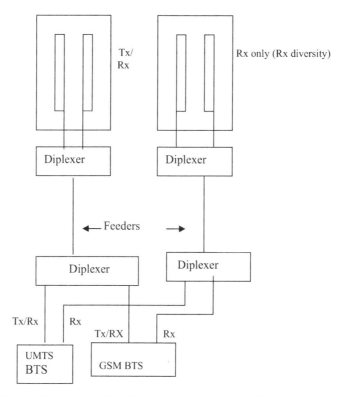

Figure 11.2 Two dual band vertically polarized antennas configured with two antenna-side and two BTS-side diplexers.

deploying single band, broadband, dual/triple band antennas, or some combination that then requires higher performance and more complex duplex or triplex transmit and receive filtering to reduce the co-site interferences. The key technical performance implications that have to be considered in these co-location situations include the effect of spurious emissions, inter-modulation, and receiver blocking, as discussed in previous sections. This section prsents various scanarios in which the antenna systems for a UMTS 1700/2100 system may be configured in GSM 1900 co-siting.

The simplest scenario would be to have a separate antenna system for the UMTS, and space it appropriately from the GSM antenna, as discussed in the previous section, to provide the necessary isolation requirement. A main disadvantage of the separate antenna system is the requirement for more feeder cables, which can be practically inappropriate and also costly depending on the situation. A main advantage is that it would allow separate control of the antenna parameters (tilt, azimuth, height) for the two systems.

An alternative would be to use dual band antennas, as shown in Figure 11.2. The dual band antenna still allows the independent control of tilting for the two systems. In this case, x-pole (cross polarized) antennas with four connectors would be preferred, as they help to reduce the number of separate physical antennas resulting in cost and space reductions. The dual cross pole antenna system can be configured as shown in Figure 11.3 without diplexers,

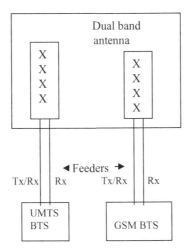

Figure 11.3 Cross pole dual band antenna configured without antenna side diplexers.

or as in Figure 11.4 with two diplexers. The second configuration helps to reduce the number of feeder cables by 2, by allowing multiplexing of the UMTS 1700 band with the GSM 1900 band on the same feeder. However, care should be taken to make sure that any losses (which can range from 0.5 dB to about 3.5 dB) are compensated for in the link budget.

Depending on which specific GSM 1900 and UMTS 1700 transmit frequencies are utilized in the site, it is possible that IM3 products could land on the 5 MHz uplink of the UMTS 1700 carrier. In this case, the use of an MHA on the UMTS antenna connector is strongly recommended to help amplify the UMTS 1700 Rx signal well above the level of IM3 products that may form by the mixing of 1900 and 2100 Tx signals downstream from the MHA in the shared feeder system

In the configuration given in Figure 11.4, the BTS side diplexers can be replaced by triplexers to allow a second UMTS carrier at 1900 MHz implemented on the same GSM 1900 band antenna, as illustrated in Figure 11.5. However that would not allow for the independent tilting and control of the UMTS 1900 band, if found necessary in certain situations. For the configuration shown in Figure 11.5, if each of the two antennas within the dual band radome can support both the GSM 1900 and the UMTS 1700/2100 band, it would be advantageous to place the GSM 1900 and the UMTS 1700/2100 bands on one cross pole and the UMTS 1900 band on the other cross pole within the radome. This would provide better isolation between the GSM 1900 and the UMTS 1900 system.

Finally, a remaining option is the use of wideband antenna covering both the GSM (1900) and UMTS bands. The advantages for a wideband antenna system are the space reduction achieved and the reduced number of feeder cables (down to 1). However, there are two main and significant disadvantages of the wideband antenna (at least with the current technology status). The first drawback is the inability to independently control the tilting for the two systems, and the second disadvantage is that it would basically provide no RF isolation between the two bands. The configuration for a wideband antenna system is illustrated in Figure 11.5.

The best specific antenna configuration for a given situation depends on several factors, including the relative placement of the frequency bands between the GSM and the UMTS,

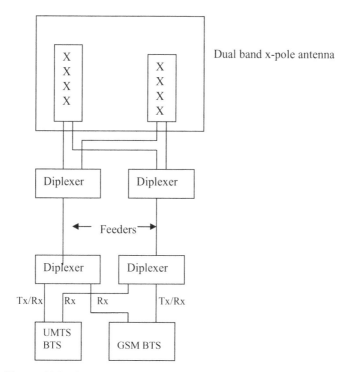

Figure 11.4 Cross pole dual band antennas configured with diplexers.

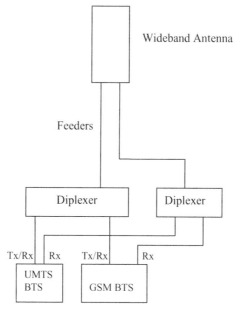

Figure 11.5 Two wideband antennas configured (one for a RX diversity). In this configuration, if each of the two antennas within the dual band radome can support both the GSM 1900 and the UMTS 1700/2100 band, it would be advantageous to place the GSM 1900 and the UMTS 1700/2100 bands on one, and the UMTS 1900 band on the other cross pole within the radome. That would provide better isolation between the GSM 1900, and the UMTS 1900.

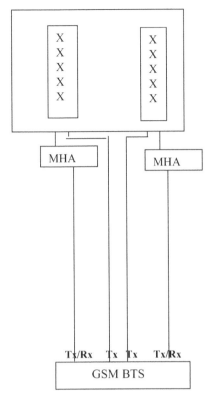

Figure 11.6 The existing GSM site configuration.

and the required isolation, the existing operator's GSM site antenna configuration, the available allowed spacing for placement of additional antennas, as well as operator constraints for replacing existing GSM antennas, adding additional feeders, etc.

Case Example
The existing GSM site configuration is shown in Figure 11.6 and consists of the following:

- One quad-pole antenna (two x-poles within one radone)
- Existing antennas are 18dBi gain, 65deg, x-pole, suitable for 1700/2100 band use; 4 ports are used for GSM transmit (4 GSM radios) and 2 ports for the GSM receive feeders
- 2 x 1900band MHAs

The operator's constraints are as follows:

- No more feeders can be added
- Antennas can be exchanged, but no additional ones can be added

A solution is to replace one of the dual band antennas within the radon with a UMTS compatible one, if it can support the UMTS band. Two combiners are then used to combine

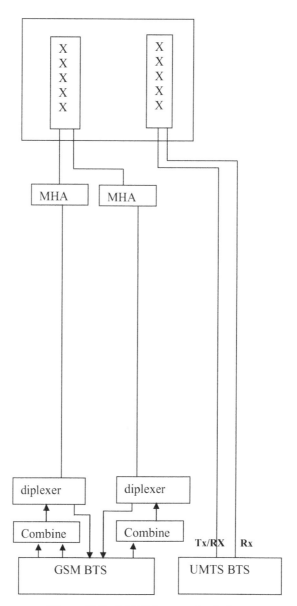

Figure 11.7 UMTS integrated solution.

each of the two GSM radios, and transmit them on one of the x-pole antennas within the radon (on two of the ports). The remaining x-pole antenna (remaining two ports on the quad-pole radone) are used for UMTS transmit and receive. In this case the effect of the combiners on the GSM link budget should be investigated to make sure that any undesirable loss in GSM coverage is compensated for by increasing the power of the GSM BTS, or taking other measures in the link budget. The solution is illustrated in Figure 11.7.

11.2 Ambient Noise Considerations

Since interference is the limiting factor in the performance of WCDMA, the ambient noise has a considerable impact to the coverage and capacity of UMTS radio network. The main sources that contribute to background noise are industrial machinery, power lines and lightening, and other radio systems. There are reports that gather measurements to quantify the background noise [4]. These reports indicate the resulting noise rise at around 4–8 dB at frequencies of 1.8–2.0 GHz, and higher at the lower frequencies. The background noise level generally depends on the environment, and it is therefore advised to measure the level and its time variations for each particular environment. This can be done with a properly calibrated spectral analyser [5, 6]. The measurement of the background noise levels is particularly critical in the base station locations close to countries that have not yet cleared the frequency band for UMTS services.

Reported measurements [6] for various urban and industrial environments place the mean background noise level in the range –105.7 dBm to –102.7 dBm in 5 MHz UMTS channels over the entire UMTS uplink band of 1920 to 1980 MHz with a standard deviation of 2.9 dB. The measurements carried out in Reference 6 also indicate that the noise floor is normally higher by about 1.9 dB at antenna locations placed at street level in the vicinity of heavy motor traffic compared to at roof-top locations in urban environment. The difference between the street level in the vicinity of heavy motor traffic and roof-top level for suburban areas is reported to be 0.5 dB.

The maximum acceptable increases in the background noise (over the thermal noise of –108 dBm in the UMTS band) are defined as the values where the coverage on the uplink and the capacity on the downlink are reduced by no more than 10%. The value at the 10% capacity reduction will depend on the ratio of the actual average path loss to maximum average path loss on the downlink. The background noise rise limit defined in this way falls in the 1.6 dB range for 10% coverage reduction, and from 2.4 to 18.1 dB for path loss ratios of –10 to –30 dB for 10% capacity reduction [6].

Since the uplink is the critical link for coverage determination, it is important to measure the background noise at the base station antenna location before validating a probable BS location from a noise level point of view. The background noise increases the effective thermal noise at the base station, and hence reduces the base station sensitivities. This effect then impacts the service coverage and capacity.

11.3 Inter-operation with GSM

UMTS is initially deployed in hot spot urban areas where there is an increasing demand for voice capacity as well as the considerable market for high-speed data services. This means that initially no continuous coverage will be provided through UMTS. Therefore proper inter-working between UMTS and GSM is critical through selective handovers between the two systems in order to provide continuous network coverage. Moreover handovers with other UMTS networks may be implemented to provide ubiquitous UMTS coverage in areas where the operator does not have continuous coverage through UMTS. Selective system camping in idle mode in areas where dual mode handsets are available, to prefer one system or another depending on the service, location, and expected load/capacity in each, is also a strategy that should be considered.

11.3.1 Handover between the Operator's GSM and UMTS Networks

The handover between the operator's UMTS and GSM (inter-system handover) should be planned based on the following considerations for each site area:

- Providing seamless service coverage
- Demand for voice versus the capacity provided from the GSM site
- Demand for UMTS high-speed data services
- Load and coverage/quality fluctuation between the two systems
- Area based service priority
- Subscriber based service differentiation

In the initial stages of UMTS deployment, UMTS coverage may be restricted primarily to dense urban and urban areas where there might be a higher level of voice congestion on the GSM networks or higher demand for the data services. This means that handover from UMTS to GSM will be required to provide continuous voice or data coverage. In the packet switched (PS) domain the term 'handover' is commonly used, but inter-system PS 'handovers' are not supported in real time due to lack of support in GSM/GPRS. However the connection (i.e., PDP context) is kept and packets are forwarded from IuPS to GPRS Gb or vice versa under control of the serving SGSN.

In areas where the operator provides ubiquitous UMTS coverage, the idle mode cell selection parameters may be set to have the mobile camp in the UMTS system offload the congested GSM network. If the area is categorized as having spare GSM capacity, the handoff parameters are set to have the mobile handover voice calls to the GSM network, offloading UMTS for data services. This means the UMTS cell neighbor list should be provided with selected neighboring cells from the GSM network in the area. This should particularly be done in areas where UMTS specific data services are expected to be in high demand.

The measurements used in the inter-system handover consist of the E_c/I_0 and RSCP of the serving cell(s) in the UMTS, the GSM neighboring cells BCCHs Rxlev, and the C/I of the serving cell when connected to GSM. The operator settable parameters, which control the handover between the cells, are the thresholds and the cell individual offsets (CIOs) and hysteresis values associated with each measurement and cell. The measurements of RSCP, and Rxlev are indicative of pilot coverage in the UMTS system, whereas the Rxlev measurements provide similar indications for the coverage in the GSM system. The E_c/I_0 is an indication of the expected quality provided by the UMTS system, and likewise the measurement of C/I when connected to GSM provides the quality information. By appropriately setting thresholds for these measurements, and the cell individual offset and hysteresis parameters, the operator is able to move the traffic between the UMTS and GSM systems through real time monitoring of the coverage and quality provided by each system.

In addition to coverage and quality criteria, the UMTS equipment is expected to support inter-system handovers based on dynamic monitoring of the load in each system, as well as service based priority. The load and service based priority can be enabled to move the traffic between the two systems based on real time local monitoring and consideration of the load, and coverage/quality by each system. In certain areas where GSM may have ample spare capacity left to handle voice, the service based priority handover feature can be used to handover voice calls to GSM, leaving UMTS to provide the higher rate data services. This feature may also be used in areas where UMTS may not provide ubiquitous coverage (for example, there are coverage holes or interference problems in UMTS from high sites, etc.) to handover the voice calls to GSM to minimize user poor service quality perceptions.

Finally, subscriber-based service differentiation can be implemented to assign priority based on the type of subscription possible. Users with gold subscriptions could by general rules be transferred to the system with the lowest load or the best available bearer services. Users with speech only subscriptions, or combined with services requiring low bandwidth data, could be kept in GSM even where UMTS coverage is provided.

11.3.2 Handover with Other UMTS Operators

The operator should consider roaming options on both the national and international level. On the national level, the operator may want to consider providing roaming with other operators' UMTS networks in areas where UMTS will not yet be provided.

The mechanism for providing the roaming will involve:

- Listing the desired partner operator in the USIMs as allowed network
- Including the neighboring cells from the partner operator in the neighbor list of the operator's border cells
- Placing the roaming subscription information and authorization in the operator's HLRs

References

[1] 3GPP TS 05.05, 'Radio Transmission and Reception', Phase 2+, Release 1999.
[2] 3GPP TS 25.104, 'BTS Radio Transmission and Reception (FDD)'.
[3] Carvalho, N.B. and Pedro, J.C., 'Compact Formulas to Related ACPR and NPR to Two-Tone IMR and IP3', *Microwave Journal*, **42**(12), December 1999.
[4] Prasad, R., Kegel, A. and de Vos, A., 'Performance of Microcellular Mobile Radio in a Cochannel Interference, Natural, and Man-made noise environment', *IEEE Trans. Vehic. Tech.*, **42**, February 1993, 33–40.
[5] Engelson, M., *Modern Spectrum Analyzer Theory and Applications*, Artech House, 1984.
[6] Laiho, J., Wacker, A. and Novosad, T., *Radio Network Planning and Optimisation for UMTS*, John Wiley & Sons, Ltd, Chichester, 2002.

12

AMR Speech Codecs: Operation and Performance

The adaptive multi-rate (AMR) speech codec was introduced at the end of 1998 as a new codec for GSM and as the mandatory speech codec for 3G wideband CDMA (W-CDMA) systems. By the beginning of 2004, AMR was the de-facto codec for existing GSM systems and is a key component for the rollout of 3G wireless designs. In GSM systems, the AMR speech coding provides the flexibility to tradeoff speech quality and capacity smoothly and flexibly by a combination of channel and codec mode adaptation, which can be controlled by the network operator on a cell-by-cell basis.

12.1 AMR Speech Codec Characteristics and Modes

The AMR coder operates on speech frames of 20 ms processing 160 samples obtained at a sampling frequency of 8000 samples per second. The coding scheme for AMR is based on the so-called Algebraic Code Excited Linear Prediction (ACELP). The speech parameters obtained are sorted according to their sensitivity to errors and are divided into three classes of importance, referred to as Classes A, B, and C. Class A is the most sensitive and the strongest channel coding is used for it on the air interface.

AMR is a multi-rate speech codec, which consists of two channel modes, FR and HR, each of which uses a number of codec modes that use a particular speech codec. AMR can operate at eight distinct bit rates (modes): 12.2 kbps; 10.2 kbps; 7.95 kbps; 7.4 kbps; 6.7kbps; 5.9 kbps; 5.15 kbps or 4.75 kbps, which are referred to as modes 7 through 0, respectively. The 12.2 kbps speech codec is equivalent to the GSM EFR codec. The highest modes, 7 and 6, are not supported in the GSM half-rate channel (AMR HR). The AMR also includes a low rate background noise encoding mode. All the eight modes are supported in GSM FR channel mode and in WCDMA. The channel mode (FR or HR) is switched to achieve the optimum balance between speech quality and capacity improvements in GSM. The uplink and downlinks use the same channel mode. The channel mode is selected by the network (probably the BSC) based on measurements of the quality of the uplink and downlinks. The lower modes within each channel mode in turn provide for more channel

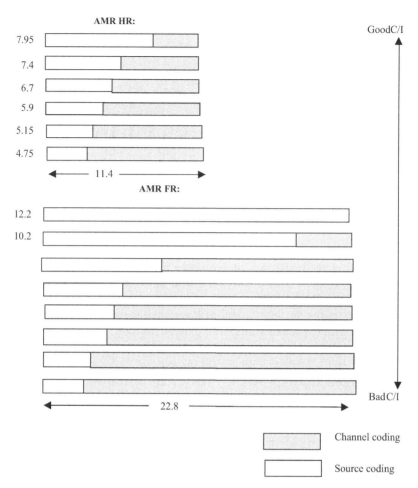

Figure 12.1 The AMR FR and HR codec modes.

coding and hence allow the use of a tighter frequency design in GSM for more capacity. The AMR speech codecs include a set of fixed rate speech codec modes for half rate (used in GSM to obtain more capacity with lesser amount of hardware) and full rate operation as shown in Figure 12.1. At low C/I, a large amount of channel coding is applied by switching to the lower codec modes (less speech coding) within each AMR codec. When the C/I increases, the speech coding is increased to provide better voice quality and the channel coding is decreased instead. The codec adaptation relies on channel quality measurements performed in the MS and the network and the signaling of this information over the air interface. The lower modes have lower inherent speech quality but provide more spare bandwidths for higher channel error correction coding capabilities The AMR FR allows for the highest channel coding modes (by using twice the total bit rate at 22.8 kbps), which is suited to the extreme bad end of channel conditions. The switching between the different codec modes are performed every other voice frame (40 ms) based on the radio channel error conditions. For compatibility with legacy systems, the 12.2 and 7.4 modes are compatible versions of the GSM EFR and the US TDMA IS-136 EFR speech codecs, respectively. The results of subjective speech quality tests for each of the AMR Codec modes are provided in Reference 1.

The GSM AMR implementation allows a selection of four (out of the eight possible) different speech and channel coding schemes. A combination of AMR modes 7, 6, 5, and 0 has been found by simulation to deliver an optimum speech quality for a broad range of channel conditions [2]. The speech coder is capable of switching its bit-rate every 40 ms per speech frame upon a signaling command, which can be transmitted in-band or on an out-of-band control channel. The AMR rate adaptation is mandatory for all UMTS equipment. The network has the option to command the enabling of the AMR in the transmitting path of the UE (uplink). The UE should be able to handle the rate adaptation in the receiving path (downlink) at any time, regardless of whether AMR in the transmitting path is enabled or not [3]. Thus the rate adaptation can be deployed in the downlink with software upgrade to existing networks without any changes in the mobile terminals.

The AMR speech codec includes a set of fixed rate speech codec modes for half rate and full rate operation, providing the possibility to switch between the different modes as a function of the radio channel error conditions. Each codec mode provides a different level of error protection through a dedicated distribution of the available gross bit rate (22.8 kbps in full rate and 11.4 kbps in half rate) between source coding and channel coding. The actual speech rate used for each speech frame depends on the radio link condition in GSM. A codec adaptation algorithm selects the optimized speech rate (or codec mode) as a function of the channel quality. By adapting the codec rate to the radio conditions the speech quality is enhanced.

The key characteristics of AMR solutions are as follows:

- Eight codec modes in full rate mode including the GSM EFR and IS136 EFR.
- Six codec modes in half rate mode (also supported in full rate), including the IS136 EFR.
- Possibility to operate on a set of up to four codec modes selected at call set up or handover.

12.2 AMR Implementation Strategies

Equipment vendors may define and implement a number of different codec sets for each channel rate in the system and let the operator select, via a command, which codec set to use for each channel rate per base station. The codec set can consist of up to a number of different codec modes (usually four), all using the same channel rate. It is not possible to have different channel rates within the same codec set. Different codec sets within each channel rate may better suit a particular radio environment such as clean or noisy speech conditions. The selection of codec mode within the codec set is based on measurements of the uplink and downlink radio channel, using threshold parameters.

The AMR rate adaptation can be implemented based on network conditions, and also based on source controlled adaptation

12.2.1 AMR Network Based Adaptation

In GSM, the rate adaptation is done instantaneously to channel conditions through the radio link adaptation function. The switching between two codec channel rates (FR or HR) can be controlled by the operator through thresholds, which can be set on a cell-by-cell basis. These thresholds may be based on measurements of hardware utilizations as well as voice quality in the cell. However, the switching between different codec modes within a certain codec set defined under a FR or HR channel is usually controlled by hard-wired thresholds that determine which codec modes should be used for a certain C/I channel quality indications. The codec mode adaptations are very fast and can be executed up to ten times per second. To avoid continuous rapid changes between two codec modes, a hysteresis, usually in the range of 2–4 dB, is used for the decision thresholds.

In WCDMA, the AMR bit rate can be controlled dynamically by the network or by the operator to provide a tradeoff between voice capacity and quality. When the AMR bit rate is adjusted dynamically by the network, the RRM algorithms within the RNC estimate the cell loading and use it to adjust the rate. The WCDMA flexible physical layer allows adaptation of the bit rate and the transmission powers for each 20 ms frame. If the cell load is too high, new voice calls are allocated a lower bit rate AMR mode or the bit rate of the existing connections is modified. The dynamic adjustment forces lower AMR rates at network busy hours to reduce voice call blocking whereas it results in maximizing the voice call quality during off-peak hours. User priority may also affect the AMR bit rates. Higher priority users may be assigned the higher AMR bit rate modes, whereas lower rates are used for the lower priority users.

The AMR bit rate can also be used to increase the cell range in WCDMA by up to 2.5 dB [4] by reducing the AMR bit rate when the mobile runs out of transmission power at the edge of the cell coverage area.

12.2.2 AMR Source Controlled Rate Adaptation

In the source controlled adaptation scheme, the AMR source controls the AMR bit rate mode based on the content and characteristics of the speech signal. The simplest version of source controlled adaptation is the discontinuous transmission (DTX) mode, which has been in common use. In DTX, the speech activity detector detects the speech silence intervals and switches to a very low rate comfort noise generation mode to save on both the mobile battery life and reduce interference in the network. However, in the more sophisticated AMR source controlled rate adaptations, various codec bit rate modes are selected during the active speech depending on the source signal characteristics. When the speech signal is rich in the information theoretic sense, the AMR source switches to higher rate AMR modes to provide the necessary quality, while it switches to the lower AMR rates otherwise to reduce the transmit power and hence the interference in the network. The AMR source controlled rate adaptation is claimed [4] to reduce the average AMR bit rate by 20–35% while maintaining the same voice quality. In the end, the gains achieved will depend on the information theoretic content and characteristics of the speech signal. The AMR codec mode rate changes can take place every other voice frame (40 ms) in GSM, and on each frame basis in WCDMA (every 20 ms).

The AMR source rate adaptation is fully compatible with the existing fixed rate AMR speech codec format with no impact to the decoding process. The AMR source rate control can be added to networks to enhance GSM and WCDMA downlink capacity without any changes to the mobile stations. The AMR source controlled rate adaptation can be implemented along with the network controlled adaptation to further reduce the transmission powers and hence the interference, further improving network capacity without decreasing quality.

12.3 Tradeoffs between AMR Source Rate and System Capacity in WCDMA

In WCDMA, the set of available codec modes (bit rates) that the UE may use are configured by UTRAN. The network selects, from the configured set of codec modes, a mode that can be supported by the system load and the TX power conditions as defined in Reference 5. The lower codec modes require less transmit power due to the lower data rate and hence result in increased capacity. The downside is the reduced inherent quality resulting from the lower source coding rate achieved. The AMR source adaptation thus helps to achieve a tradeoff between the AMR bit rate (quality) and the required network capacity. The estimated capacity gain is about 15–25% [4]. The AMR speech codec also helps to increase the battery life by transmitting at lower power with the lower source rates.

The tradeoff between capacity/coverage and AMR source rate (quality) in WCDMA can be seen from the following equation [6]:

$$\begin{aligned}\text{Coverage or capacity gain} &= 10.\text{Log}_{10} \frac{DPDCH(at12.2kbps) + DPCCH}{DPDCH(atAMRbitrate(kbps)) + DPCCH} \\ &= 10.\text{Log}10 \frac{12.2 + 12.2 * 10^{-3dB/10}}{AMR\_Bi\_Rate(kbps) + 12.2 * 10^{-3dB/10}}\end{aligned} \quad (1)$$

where the power difference between the DPCCH and DPDCH channels was assumed to be -3 dB for 12.2 kbps AMR speech. For lower AMR bit rate modes, the DPCCH power is kept the same while the power of the DPDCH is changed according to the bit rate. The reduction of the total transmission power calculated from the above equation can translate into either increased cell range (on the uplink for instance) or more capacity. The trends calculated from the equation are shown in Figure 12.2.

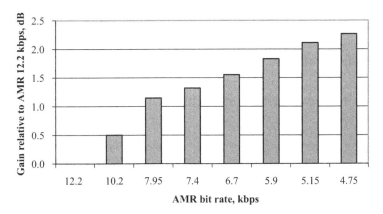

Figure 12.2 Capacity gains achieved relative to AMR 12.2.

The increased gain achieved in the transmission power (power saving) results in increased coverage range as the AMR bit rate is reduced.

The AMR bit rate adaptation also helps to increase the cell range in WCDMA by reducing the bit rate (switching to a lower AMR mode) when the mobile runs out of transmission power at the cell edge.

12.4 AMR Performance under Clean Speech Conditions

Experiments performed in clean speech condition, that is under no background noise of any kind, have shown [7] that the combination of all eight speech codec modes used by the AMR FR provide the performance equivalent of the enhanced full rate speech (EFR) down to 4 dB C/I in mean opinion score. It is to be noted that mean opinion scores can only be representative of the test conditions in which they were recorded (speech material, speech processing, listening conditions, language, and cultural background of the listening subject). However, the relative performances of different codecs in test situations can be considered more reliable and less impacted by cultural differences between listening subjects. These results have also shown that the four highest codec modes (12.2, 10.2, 7.95, and 7.4) are equivalent to EFR down to about 10 dB, and are barely affected by propagation errors over wide range channel conditions down to about 7 dB of C/I. The four lowest codec modes (6.7, 5.9, 5.15, and 4,75) are all judged to be effectively equivalent to EFR down to 10 dB C/I. The performance of the three lowest codec modes are statistically unaffected by propagation errors down to 4 dB C/I.

Similar experiments carried out by the ITU with AMR HR have shown that the combination of all six speech codec modes provide the performance equivalence of the ITU Recommendation G.728 [1] (16 kbps) speech codec down to 16 dB C/I. Furthermore, the results have shown that AMR can provide significantly better performances than GSM FR and GSM HR codec down to 7 dB C/I. AMR HR optimizes the usage of capacity when the traffic load is high, whereas it offers the best possible speech quality when the traffic load is low. This is done by dynamically varying the amount of channel coding based on channel conditions.

The four highest codec modes (7.95, 7.4, 6.7, and 5.9) were found in these experiments to be significantly better than the GSM FR in error free conditions down to 13 dB C/I and at least equivalent to the EFR at 10 dB C/I. The three highest modes (7.95, 7.4, and 6.7) were found to be equivalent to the error free EFR in very low error conditions. The two lowest modes were found to be at least equivalent to the GSM FR over the full range of test conditions down to about 4 dB.

12.5 AMR Performance under Background Noise and Error Conditions

The codec performances under background noise and error conditions have been measured in six different experiments [7] of the GSM characterization phase of testing for AMR FR and AMR HR. The following background noise types were included in the tests: Street noise at 15 dB SNR, Car noise at 15 dB SNR, and Office noise at 20 dB SNR.

At high C/I (down to 13 dB) the three highest codec modes (12.2, 10.2, and 7.95) were found to be equivalent to EFR in error free condition. All codec modes down to AMR 5.9

performed better than the GSM FR across all test conditions. A couple of codecs (6.7 and 5.9) still provided at 4 dB C/I, a quality equivalent to the GSM FR at 10 dB C/I. The two lowest modes (5.15 and 4.75) were usually found to be worse than the GSM FR at 10 dB C/I across the range of test conditions[1].

The above experiments have shown that both the AMR FR and AMR HR provide significantly better performance (in MOS) than GSM FR and GSM EFR under all the stated test conditions below C/I of 7 dB, at which point the quality performance of the GSM FR and EFR begins to sharply drop.The EFR codec is roughly 3 dB more robust, that is, for the same MOS value, the EFR codec can support double the level of interference [8]. On the other hand, HR connections generate half the amount of interference (3 dB better theoretical performance) because two can be placed on one frequency in GSM. From this consideration it is expected that HR has a network performance similar to that of EFR, in terms of speech quality outage. This means that with the criteria considered HR and EFR have a similar performance in terms of spectral efficiency. The use of HR results in speech quality degradation because the HR speech codec has a worse error-free MOS, but the HW utilization improves so the number of TRXs required for the same traffic is lower in GSM.

12.6 Codec Mode Parameters

The codec mode parameters consist of thresholds to control the switching between half and full rate, and adaptation parameters within each to control the rate. These are referred to as the compression handover and AMR adaptation parameters, respectively.

12.6.1 Compression Handover Threshold

The compression threshold is used to control the switching from full rate codec mode to half rate. This parameter is set by the operator to control the triggering to half rate assignment based on considerations of radio conditions and traffic load. Full rate channels are allocated if sufficient radio resources are available thus offering highest quality whenever possible. Once the cell traffic exceeds a specific threshold, additional cell capacity is provided by switching from full rate to half rate. This compression handover is triggered for particular calls having good radio conditions in terms of C/I. If the radio conditions of a particular call fall below a specific threshold, a switch back from half rate to full rate is triggered. In addition, cell load dependent decompression handover is executed in the case of relaxing traffic load below a certain threshold.

12.6.2 AMR Adaptation Parameters

After a handover to half rate mode, the AMR call starts in the initial codec mode, for example, the robust AMR HR 5.15. Then the fast codec mode adaptation (CMA) switches the rate to that of the best-suited codec mode according to the prevailing radio link quality. In full rate mode the call starts in the initial codec mode, for example, AMR FR 5.9, and the codec mode adaptation algorithm chooses the proper codec mode. The AMR codec mode adaptation process is very fast and can be executed up to ten times per second. The codec

---

[1]The support of the two lowest modes in full rate is required to allow Tandem Free Operation between a half rate MS and a full rate MS. They should not be the primary choice for operation in full rate mode only.

mode adaptations are controlled by thresholds that determine which codec mode should be used for a certain C/I (channel quality). To avoid continuous rapid changes between two codec modes, a hysteresis, usually in the range 2–4 dB, is used for the decision thresholds. These parameters is usually hardwired by the equipment vendors based on the results of their field trial tests and measurements.

12.7 The AMR-Wideband (WB)

The AMR-WB speech coder was developed by 3GPP and consists of the multi-rate speech coder, a source controlled rate scheme including a voice activity detector and a comfort noise generation system, and an error concealment mechanism to combat the effects of transmission errors and lost packets [9]. The AMR-WB uses a sampling frequency of 16,000 Hz and the processing is performed on 20 ms frames, i.e., 320 speech samples per frame. The speech coder is capable of switching its bit-rate every 20 ms speech frame upon command [10]. The wideband AMR codec has also been selected by the ITU-T in the standardization activities for a wideband codec around 16 kbps, marking the first time that the same codec is adopted for wireless and wireline services. This will eliminate the need for transcoding and facilitate the implementation of wideband voice services across a wide range of communication systems. The wide band codec has also been standardized as the default codec for '16 KHz speech' media type for packet switched streaming services (PSS) [11] and multimedia messaging services [12].

The multi-rate speech coder is a single integrated speech codec with nine source rates from 6.60–23.85 kbps and a low rate background noise encoding mode (sending silence descriptor frames at regular intervals amounting to 1.75 kbps). The AMR-WB modes are closely related to each other and employ the same coding framework.

The AMR-WB codec provides superior speech quality and natural voice compared to existing 2G and 3G systems. In typical operating conditions (C/I >10 dB), the AMR-WB offers superior quality over all other GSM codecs. Under poor radio conditions (C/I < 7 dB), AMR-WB still offers quality comparable to AMR-NB and far exceeds the quality of the fixed rate GSM codecs [1].

12.8 AMR Bearer QoS Requirements

3GPP TS 3GPP TS 23.107 has specified the following requirements for the transmitting of AMR voice over the UMTS network:

- **Delay:** end-to-end delay not to exceed 100 ms (codec frame length is 20 ms)
- **BER:** 10^{-4} for Class A bits; 10^{-3} for Class B bits

For some applications, a higher BER class ($\sim 10^{-2}$) might be feasible:

$$FER < 0.5\% \text{(with graceful degradation for higher erasure)}$$

Generally, the AMR speech codec can tolerate a frame error rate (FER) of 1% for class A bits without any degradation of the speech quality. For class B and C bits, a higher FER is allowed.

References

[1] Cox, R. and Perkins, R., 'Results of a Subjective Listening Test for G.711 with Frame Erasure Concealment', Committee Contribution TIA1.7/99-016, May 1999.
[2] Werner, M., Junge, T. and Vary, P., 'Quality Control for AMR Speech Channels in GSM Networks', IEEE, 2004.
[3] 3GPP TS 26.093, 'AMR Speech Codec; Source Controlled Rate Operation'.
[4] Holma, H., Melero, J., Mainio, J., Jalonan, T. and Makinen, J., 'Performance of AMR Speech in GSM and WCDMA', *IEEE Vehicular Technology Conference*, VTC, 2003.
[5] 3GPP TS 25.133, 'Requirements for Support of Radio Resource Management'.
[6] Holma, H. and Toskala, A., *WCDMA for UMTS*, John Wiley & Sons, Ltd, Chichester, 2000.
[7] 3GPP TS 26.975-600, 'Performance Characterization of the Adaptive Multi-Rate (AMR) Speech Codec'.
[8] Halonen, T., Romero, J. and Melero, J., *GSM, GPRS and Edge Performance*, 2nd Edition, John Wiley & Sons, Ltd, Chichester, 2003.
[9] 3GPP TS 26.193, 'AMR Wideband Speech Codec Source Controlled Rate Operation'.
[10] 3GPP TS 26.171-500, 'Adaptive Multi-Rate – Wideband (AMR-WB) Speech Codec; General Description'.
[11] 3GPP TS 26.234, 'Packet Switched Streaming Services (PSS); Protocols and Codecs'.
[12] 3GPP TS 26.140, 'Multimedia Messaging Service (MMS); Media Formats and Codecs'.

13

The Terrestrial Radio Access Network Design

The terrestrial radio access network design is concerned with RNC planning and dimensioning, the interconnection topology, and the sizing of the transmission links connecting the Node Bs to the RNCs. The RNC (radio network controller) planning owns and controls the radio resources in its domain. The RNC is the service access point for all the services that the radio access network (RAN) provides to the core network (CN) (e.g., management of connections to the UE) and provides a traffic concentration point between Node Bs and the core nodes. Node B planning (radio sites) was addressed in Chapter 5 and was determined based on considerations of service mix coverage, QoS requirements, and radio link budgeting. The RNC planning is concerned with the determination of the number and location of RNCs and the homing of Node B clusters to each RNC.

13.1 RNC Planning and Dimensioning

RNC planning and dimensioning involves the determination of the number of RNCs to support the estimated traffic mix, the location of each, and the homing of Node Bs to each RNC. Since the network coverage normally expands an area that is too large to be handled by one RNC, it is divided into regions each of which is handled by one RNC. The number of RNCs required is determined by the estimated network traffic, the number and distribution of Node Bs, and the capacity limitations of the RNCs. The distribution of Node Bs impacts the interconnect topology and clustering options [1] that can be performed to minimize the number of RNCs, and transmission costs. The capacity of RNCs is constrained by a number of factors, the most demanding of which is selected for dimensioning. The RNC capacity limiting factors include the maximum number of cells and Node Bs, the maximum Iub throughput, and the number and type of interfaces (such as E1, ATMs, STm, etc.), and these factors are vendor dependent. In calculating the number of RNCs required to handle the Iub throughput, the estimated traffic should be scaled up by a soft handover overhead percentage of about 30–40%.

The next step in the design process is where to place the RNCs and which subset of the Node Bs to home into each of the RNCs. The location of the RNCs is determined by a number of criteria to include:

- Traffic distribution as determined by the Node B locations
- RNCs traffic handling capacity
- Least cost transmission
- Reducing inter-RNC handovers
- Co-location with Node Bs or existing MSC sites (based on reducing site acquisition costs)
- Other site constraints as determined by the client

Since from the transmission point of view, the RNCs provide the function of traffic concentration and hubing for the radio sites (Node Bs), it is desirable to place RNCs in areas of heavy traffic distribution. This will provide optimum hub location for the busy Node Bs and help to reduce the transmission costs. The optimum hub location criteria attempt to form clusters of Node Bs that are geographically near one another and home them to an RNC, which is placed at the cluster 'center of mass'. The cluster center of mass may be computed using the distance to each Node B in the cluster (from a reference point) weighted by the aggregate peak hour traffic to and from each Node B.

The center of mass criteria will help to minimize inter-RNC handovers, due to the fact that it is computed over clusters of Node Bs that are geographically closely located.

There is also the option of placing the RNCs at the core node locations such as at the MSC sites. The benefits offered in MSC co-location are reduced site costs, elimination of RNC-core node interconnect transmission facilities, and lower costs on site maintenance operations. However these benefits should be carefully traded off against the benefits achieved by placing the RNCs close to Node B concentrations to achieve higher multiplexing gains on the Iub links, and with consideration of available transmission facilities and options in the area.

The number of Node Bs that can be clustered and homed into an RNC will depend on the traffic volume to/from the Node B and the throughput that can be supported by the RNC, as mentioned earlier. The required sizing for a given RNC is dictated by the aggregate peak hour packet and circuit switched traffic to/from the cluster of Node Bs homing into the RNC. The required RNC's capacity should be expressed in terms of packet and circuit switched throughput components.

The steps proposed for the RNC planning process are given in Figure 13.1.

13.2 Node Interconnect Transmission
13.2.1 Node B to RNC

The trunking between the cell sites (i.e., the Node Bs) and the RNCs can be obtained through one of several different approaches: ATM virtual paths provided by an ATM network, microwave line of sight links, and leased lines. These approaches are discussed in the following sections.

The Terrestrial Radio Access Network Design

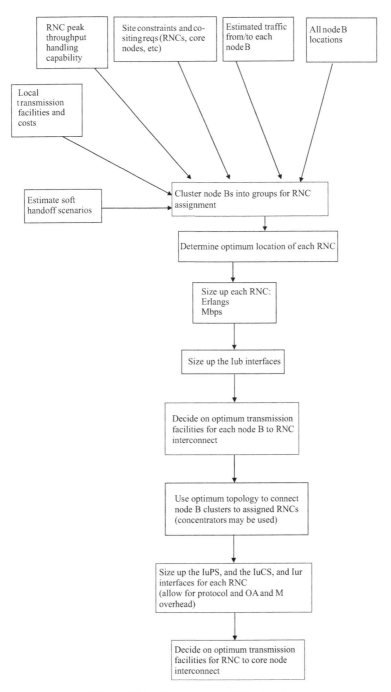

Figure 13.1 Flowchart for RNC planning.

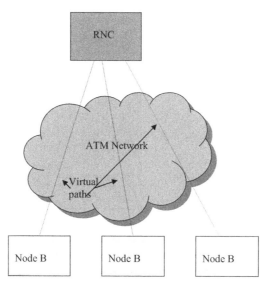

Figure 13.2 Using an ATM network to interconnect Node Bs and RNCs.

13.2.1.1 Using ATM Virtual Paths

The cell sites (Node Bs) connect to the RNCs through the ATM based Iub interface (Node B is the termination point for ATM). Since this interface is ATM based, ATM virtual paths [2] can be used to provide the interconnection between the cell sites and the RNCs as illustrated in Figure 13.2. The ATM virtual paths [3, 4] can be provided through either a public ATM network (a network perhaps owned by the operator) or a private ATM network set up for this purpose by the operator. Whatever option is used, the ATM network provides enhanced reliability and availability through the inherent redundancies that are built, particularly in public ATM networks, for routing around failures in the network. Whereas, in the case of leased lines, it will not be economical to have the same amount of redundancy because it incurs a very high operational cost.

Use of an ATM network can also result in reduced transmission costs due to gains that are achieved through statistical multiplexing and bandwidth on demand usage.

13.2.1.2 Using Microwave Links

It is possible to use microwave transmission facilities, depending on the demography, for carrying the ATM formatted cells between the cell sites and the RNCs. Microwave transmission design usually employs hot-standby or $(1 + 1)$ redundancy configuration. Hot standby redundancy provides the most reliable interconnection links.

13.2.1.3 Using Leased Lines

The leased line approach makes use of private lines such as T1, E1, E3, T3, etc. as leased from a common carrier. Since, the cell site to RNC interface is ATM based, the un-channelized

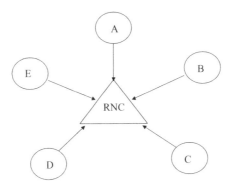

Figure 13.3 Star topology.

versions of T1, E1, T3, E3, etc. are used, which also help to multiplex a large number of ATM virtual circuits on the same link and reduce the required transmission bandwidth through statistical gains. In the case of using leased lines, various topology options [2, 5] can be used to architect the interconnection. These topologies for connecting clusters of cell sites to the RNC will be examined for the evaluation of economic and reliability in the following sections.

The Star Topology
In star topology, illustrated in Figure 13.3, the cell sites are directly connected to the RNC by trunk circuits. To avoid complete isolation of a site by a single failure of a trunk circuit, two links can be provided for each site to the RNC. The second link can be a standby whenever the first one fails. Star topology is generally uneconomical for small and remote cell sites that use only a small fraction of the full capacity. However, in practice it is the most simple and common topology used, and it is easy to dimension.

Daisy Chain Topology
In daisy chain topology, illustrated in Figure 13.4, several sites can be interconnected in a chain configuration via trunk circuits. The number of sites interconnected in the chain should be limited to prevent isolating a large number of sites. The chain topology requires about the same total number of circuits as the star topology (more if there are heavily loaded cell sites at the distant end of the chain); however, the effects of circuits or Node B failures are far more severe.

Ring Topology
In ring topology, cell sites A, B, C, and D are connected in a ring [6], where drop and insert techniques are applied at sites B, C, D, and E (see Figure 13.5). The failure of one link, for example AE, will not affect cell A since a standby link is available from cell sites A to the RNC. Although ring connection is highly reliable it is expensive because each of the two links branching from the RNC will have to be sized to carry all the traffic requirements of all the cell sites on the ring. In this way, the worst case failure is avoided.

Ring topology requires more circuits than the chain, but has the ability to withstand the effects of a circuit or base station failure due to the presence of two separate routes between each cell site and RNC in both directions.

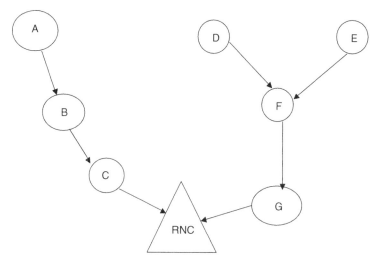

Figure 13.4 Daisy chain topology.

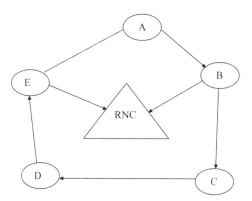

Figure 13.5 Ring topology.

Hub Topology
In hub topology, illustrated in Figure 13.6, several cell sites can be interconnected via trunk circuits to a hub (limited number of sites should be interconnected to a hub) that serves as a concentration point. This will improve the traffic efficiency from hub to RNC. When available, and there is substantial amount of traffic, a hub could be interconnected to RNC via high speed trunk circuits. To increase the reliability of hubs to RNC, $r + 1$ redundancy should be provided such that the failure of one link will not affect the service quality.

Evaluation of Leased Line Alternative Interconnect Topologies
Table 13.1 evaluates the alternative topology options that can be used when leased lines are used for the interconnection of Node Bs to RNCs.

The reliability and cost evaluation of 3G wireless access network topologies are further discussed in Reference 2.

The Terrestrial Radio Access Network Design 221

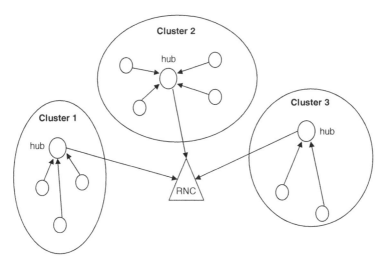

Figure 13.6 Hub topology.

13.2.1.4 Sharing GSM Transmission Facilities

In a scenario where the UMTS operator already owns a GSM network, the UMTS access links can be integrated with the existing GSM transmission infrastructure to save on costs. This means that the UMTS radio sites and RNCs would be co-located as much as possible with the GSM sites and base stations. Then there are two options for integrating the access traffic from UMTS with GSM as follows:

- *Option 1*: In this option when the UMTS traffic is still in the growing stages, the UMTS ATM traffic is mapped over fractional E1s (64 kbps time slots) provided by the GSM access transmission facilities. This is called ATM over fractional E1 approach. This requires placing digital cross connects (DXUs) at the co-located UMTS-GSM radio sites, and also at the co-located UMTS RAN and GSM base station controllers to multiplex and de-multiplex the traffic from the two networks to and from corresponding elements.
- *Option 2*: This option is more applicable when large traffic is projected for UMTS traffic. In this approach all of the access transmission network is converted to a common ATM based transmission with the GSM traffic carried over ATM circuit emulations. The GSM BTS are connected to the co-located UMTS Node Bs. Then in Node B, the GSM fractional E1s are mapped to ATM emulated circuits and transmitted towards the RNC together with the UMTS ATM traffic over common ATM transmission facilities. In the RNCs, the ATM emulated GSM traffic is filtered (using an ATM switch) and converted to E1s and delivered to the GSM base station controller.

13.2.2 RNC to Core Network Nodes

The RNCs connect to the core packet switching network (the SGSN nodes) via the ATM/ AAL5 based IuPS interface and to the circuit switching core network (MSC nodes) via the

Table 13.1 Evaluation of leased line alternative topologies.

Criteria	Star	Daisy chain	Ring	Hub
Reliability	High: Only one cell site is isolated.	High: Failure of any one link would isolate, at most, one cell site.	Very high: Failure of one link does not isolate any one cell.	High: Failure of one link does not isolate any one cell. More than one cell may be affected when two links fail, but re-routing is possible in may cases to avoid complete isolation of more than two cells.
	Very high with use of redundant link as a back-up.	Very high with use of redundant link as a back-up.	Re-routing of trafficis required through an alternate link.	
Expansion	Major: Reconfiguration is needed when network expands to include more core nodes.	Little: Links are added as cells are installed. Sections are spilt to accommodate more core nodes.	Some: Reconfiguration of trunk routing is required when new cells and core nodes are added.	Little: Easy expansion as new cells are homed to the hub.
Maintenance	Easy	Easy	Difficult when there are many cells and multiple loops.	Easy
Application	Best when cells are close to the core node and leased lines are used.	Best when there are not many cells to interconnect. Microwave system would be based on this topology.	Best when cells are close to the core and highest reliability is required. This topology would require a high capacity transmission system such as microwave or fiber optic link.	Best for leased line application where the cells to be interconnected are clustered and far away.

ATM/AAL2 based IuCS interface [3, 4]. Since, the information is transmitted on both of these interfaces in the form of ATM cells, the interconnection between the RNC and the core network can generally be made through ATM virtual paths. The ATM virtual paths can be provided through either a public ATM network (a network perhaps owned by the operator) or a private ATM network set up for this purpose by the operator. Whichever option is used, the

ATM network will provide route redundancy and reliability by offering to route around failures in the network and provide bandwidth on demand.

It is also possible to use point-to-point leased lines carrying the ATM formatted cells between the RNC and the core network nodes. In that case, the same topology options as discussed in the previous sections can be used for the interconnection.

13.3 Link Dimensioning

Link dimensioning is concerned with sizing of the transmission links that are used to interconnect the cell sites (Node Bs) to the RNC, and the RNC to the core nodes (MSC, SGSN, or IGSN). These transmission links may consist of leased lines or ATM virtual paths provided by an ATM network (from the dimensioning point of view, microwave links are also classified as leased lines).

Since the air interface is always the limiting factor in traffic handling capacity for a cell site, it is important to dimension the terrestrial access links sufficiently in order to handle easily the maximum traffic that can be handled by a cell site in both the forward and the reverse direction. In other words, the terrestrial access links should not add any more limitations to the system traffic handling capacity (not more than 0.1% blocking, say) than what already is placed by the more limited and expensive resources of the air interface.

13.3.1 Protocol Overhead

In dimensioning the links connecting various network element components, the additional traffic caused by the protocol header overhead should be added to the user traffic on the link. Each interface uses certain protocols for transmitting the user traffic. Each protocol adds a certain amount of header overhead to the traffic. The percentage protocol overhead is estimated by the ratio of the packet size used on the interface, which will vary according to user and application, and the total header overhead from the various protocols used by the interface.

Table 13.2 lists the protocols used on the various interfaces involved in transmitting user traffic, along with the header size of each protocol. These data can be used to calculate the overhead added to each packet of the user traffic.

Table 13.2 Protocol overhead on UMTS internal interfaces.

Protocol	Header overhead, bytes	SGSN–GGSN	RNC–SGSN	Node B–RNC, RNC–RNC
UDP	8	x (or TCP)	x (or TCP)	
TCP	20	x (or UDP)	x (or UDP)	
GTP	12	x	x	
IP	20	x	x	
ATM	5	option	x	x
AAL2	3	option (voice)		x
AAL5	8	option (data)	x	
PPP or LLC	8	if private lines used		
RLC	8% on user encoded data			x
RTP	12	if packet voice used		

Figure 13.7 Protocol layers involved in transmitting IP/UDP data over Iu interface.

AAL2 (ATM adaptation layer 2) [1, 2] is used for transmitting packetized voice and AAL5 is used for transmitting IP data. To adapt IP to ATM, LLC encapsulation is used. This calls for an overhead amount of 8 bytes per datagram. We can denote IP + Protocol Encapsulation as Classical IP (CLIP). Alternatively, the PPP protocol can be used, which would involve the same amount of overhead.

Packetized voice (VoIP) will also require another level of protocol, which is referred to as the RTP protocol as listed in Table 13.2.

On the RNC–SGSN interface, the transport protocol used is TCP or UDP for IP data (TCP is used for the reliable transport of data such as email, web applications, and FTP). Note that the GTP protocol used over the GGSN–SGSN link is extended from the SGSN to RNC in UMTS, whereas in GPRS it is terminated at the SGSN.

The case for transmitting UDP/IP data with user packet data sizes of 512 bytes over the Iu interface is illustrated in Figure 13.7. The proper overhead can be calculated through the sheet given in Table 13.3

The transmission overhead from RNC to MSC is similar to the overhead requirement for CS traffic in Iub. Therefore the calculation of Iu traffic from RNC to MSC is similar to Iub traffic calculation. The only difference is that all subscribers under one RNC are pooled together and the calculations are based on that. No SHO should be included because all connections are terminated at the RNC.

13.3.2 Dimensioning of Node B–RNC Link (Iub)

On the Iub link, the traffic that is transported consists of compressed circuit mode voice (usually at 12.2 kbps with AMR) and various data applications with varied QoS requirements. We will assume that voice is transmitted in circuit mode in these derivations. In the next generation all-IP access, VoIP is supported, in which case the formulation derived here for sizing the data link capacity is easily applicable by treating voice as data of the highest priority.

Table 13.3 Example protocol overhead calculation for UDP/IP data.

User Data	$= 512$ Bytes (TCP session)
AAL5 PDU	$= (512 + 8\text{Encap} + 20\text{IP} + 8\text{UDP} + 12\text{GTP})$
	$= 560$ bytes
AAL5 PDU + trailer	$= (560 + 8) = 568$
AAL5 PDU + trailer + padding	$= \text{roundup}(568/48) \times 48$
	12 cells needed, 20 bytes of padding
Including ATM headers	$= 568/48 \times 53 = 636$
Total OH required	$= (636 - 512)/512$
	$= 25\%$ OH

The voice traffic will be symmetrical requiring the same amount of bandwidth in both directions on the link, whereas the data applications such as web queries will normally be highly asymmetrical requiring a much larger bandwidth in the RNC to cell site direction. Moreover, the voice traffic is highly delay sensitive and also has rather known statistics. Therefore, to make the approach more general, we will treat the voice and data traffic as two different streams and show how to estimate the link bandwidth required for each separately. With using an ATM network to implement the link, it is quite easy to set up two separate ATM virtual paths on the same physical port and use one for voice and the other for transmitting the data packets. Once the network is migrated to all-IP allowing VoIP, we can use only one link to transmit both voice and data, and the methodology is easily extended.

To start, we define the following notation:

- $E =$ the busy hour expected cell voice call handling capacity in erlangs (including soft handover connections).
- $\lambda =$ the average rate of each voice call bit rate (kbps). This is 16 kbps on the Iub and in most designs on the Iu interface as well.
- $\lambda_i^F =$ the busy hour expected traffic for the cell site in the direction of RNC to Node B for data application type i, $I = 1, 2, \ldots, P$, in average kbps.
- $\lambda_i^R =$ the busy hour expected traffic for the cell site in the direction of Node B to RNC for data application type i, $i = 1, 2, \ldots, P$, in average kbps.
- $P =$ the number of different data applications.
- $OHdata_{prot} = \%$Data overhead protocol (this includes IP, TCP, or UDP, PPP, ATM, and AAL5 for IP based data). This would depend on the packet size used. But sample calculations are given in Section 13.3.1, which result in 25% for packet sizes of 512 bytes.
- $OHvoice_{prot} = \%$Voice protocol overhead (includes RLC, ATM, and AAL2). This is estimated at 20%.
- $SHO_{voice} = \%$soft handover overhead for voice connections.
- $SHO_{data} = \%$soft handover overhead for data connections.
- $OH_{sig} =$ signaling overhead, about 10%.
- $\rho_v =$ voice activity factor (usually set to 50 or 60%).
- $\rho_d =$ utilization factor specified for the data link in each direction (link load factor).
- $R_v =$ estimated required voice link bandwidth in bps.
- $R_d^F =$ estimated required data link bandwidth in the RNC to cell site direction (kbps)
- $R_d^R =$ estimated required data link bandwidth in the cell site to RNC direction (kbps)

13.3.2.1 Sizing the Voice Links

Assuming that quantity E is large enough (a few 10s) to allow for the statistical multiplexing to take effect (that is, to make the random synchronization of all voice call activity periods highly unlikely), we will have simply

$$R_v = \rho_v \times E \times \lambda (1 + OH_{sig} + SHO_{voice})(1 + OHvoice_{prot}) \qquad (1)$$

The protocol overhead $OHvoice_{prot}$ is due to ATM, AAL2, and can be estimated from Table 13.3 to be about 20%. The overheads due to handovers is estimated at about 30–40%. A signaling overhead of 10% is assumed ($OH_{sig} = 0.10$).

The value calculated by Equation (1) gives the required sizing for a leased line kbps, which can be converted into the number of E1s required or expressed simply as the bandwidth required if a constant bit rate (CBR) type ATM virtual path is used for the transmission. Alternatively, an ATM VBR-rt (variable bit rate, real time) type of virtual path may be used with a peak bit rate given by Equation (1), a sustained bit rate given by $E \times \lambda(1 + OH_{sig} + SHO_{voice})(1 + OHvoice_{prot})$, and a maximum burst size determined by the worse case active voice period, as a general rule. Depending on costs and availability, either of these three transmission choices is expected to provide the needed bandwidth to handle the voice traffic.

13.3.2.2 Sizing Data Links

The link dimensioning for data in each direction is obtained by estimating the busy hour aggregated throughput for all the applications and scaling up to provide the specified percent link utilization.

The downlink (RNC to cell site) dimension is then given by

$$R_d^F = (1/\rho_d) \sum_{i=1}^{i=P} \lambda_i^F (1 + OH_{sig} + SHO_{overhad})(1 + OHdata_{prot}) (kbps) \qquad (2)$$

and the uplink (cell site to RNC) dimension is given by

$$R_d^R = (1/\rho_d) \sum_{i=1}^{i=P} \lambda_i^R (1 + OH_{sig} + SHO_{overhad})(1 + OHdata_{prot}) (kbps) \qquad (3)$$

The protocol overhead, $OHdata_{prot}$, for IP data is due to ATM, AAL5, PPP, IP, and TCP or UDP and can be calculated from Table 13.2 depending on what IP packet size is used. This overhead will be about 25% for a packet size of 512 bytes. The handover overhead is typically assumed to be about 20%.

The overhead adjusted quantities calculated by Equations (2) and (3) are then used to specify the sizing for a leased line or a constant bit rate ATM virtual path in each case. Alternatively, an ATM virtual path of the ABR class may be used, in which case the minimum and the maximum bit rates for it would be given as a general rule by $\rho_d \cdot R_d^R$, and R_d^R, for the uplink, respectively and by $\rho_d \cdot R_d^F$, and R_d^F, respectively, for the downlink, or the larger of the two in the case where no asymmetrical circuits are available.

13.3.3 RNC–MSC Link Dimensioning

The RNC–MSC link is used for the transmission circuit mode voice between the RNC and the 3G MSC. The dimensioning of the symmetrical RNC–MSC link is done as in the case of the Node B–RNC, except that here the aggregate traffic from/to all the Node Bs (cell sites) connected to the RNC should be considered.

With K_v denoting the dimension of the RNC–MSC link in each direction, we then have

$$K_v = \rho_v \sum_{cells} \lambda . E(1 + OH_{sig} + OH_{prot}) (bps) \qquad (4)$$

in which the quantities ρ_v, E and λ are defined in Section 13.3.2 and the summation is performed over all the cell sites served by the RNC. The quantity calculated by Equation (4) is then used to specify the sizing for a leased line or a constant bit rate ATM virtual path. Alternatively, an ATM virtual path of the VBR-rt class may be used, in which case the sustained cell rate and the peak cell rate for it would be given as a general rule by ρ_v, K_v, and K_v respectively. The maximum burst size for the ATM VBR-rt cicuit is given by the typical voice activity period.

13.3.4 RNC to SGSN Link Dimensioning

The RNC to SGSN link is used for the transfer of packet switched data service, mostly IP based applications from the RNC to SGSN. Since the major capacity limitation is over the radio interface, the RNC to SGSN terrestrial link should be sized sufficiently to handle the maximum possible traffic originating from the cells and avoid any further blocking or queuing in the RNC. This link can be dimensioned as in the case of cell site to RNC link for data, except that here the traffic aggregated over all the cell sites served by the RNC should be considered. Therefore, using same notation as before, and using K_d^R to denote the RNC to SGSN link dimension, we have

$$K_d^R = (1/\rho_d) \sum_{cells} \sum_{i=1}^{P} \lambda_i^R (1 + OH_{sig})(1 + OH_{prot}) \qquad (5)$$

in which the outer summation is performed over all the cell sites served by the RNC. The quantity calculated by Equation (5) is then used to specify the sizing for a leased line or a constant bit rate ATM virtual path. Alternatively, an ATM virtual path of the ABR class may be used in which case the minimum and the maximum cell rates for it would be given as a general rule by ρ_d, K_d^R, and K_d^R, respectively.

13.3.5 SGSN to RNC Link Dimensioning

The SGSN to RNC link is expected to handle much higher traffic than in the reverse direction due to web application downloads. The SGSN to RNC link must be sized and engineered to handle heavy traffic for a variety of high bit rate applications as well as low bit rate applications with varied QoS requirements to all the cell sites served by the RNC. Because of the traffic handling limitations of the air interface, queuing and hence the need for buffering is expected within the SGSN over the interface with RNC. In order to economically size up the link, and in the mean time allow the QoS requirements in terms of transport delay to be met for the various applications using the link, we will provide the formulation for the cases when no priority is implemented among service classes, and for the case when a priority scheme is implemented. These are discussed in the following two sections.

13.3.5.1 No Service Priorities Implemented

In this case, the packet flows from different services are treated with the same priority. We will assume that the maximum buffering delay allowed at the SGSN-RNC interface is D seconds for all the services. Without loss of the generality of the basic approach we will assume that the aggregate packet flow arriving from all the services for transmission over the SGSN-RNC link is Poisson distributed. Once dynamic equilibrium is established over the maximum tolerable delay window D, we can expect that the number of packets transmitted over the link within the time window D, $N_D(N, D)$ is equal to the number of packet arrivals in time interval D. That is

$$N_D(N, D) = n_a(\lambda, D) \qquad (6)$$

where $n_a(\lambda,D)$ is the number of packets that arrive in a time window D given a rate of arrival λ. The number of arrivals $n_a(\lambda,D)$ is a random number with an average λD. However the actual number of arrivals in a given time window, D, can be larger than the average. If the system is dimensioned such that the link resource is just enough to transmit the λD packets in time D, then the system will fail to meet the delay criteria whenever the number of arrivals is larger than the average. To guarantee that delay criteria will be met 99.9% of the time the link should be sized so that

$$P[n_a > n_a(\lambda, D)] \leq 0.001 \qquad (7)$$

By imposing this condition it can be assured that the link is sized to serve the arriving packets say 95% of the time. Dimensioning the system in this fashion guarantees that the delay criterion for each service class will be met 95% of the time or whatever other value is chosen as desired.

To find out the link size that will meet the criteria stated by Equation (7), the mean packet arrival P in the time interval D is calculated by dividing the product of the aggregate data rate, R, and the time delay D by the average packet size, μ, first. So, the average number of packet arrivals in the time window D is

$$P = R.D/\mu \qquad (8)$$

where μ is the packet size. Then, with the Poisson assumption for the aggregate packet arrival and a mean rate given by P, we obtain the cumulative probability density function plot (CDF), from which we read the peak packet arrivals rate, h, which will not exceed a probability of 95%. Then the link is sized by

$$\text{Link bandwidth required} = h.\mu/D \qquad (9)$$

The CDF for the case of $D = 500$ ms, $\mu = 512$ bytes (512×8 bits), and $R = 100$ kbps is given in Figure 13.8. From this figure we read that the peak packet arrival rate, h, not to exceed 95% of the time is 18 packets per 500 ms. Therefore, the link bandwidth required is (using Equation (9))

$$\begin{aligned}\text{Link bandwidth required} &= h \times 512 \times 8/0.500 \\ &= 18 \times 512 \times 8/0.500 \\ &= 147.456 \text{ kbps}\end{aligned}$$

The Terrestrial Radio Access Network Design

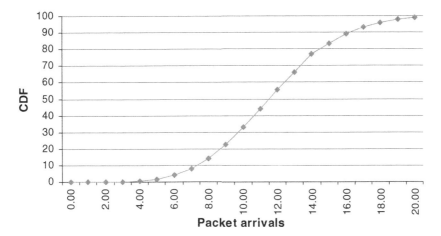

Figure 13.8 CDF for the Poisson packet arrivals in 500 ms time window.

As the observation time D (maximum allowable delay) becomes smaller, the bandwidth required for the design exceeds the average requirements; that is, the ratio (bandwidth required)/R becomes larger and larger (R is the busy hour average aggregate data rate of the services). Thus for data services with no delay criteria (D large) the link can be sized based solely on the average requirement.

13.3.5.2 Service Priorities Implemented

We will assume some form of priority queuing implemented within the SGSN on this link. This is also commensurate with the fact that the SGSN is an edge node within the core network and vendors may consider implementing some priority queuing mechanism on its downlink access links.

We will assume that there are P levels of priority, with one associated with each of the P application services types (assumed previously). We will also assume that the arrival processes to each of the queue will be Poisson distributed. The Poisson assumption has been used frequently in analyzing queuing in packet data networks, although it is now being challenged by the recent self-similar theories in traffic engineering. However, to render the analysis to closed form solution and illustrate the basic approach, we will use the Poisson assumption here. We also need to assume some form of queue scheduling algorithm for serving the P queues. Various scheduling algorithms have been proposed and studied, such as HOL (head of the line) and round robin servicing. Since, the HOL algorithm lends itself to a more tractable analysis, and is the simplest to implement in practice, it will be used here. In HOL scheduling, a lower priority queue is served only when there are no packets in a higher priority queue.

We use the following notation:

- λ_i = the expected busy hour averaged aggregate data rate (kbps) from the ith application service class on the SGSN to RNC link (this should include the protocol overheads).
- λ^T = the total expected busy hour averaged aggregate data rate (kbps) from all application services on the SGSN to RNC link (this includes the protocol overheads).

- ρ = the overall utilization of the SGSN to RNC link.
- K_d^F = the dimension of the SGSN to RNC link in units of kbps.
- W_p = the dimensionless mean queuing delay (waiting time) for cells from application class p (p = 1, 2, ..., P) in units of the ATM cell transmission time on the SGSN to RNC link.

Using the results given in Reference 7], we have

$$W_p = \frac{\rho/2}{\left[1 - \rho \sum_{i=p}^{P} \frac{\lambda_i}{\lambda^T}\right]\left[1 - \rho \sum_{i=p+1}^{P} \frac{\lambda_i}{\lambda^T}\right]} \tag{10}$$

in which

$$\lambda^T = \sum_{i=1}^{P} \lambda_i \quad \text{(the total busy hour cell data flow in the downlink direction)} \tag{11}$$

and

$$\rho = \lambda^T / K_d^F \quad \text{(the total utilization factor)}$$

In Equation (11), the quantities λ_i, $I = 1, 2, \ldots$, are obtained by summing the busy hour averaged cell flow from application service class i to all the cell sites served by the RNC. The dimensionless cell waiting time (queuing time) for cells from application service type p (of priority p) can be plotted against a feasible range of values (0–1) for the overall link utilization ρ for each priority class p in the system. The plots will then give an idea of the highest link utilization, which can satisfy the *average* delay constraints for the most sensitive (in this terminology, the Pth priority class that is the highest priority level) traffic class. Note that the delay variances can also be calculated but with a much more complicated expression [7]. However, for simplicity the allowed utilization may be estimated based on the average delay constraint, which can be set to slightly higher than the maximum service delay constraint to provide a safety delay margin. Then, the link dimension is simply calculated from Equation (7).

References

[1] Lauther, U., Winter, T. and Zeigelmann, M., *Proximity Graph Based Clustering Algorithms For Optimized planning of UMTS Access Network Topologies*, ICT 2003.
[2] Houeto, F., Pierre, S., Beaubrun, R., Lemieux, Y., 'Reliability and Cost Evaluation of Third-Generation Wireless Access Network Topologies: A Case Study', *IEEE Transactions on Reliability*, **51**(2), June 2002, 229–239.
[3] Wu, T. Y. and Yoshikai, N., *ATM Transport & Network Integrity*, Academic Press, San Diego, 1997.
[4] Ibe, O.C., *Essential of ATM networks and Services*, Addison-Wesley, Reading MA, 1997.
[5] Schwartz, M., *Telecommunication Networks*, Addison Wesley, Reading MA, 1987.
[6] Otsu, T., Okajima, I., Umeda, N. and Yamao, Y., 'Network Architecture for Mobile Communications Systems Beyond IMT2000', *IEEE Wireless Communications*, **8**(5), Oct. 2001, 31–37.
[7] Rahnema, M., 'Smart Trunk Scheduling Strategies for Future Integrated Services Packet Switched Networks', *IEEE Communications*, February 1988, **26**(2), 43–50.

14

The Core Network Technologies, Design, and Dimensioning

The core network interfaces the radio access network to external networks such as the Internet and the public fixed telephony network. It performs the switching and routing of the traffic between the radio access and the fixed external networks. Therefore understanding its function, the nodes that make up the core network, databases, and service management platforms, the protocols and the transport technologies that are commonly used, the architectural options such as the distributed implementation of SGSN and HLRs, the transport technology options, and the dimensioning of the core network elements and links is important for the planning and optimization of the UMTS network. We will discuss all these aspects in this chapter and provide guidelines for dimensioning, as well as a comprehensive coverage of the evolution to soft switching and all IP core networks. Soft switching, which is already being implemented by an increasing number of networks, provides the framework for the long-term growth of 3G (and higher) networks.

14.1 The Core Network Function

The core network in UMTS and GPRS (General Packet Data Services) or Edge provides the interoperability between the radio access network (RANs) and the external networks. The external networks include the Internet and the public switched telephony network (PSTN). The interface to the PSTN is provided through the gateway MSCs. The signaling for the circuit switched calls uses the same signaling transport infrastructure as in GSM [1], which is based on the CCITT Signaling System No 7 (SS7), and a mobility specific application part (MAP) [2]. The SS7 signaling transport infrastructure is based on packet switching of call signaling messages between the switches and the databases, and is a highly redundant and reliable network. The SS7 network uses multiple 56 or 64 kbps TDM circuits for connecting the signaling nodes.

The RAN interfaces with the Internet and any other packet switched public networks through the IP based packet core network, which is specific to GPRS [3] and 3G networks. The IP core network also provides and supports the feature servers that operators deploy to

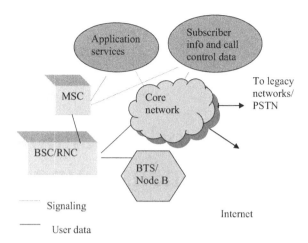

Figure 14.1 Basic core network connectivity to other network elements and mediums.

increase the number and variety of subscriber services for internally managed content. The core network extends the Internet protocol based packet data services to the radio access infrastructure and hence is capable of routing IP packets within the context of mobility. The basic core network infrastructure and elements in UMTS are the same as in GPRS with the exception of required higher capacity and different authentication and ciphering algorithms [4]. The basic core network connectivity to various other network elements is shown in Figure 14.1.

14.2 The IP Core Network Architecture

The packet domain core network functionality is logically implemented on two network nodes: the Serving GPRS Support Node (SGSN) and the Gateway GPRS Support Node (GGSN). The GPRS support nodes originating from GPRS evolve to UMTS network nodes, most often with simple software upgrades particularly if GPRS Release 99 is used. Therefore, the name GPRS Support Node is used even if the node provides UMTS functionality only.

The implementation of the packet core network architectures begins most often with deploying the GPRS for enabling packet data services in GSM and CDMA One networks with incumbent operators [5, 6]. GPRS overlays a wide-area IP based data network on the existing GSM network infrastructure to achieve interconnectivity with packet data networks and the Internet. This data overlay network provides packet data transport at rates from 9.6 to 171 kbps, theoretically with the commercial ones up to 115 kbps. Additionally, multiple users can share the same air-interface resources.

The UMTS 3G networks, which are based on the European wideband CDMA air interface standard, are based on the same core network infrastructures used in GPRS (the 2.5G networks). Therefore, from the core network point of view, there is basically no functional differences between GPRS and the UMTS 3G networks, other than the fact that the 3G core networks will have to be sized to accommodate higher traffic handling capacity, and might use

The Core Network Technologies, Design, and Dimensioning

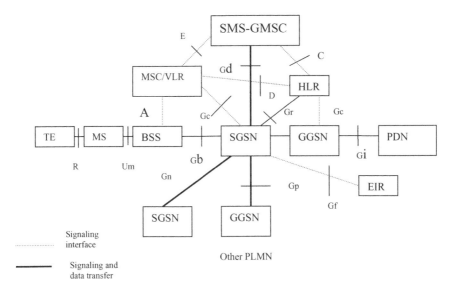

Figure 14.2 The GPRS core network architecture.

different algorithms for authentication and/or ciphering [4]. Release 1998 is the earliest release of the GPRS core network followed by Release 1999. The main difference between GPRS Releases 1998 and 1999 is that Release 1999 is capable of supporting UMTS RANs. Also, the QoS management principles used for Release 1998 are similar to the ones used for UMTS Release 1999, except that with Release 1999 it is possible for a subscriber to use more than one PDP (packet data protocol) context with different QoS parameters and share the same PDP address. This is aimed at the support of multimedia connections (a UMTS service).

GPRS attempts to reuse the existing GSM network elements as much as possible, but in order to effectively build a packet-based mobile cellular network, some new network elements, interfaces, and protocols that handle packet traffic are required. Therefore, GPRS requires modifications to numerous network elements to include the core network elements such as the HLR databases. The GPRS system architecture and the various network elements and interfaces are shown in Figure 14.2. The Gb interface in UMTS core network is renamed to Iu [7]. The core network elements that are specifically developed for the GPRS network architecture consist of the serving GPRS support node (SGSN), gateway GPRS support node (GGSN), and upgraded location registers. These elements, which are also used in the UMTS core network with some variations are discussed in detail in the following sections.

14.2.1 The Serving GPRS Support Node (SGSN)

A GPRS Support Node (GSN) contains functionality required to support GPRS and/or UMTS packet domain functionality. In one public land mobile network (PLMN), there may be more than one GSN. The SGSN plays the important role of interfacing with the RAN and mobility management elements of the GSM network – all of which have been upgraded in some respect for enabling packet data services in a circuit-based RAN network. SGSNs contain visitor location register or VLR-like functionality, which detects new GPRS or UMTS mobile stations

(MSs) in a given service area, processes registration of new mobile subscribers, and keeps a record of their location inside a given area. Therefore, the SGSN performs mobility management functions such as mobile subscriber attach/detach and location management. The SGSN is connected to the base-station subsystem via a Frame Relay connection to the PCU (packet control unit) in the BSC in GPRS, and via ATM connections to the RNC (radio network controller) in UMTS. The PCU is a GPRS add-on to the BSC.

A mobile station (GPRS or UMTS) can initiate a packet data session via the RAN to the SGSN. The SGSN can route the data between it and the MS to another SGSN or to the GGSN for accessing an external data network. Likewise, it is also responsible for delivery of packet data to the mobile's within its service area.

The SGSN represents the mobile's point of attachment to the core network and handles and provides the following specific functions for the data services:

- call control signaling with data services location registers
- mobility management such as tracking of mobile's IP routing area and serving cell
- user authentication and verification
- billing data collection
- handling of the actual user's traffic and conversion between the IP core and radio network
- standard interfaces to the HLR for management of end-user subscriber data

Each SGSN in the network serves mobiles in a limited area, referred to as the SGSN area. An SGSN area is thus the part of the radio access network served by an SGSN, and may consist of one or several routing areas. An SGSN area may also consist of one or several BSC areas. There is not necessarily a one-to-one relationship between an SGSN area and a MSC/VLR area.

14.2.1.1 SGSN Node Architectures

Vendors are implementing the SGSN hardware through two different architectures. These are referred to as the integrated and the split (or distributed) architecture.

In the integrated SGSN architecture, the functions of control signaling and handling of the actual traffic data are provided within the same hardware unit.

In the split SGSN architecture, the functions of control signaling and handling of the actual user's traffic are separated into an SGSN server and the so-called Media Gateway (MGW) unit, respectively. By putting the traffic handling functionality in a separate unit (i.e., the MGW), the number of SGSN servers can remain the same when the traffic per subscriber increases. This results in a more scaleable design.

The split SGSN architecture also makes the transport network completely transparent to the control and services planes. This means that changes in the transport network or user plane will hardly affect the higher layers in the network. For instance, migrating from an ATM based transport to an IP based transport in the core network could be done without any impact to the servers and other higher layer network elements.

14.2.2 Gateway GPRS Support Node (GGSN)

GGSNs are used as interfaces to external IP networks such as the public Internet, other mobile service providers' GPRS services, X.25 PDNs, and enterprise intranets. The GGSN is the node that is accessed by the packet data network due to evaluation of the PDP address. It contains routing information for attached GPRS or UMTS users. The routing information is used to tunnel the protocol data units (PDUs) to the MS's current point of attachment, i.e., the Serving GPRS Support Node. The GGSN may request location information from the HLR via the optional Gc interface.

Other functions include network and subscriber screening and address mapping. One or more GGSNs may be provided to support multiple SGSNs. The GGSN can deliver to the MS IP-based data services that are consistent with the look and feel of the Internet or other IP-based networks that the subscriber may be accustomed to accessing via a wireline environment. As regards the external IP networks, GGSN is viewed as performing common IP router functions. This may include firewall and packet filtering functions.

A GGSN may serve more than one SGSN nodes, just as an SGSN may be served by more than one GGSN for redundancy purposes or for connecting to multiple ISPs and external data networks.

14.2.3 The HLR

The HLR stores all the subscriber related data, such as subscriber service profile, user's location/serving area, etc., which are needed by the SGSN to perform routing and data transfer functionality. The GPRS HLR is a software-upgraded version of the GSM HLR to handle the new call models and functions introduced by GPRS. The UMTS HLR require software upgrade of the GPRS HLR. The communication between the HLR and the SGSN takes place through the standardized interface referred to as the Gr interface. The so-called Ge interface provides the communication interface with the GGSN. The higher layer signaling protocols between the HLR-SGSN and the HLR-GGSN are based on the SS7 MAP/TCAP protocol suite.

14.2.3.1 HLR Implementation Architecture

The HLR can be implemented as one element (central) in the network or via a distributed architecture depending on the size of the network. For an initially small network, one HLR may be all that is required. However, this means that all signaling traffic for subscriber service profile verification and location determination would have to be routed to one central place, resulting in costly transmissions. The distributed HLR architecture allows operators to make considerable operational expenditure savings in transport cost because not all HLR traffic has to travel to a central point but can be kept local. In the distributed implementation, one single HLR can be implemented in a physically distributed architecture. The distributed architecture limits the impact when system outage occurs, because a different node is able to assume the functionality of the effected equipment. This helps operators to build more resilient networks and also helps them develop their 'disaster recovery' strategy.

14.2.4 The Core Network Protocol Architecture in GPRS

The core network in GPRS and UMTS uses the Internet protocol, IP, as the protocol in the network layer [8]. Options for the underlying data link layer to transport the IP protocol data units between the core network elements include ATM, Ethernet, or protocols running on private lines such as PPP and HDLC. The protocols used within the transport layer are UDP for IP services and TCP for services that require delivery guarantee such as X.25 services.

Data transfer between two GPRS support nodes (GSN) such as a SGSN and a GGSN, or between two SGSNs, takes place through a tunneling protocol referred to as the GPRS Tunneling Protocol (GTP). The GTP [9] encapsulates and carries the protocol data units (PDU) that need to be transported between two GSNs. GTP uses a Tunneling ID (TID) within its header to indicate to which tunnel a particular PDU belongs. In this manner, packets from different mobiles are multiplexed and de-multiplexed by GTP between a given GSN pair. The UDP/TCP destination port reserved for GTP is 3386.

The GTP tunnel ID value is established by a procedure that is referred to as the Packet Data protocol (PDP) context establishment, which takes place on the signaling plane. PDP context activation specifies a two way GTP tunnel between the SGSN and the GGSN used to carry encapsulated user data packets between a mobile and the external packet data network. The PDP context activation process also leads to the establishment of an SNDCP (Sub-Network Data Convergence Protocol) tunnel between the mobile station (MS) and the SGSN serving the mobile. The SNDCP is used to transfer the PDP data between the SGSN and the MS.

The PDP context specifies two important parameters, that are defined as follows:

- **The PDP Type** defines the end-user protocol used between the external packet data network and the MS. The PDP type is divided into an organization and a number field. The organization field specifies the organization responsible for the PDP type number field and the PDP address format. For X.25, for instance, the organization is ETSI, and the PDP Type number is 0. The PDP address is the X.121 format.
- **The PDP Address:** is the address that the PDP context of the MS is identified with from the external packet data network. This address is assigned dynamically to the MS at PDP context activation by the GGSN for that service. When the MS is assigned a permanent IP address at subscription time, the address is activated by the PDP context establishment.

The PDP context activation will also specify the quality of service profile information elements that represent the QoS values negotiated between the MS and the SGSN.

The protocol architecture for GPRS is shown in Figure 14.3.

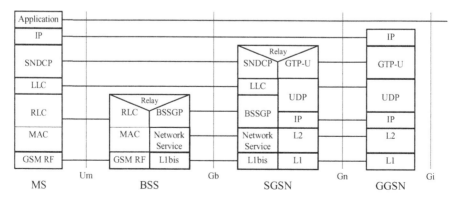

Figure 14.3 GPRS protocol architecture.

14.2.5 SS7 Over IP Transport Option (SS7oIP)

With the increasing customer demand in intelligent network services and supplementary services (calling ID, call return, etc.), and more roaming services in the wireless networks, the SS7 networks continue to grow at an unprecedented rate. In Europe and Asia, more and more customers are using Short Message Service (SMS) for 'instant messaging'. All these applications use SS7 and are contributing to double-digit overall annual growth for SS7 applications.

The large growth in SS7 networks necessitates the provisioning of additional dedicated 56 or 64 kbps circuits. Most network operators add these circuits in redundant pairs to provide the needed reliability. This sustained growth can make forecasting challenging. Determining the link requirements 3–6 months ahead is very difficult, particularly when usage has been increasing at over 400% annually. Over-provisioning may actually take place simply to ensure that network blocking will not occur. But as more SS7 links are terminated on the SS7 network, core devices like STPs or HLRs also increase in size and complexity, with a corresponding increase in the number of ports and the need for higher processing speeds just to handle the load. Each STP or HLR augment requires network reconfiguration, resulting in additional network management costs.

A cost effective solution that provides flexible capacity growths to handle the increasing demand for SS7 signaling load is to integrate the signaling transport with the same IP core infrastructure used in GPRS or UMTS to handle the IP traffic and related signaling. This is done by the use of the Streamed Control Transmission Protocol (SCTP) [10] defined by the IETF. The IETF has defined a set of standards under SigTran (signaling transport) for the transport of SS7 signaling over IP, which are based on the SCTP protocol stack. SCTP is designed to transport SS7 signaling messages over IP networks. It operates directly on top of IP at the same level as TCP. SCTP's basic service is connection oriented reliable transfer of messages between peer SCTP users. With SCTP, the SS7 application layer signaling messages (such as MAP) are transported over IP using the same physical links that are used to transmit the user IP packets and control messages within the IP core network. By using the lower costs shared IP links, statistical multiplexing gains are also achieved saving further on transmission costs. The SS7oIP approach also eliminates the need for dedicated 56 or 64 kbps TDM circuits for SS7 messages, and moreover reduces the number of ports on HLRs and switches, saving on equipment size and costs.

14.3 Mobility Management in GPRS

Mobility management (MM) keeps track of the location of a mobile station (MS). Because a MS is often moving from one place to another, the network must be aware of the MS's position in order to maintain connectivity. Mobility management refers to the range of procedures that make this possible. These include identification and authentication of the mobile subscriber, security, access to wireless services, transfer of subscriber data among network nodes, location updating, and registration. Mobility management involves the exchange of signaling and control messages between the MS and the network node with which the MS communicates. A subscriber can access network services only if the MM functions are successfully completed, the network has granted permission, and (in some cases) the network has authenticated the user.

14.3.1 Location and Routing Area Concepts

Location areas (LAs) and routing areas (Ras) are concepts defined within the core network for the implementation of mobility management functions. The LA and RA together define an area in which the MS can access CS and PS services, respectively, subject to the access granted by the network. One LA is handled by one CN node. Similarly, one RA is handled by one CN node. LA is used by the MSC/VLR to page the MS for circuit switched calls (such as voice calls). RA is used by the SGSN to page the MS for packet data calls.

When a user moves from one location or routing area to another, it has to update the information with the network. The LA/RA update procedure is always initiated by the MS. There are two triggers that initiate the update procedure: periodic and normal. A periodic update occurs when the update timer expires. A normal update is triggered when the RAI changes or when a change in the cell ID has been observed. LA is defined by the network operator and can be a group of cells, all of which fall within one Location Area Identity (LAI, where LAI = MCC + MNC + LAC), where MCC denotes the mobile country code, MNC denotes the mobile network code, and LAC denotes the location areas code assigned by the operator. RA is defined by its routing area indicator RAI (RAI = MCC + MNC + LAC + RAC), where RAC defines the routing area code assigned by the operator, and is a subset of one, and only one, location area. This means that a RA cannot span location areas, because it is served by only one SGSN. Those cells that do not support packet-domain services within a LA are grouped into a null RA in which there is no SGSN available to support packet services.

14.3.2 User States in Mobility Management

GPRS mobility management consists of two levels: macromobility to track the current serving SGSN and micromobility to track the mobile station location down to the routing area or cell level. Macromobility management results in the downloading of the user's service profile from the HLR to the serving SGSN and GTP tunneling and packet data protocol (PDP) context management.

At the micromobility level, the MS is tracked depending on its mobility management state. In the Idle state, the user cannot be reached by the GPRS network. In the Idle state, the MS does not have a logical GPRS context activated or any Packet-Switched Public Data Network (PSPDN) addresses allocated. In this state, the MS can receive only those multicast messages that can be received by any GPRS MS. Because the GPRS network infrastructure does not know the location of the MS, it is not possible to send messages to the MS from external data networks.

The user can move from the Idle state to the Ready state through the GPRS Attach Procedure. In the Ready state, the mobile station's location will be known down to the cell level. The user slips to the Standby state when there is no active transmission. In the Standby state, the MS is tracked down to a routing area, which is a subset of a GSM location area. The main reasons for the standby state are to reduce the load in the GPRS network caused by cell-based routing update messages and to conserve the MS battery. When a MS is in the standby state, there is no need to inform the SGSN of every cell change – only of every routing area change. The operator can define the size of the routing area and, in this way, adjust the number of routing update messages.

A cell based routing update procedure is invoked by the MS when it enters a new cell. When a MS in an active or a standby state moves from one routing area to another in the service area of one SGSN, it must again perform a routing update. The routing area information in the SGSN is updated and the success of the procedure is indicated in the response message.

The inter-SGSN routing update is the most complicated of the three routing updates. In this case, the MS changes from one SGSN area to another, and it must establish a new connection to a new SGSN. This means creating a new logical link context between the MS and the new SGSN, as well as informing the GGSN about the new location of the MS.

14.3.3 MS Modes of Operation

A GPRS MS can operate in one of three modes of operation as discussed below:

- **Class A:** The MS can attach to both GPRS and other GSM services, and simultaneously support both services.
- **Class B:** The MS can attach to both GPRS and other GSM services, but the MS can operate only one set of services at a time, and is capable of monitoring one paging channel at a time.
- **Class C:** The MS can exclusively attach to GPRS services.

The mobility management coordination issue here is how to page for GPRS and GSM services when the MS only listens to one paging channel, as in the class B mode of operation. This is facilitated through the GPRS network mode of operation whereby mobility management is coordinated between GPRS and GSM through the Gs interface (the standardized interface between the MSC and the SGSN) so that class B MSs can still be paged for voice or circuit switched calls when camping on the packet side (i.e., when listening to the GPRS paging channel). In that mode, all MSC originated paging of GPRS attached MSs shall go via the SGSN, thus allowing network coordination of paging by the SGSN based on the IMSI. The SGSN converts the MSC paging message into an SGSN paging message. Moreover, paging via the SGSN will be more efficient because of the fact that the SGSN pages in either a cell or routing area that is always contained in a GSM location area.

Another feature that saves radio resources is the combined circuit switched and packet switched mobility management procedures. For instance, a combined RA/LA update procedure takes place when the MS enters a new LA while the MS is IMSI-and GPRS attached. The MS sends a single RA update request, indicating that a LA update shall also be performed. The SGSN interprets the request, executes the RA update, and forwards the LA update to the MSC/VLR. Complete mobility management and location management procedures are outlined in References 11 and 12.

14.4 IP Address Allocation

To begin a packet data session, the mobile station must request a PDP context establishment from the SGSN, which passes this request along to the GGSN. This process results in the allocation of an IP address by the GGSN, which can be statically or dynamically allocated. In the dynamic allocation approach, IP addresses are allocated by the visited SGSN or the home network GGSN when roaming. Dynamic IP address allocation enables the operator to use and re-use IP addresses, from a pool of IP addresses, which significantly reduces the total number of IP addresses required per PLMN. There is an IP address pool per APN. The IP addresses can also be allocated dynamically, by the ISP/corporate at the time of authentication through for instance a RADIUS server.

In static IP address allocation, IP addresses are defined for the subscriber in the HLR. Static IP addresses allocation allows the operator to offer subscribers the ability to use an IP address provided by the subscriber. This can be used when accessing secure networks that use the calling IP address as a security check.

14.5 Core Network in WCDMA

The core network in Wideband-CDMA [13] Release 99 is based on the GPRS protocols discussed in previous sections and uses the same core network elements. Mobility management is also handled by specific GPRS protocols. However, there are enhancements that have been made in later releases (such as Release 4 and up) to handle efficient handling of multimedia traffic. For instance, once an IP address has been allocated to a MS, it can establish multiple packet data sessions with the 3G-SGSN, each identified with its own PDP context, and have its own QoS depending on the PDP context. In addition, the 3G core network Release 6 has defined IP multimedia subsystem (IMS) and session initiation protocols for the efficient configuration of multimedia and VoIP services, as will be discussed in the next section. Since the data rates involved in WCDMA can be much higher, the core network in this case should be sized to handle a much higher throughput.

With respect to packet services in Iu mode, a MS attached to a packet switched domain may operate in one of the following MS operation modes [8]:

- PS/CS mode of operation or
- PS mode of operation.

The terms 'PS/CS mode of operation' and 'PS mode of operation' are also referred to as 'MS operation mode A' and 'MS operation mode C', respectively.

In network operation mode I and II [8], a MS in PS/CS mode of operation uses the same procedures as for a GPRS MS operating in MS operation mode A, unless it is explicitly stated for A/Gb mode only or Iu mode only.

In network operation mode I and II, a MS in PS mode of operation uses the same procedures as for a GPRS MS operating in MS operation mode C, unless it is explicitly stated for A/Gb mode only or Iu mode only.

The network operation mode III is not applicable for Iu mode [8].

14.6 IP Multimedia Subsystem (IMS)

IMS is a Release 5 and up 3GPP defined mobile core network subsystem [14–16], which enables the convergence of data, speech, and mobile network technology over an IP-based infrastructure. The Open Mobile Alliance (OMA) is then the organization that is focused on defining applications and services to run on top of the IMS infrastructure. IMS specifications provide a clear evolution path from 2G to 3G networks and also from

3GPP Release 99 to Release 6 networks. A basic advantage of IMS is that it offers the benefits of incremental add-on to the existing network rather than a radical replacement [17, 18]. IMS allows the service creator the ability to combine services from CS and PS domains in the same session, and for sessions to be dynamically modified 'on the fly' (e.g., adding a video component to an existing voice session, or adding or dropping a user in a conference calling session, etc.). The video-conferencing service extends the point-to-point video call to a multi-point service and requires an IMS conference bridge service, which links the multiple point-to-point video calls together, and implements the associated service logic. Other services made possible by IMS include instant messaging, messaging between a mobile terminal and the fixed network (a PC, etc.), and presence services [19]. Messages can contain any media type content such as text, image, audio or video clips, application data, or a combination of these. The message is sent through the packet core network to the IMS, which locates the terminating IP client and routes the message to the recipient.

3GPP has done most of the work in defining the core elements of the IMS infrastructure. The Session Initiation Protocol (SIP) has been specified as the protocol for the communication between the core elements and the application servers. The SIP protocol operates on top of the IP layer, which facilitates the inter-networking of the fixed and wireless networks. IMS uses the SIP protocol for multimedia session negotiation and session management. IMS is essentially a mobile SIP network designed to support this functionality, where IMS provides routing, network location, and addressing functionalities. The SIP protocol is well defined and can be easily extended so as to develop new differentiated services.

The core network in an IMS infrastructure consists of a set of SIP proxies called Call Service Control Function (CSCFs) that together with Home Subscriber Server (HSS) are capable of routing any SIP request/message to devices and other networks. The interfaces and protocols used by these core elements are well standardized, which enables for the rapid development and deployment of any new service in such a framework. Another component of the IMS architecture is the Media Control Gateway Function (MCGF), which is responsible for transferring and distributing media between participants in multimedia sessions. Such media include audio, video, and messages. The core elements consisting of the CSCFs and HSS form a network that enables the discovery and communication between end-users and services and between services themselves. In addition, it describes common services, for example, related with charging, that can be used uniformly by all the applications. IMS deployment does not require a 3G RAN. IMS-like services can be and are deployed through GPRS and EDGE.

The standard interfaces and the common service components and functions provided by IMS helps to significantly reduce the time-to-market for rolling out new IMS Services and selectively delegate the development of new applications to third parties. The IMS entities and infrastructure provide a reusable service platform. This means that in most cases new services can be created on a SIP application server and client plug-in, by re-using existing features, such as Presence, conferencing, and instant messaging.

14.7 Roaming in Mobile Networks

Roaming is defined as the use by a customer of one mobile operator of another mobile operator's network to make or receive a call, usually because the customer is out of range of his or her own operator's base stations. There are generally two aspects to the requirements for facilitating roaming between networks: the business aspects for charging and billing arrangements between the operators and the technical capabilities and interfaces required between the networks involved to make roaming possible.

14.7.1 Mobility Handling Mechanisms in Roaming

Mobility management between different operator networks is handled by the Mobile IP protocol [20]. This enables users to connect to different IP subnetworks, while always using their permanent or long-term IP address (registered in its HLR). Mobile IP is the protocol designed to support macromobility, i.e., mobility between different subnetworks. Mobility within the subnetworks, e.g., handovers between base stations, is handled by the underlying network.

When the mobile terminal is located in its home subnetwork, it operates like a regular terminal, i.e., no Mobile IP signaling occurs. When the mobile terminal arrives in a foreign subnetwork, it obtains a temporary address (a care-of address), which it registers with its home agent (HA), a node in its home network. When datagrams arrive in the home network, the HA intercepts them and tunnels them to the current care-of address. In IPv4, the mobile terminal has to register with the Foreign Agent (FA), a router in the foreign network, which provides routing services to the mobile node while registered. The care-of address in the foreign network can either be determined from a FA's advertisements (a foreign agent care-of address) or by some external assignment mechanism such as DHCP (a co-located care-of address). In the former case, the IP tunnel from the HA is terminated in the FA and in the latter case in the terminal.

Extensions to the base Mobile IP protocol have been defined by IETF in Mobile IP6 to allow for optimization of datagram routing from a correspondent node to a mobile node. Without Route Optimization, all datagrams destined to a mobile node are routed through that mobile node's home agent, which then tunnels each datagram to the mobile node's current location. The extensions provide a means for correspondent nodes to cache the binding of a mobile node and then tunnel their own datagrams for the mobile node directly to that location, bypassing the route for each datagram through the mobile node's HA.

In IPv6 (RFC 2460), the FA has been eliminated and the terminal will always use a co-located care-of address. It is mandatory for all nodes to support caching of bindings for mobile nodes, which means that routing is always optimized, unless the terminal chooses to communicate via its HA. Route optimization is an integral part of mobile IP6, whereas it is an add-on in mobile IP4.

Mobile IP4 requires reverse tunneling to avoid ingress filtering problems (where firewalls drop the mobile's outgoing packets) because the packets are sent with the home address as the source. In mobile IP6, packets may be sent with the care-of address as the source address, hence eliminating problems with ingress filtering. Mobile IP6 uses the IPSEC protocol suite for implementing security mechanisms, whereas Mobile IP4 provides its own security measures.

14.8 Soft Switching

Similar to the split architecture option for SGSN, the MSC (mobile switching center), which handles the circuit switched voice calls, can also be split into call control and media switching modules. The approach is then referred to as soft switching.

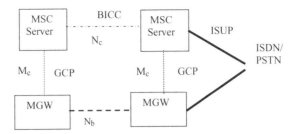

Figure 14.4 Soft switching architectural elements and interfaces.

Soft switching separates the MSC's call/mobility control and the switching functions into separate nodes. These are called the MSC server (MSC-S), which handles call control and mobility handling, and the mobile media gateway (M-MGW), which handles the switching, trans-coding, and inter-working functions for the actual user traffic. These elements are also referred to as the media gateway controller (MGC) and media gateway (MG), respectively. The media gateways basically provide the capability to cross-connect various kinds of media streams under the control of the MGC, and provide the resources to process voice and data and support widely used transmission interface protocols. The MSC servers (the media gateway controllers) communicate with the media gateways over the M_c interface through the so-called Megaco protocol defined by 3GPP Release 4 Specifications. The elements and their interfaces in the soft switching architecture are illustrated in Figure 13.4. The functions of the MSC server are call control and routing, accounting, authorization, billing, provisioning, fault monitoring, and recovery. The MSC server terminates the user-network signaling (for call control and mobility management) and translates it into the relevant network–network signaling. The MSC servers use the BICC (Bearer independent call control) protocol to communicate signaling information with each other over the N_c interface. The BICC is independent of the transport technologies used to transfer the actual user traffic between the M-MGW, as well as for the transfer of the signaling information. The signaling exchange between the media gateway controllers takes place over the N_b interface, which uses protocols such as SS7 and H.323 or the session initiation protocol (SIP) depending on whether the network that connects the media gateways uses circuit or packet mode communication, respectively.

14.8.1 Benefits of Soft Switching

By using the strategy of distributing the switching function (performed by the media gateways) to the high radio access concentration periphery points of the network, such as within the BSCs and RNCs, local switching of mobile-to-mobile calls and the breakout to PSTN is made possible. This helps to off load traffic from backhaul transmission facilities and save significant operating costs. In the meantime, the more complex call control functions, performed by the media gateway servers, can be centralized to only a few locations to save on site and equipment costs while providing enhanced reliability through redundant server configurations. This strategy also enhances the flexibility for network expansion and evolution towards the efficient provisioning of multimedia services through an all IP core network [21].

Soft switching provides the opportunity for more efficient network design through optimizing equipment locations, improving scalability and simplifying the O&M. Furthermore, it provides the flexibility for easy migration to VoIP for converged service offerings and multimedia support. The flexibility and scalability of physically separated call control and signaling servers and gateways based on latest technology hardware provides for the efficient handling of traffic growth.

Since most mobile networks have grown quickly in response to the market dynamics, large cost savings are possible by redesign and restructuring of the network. This involves re-examining and rationalizing sites, nodes, technology, and operations. This is facilitated in a migration to soft switching, which provides a layered architecture in which the application, control, and connectivity functions are moved into separate layers, allowing for each to be independently optimized in terms of technology, location, capacity, etc.

The pooling of servers and media gateways that is made possible in soft switching provides for redundant hardware configurations and enhanced reliability. The server pooling also allows for easy addition of nodes to absorb growing traffic without the need for the re-parenting of equipment or reconfiguration of the network. Moreover, new coverage areas can be built with transport efficient solutions while utilizing MSC-Server pools in the network. Soft switching can reduce the transmission costs, by allowing the distribution of the M-MGWs to radio network concentration points, enabling local switching of mobile-to-mobile calls. It eliminates unnecessary trans-coding (for mobile-to-mobile calls, for instance) by performing it only when necessary, and reduces equipment footprint and site rentals by more centralization of the call control layer equipment to low rental areas.

The pooling of the MSC-S provides further benefits through increased capacity, service availability, and reduced OPEX. Capacity benefits result from the traffic smoothing effects of enlarged service areas, which helps to handle abnormal traffic peaks from individual MSCs. Additionally, server pools can provide both local and wide area back up services through geographical redundancy, which ensures service continuity if a major site goes down.

14.8.2 Transition to Soft Switching

A migration to soft switching can be done by, for instance, moving some of the user traffic from the MSCs to media gateways placed at strategic places while maintaining the call processing functions within the MSCs. Over time, more and more traffic can be transferred to the M-MGW, providing a smooth transition to a common GSM/WCDMA soft switch architecture. This approach is highly attractive in terms of protecting existing investment, but increases network complexity. Alternatively, complete BSCs can be transferred one at a time to the overlay soft switch network to effectively create a common GSM/WCDMA network. This allows the traditional MSCs in the GSM network to be taken out of service and, where appropriate, re-configured as pure MSC-S. This approach can still protect existing investment and provides a more simplified migration. The pooling of servers in a common GSM/WCDMA call control platform helps scalability and removes the need for re-parenting of BSCs and RNCs to accommodate growth. The efficient handling and management of inter-system handovers are also facilitated by equipping all the servers and gateways with simultaneous GSM/WCDMA mobile service capabilities and support for standardized network interfaces.

The deployment of soft switching with WCDMA and moving the GSM MSCs to a common platform can also facilitate the transition to an all-IP common core network. A common core, handling GSM and WCDMA traffic in the same server pools and gateways, simplifies network planning by removing the need to forecast the split of GSM and WCDMA traffic or to plan and dimension two separate core networks. Moreover, using a single unified transport infrastructure for all services (signaling, user plane payload, billing data, O&M, etc.) simplifies planning and reduces operating costs. Additionally, using the unified IP based transport greatly reduces the complexity of provisioning and dimensioning the many point-to-point TDM links.

Greenfield operators in particular have an excellent opportunity to take advantage of the latest proven technologies to build a scalable, flexible, and efficient network from the start. A properly designed and implemented soft switching network improves service flexibility and provides more scaleable capacity at reduced operating costs. Soft switching is currently being deployed in an increasing number of networks.

14.9 Core Network Design and Dimensioning

The design of the core network involves several considerations and decisions that need to be made to ensure a design that is scaleable, reliable, cost effective, and meets the QoS objectives for delay-tolerant data traffic, as well as real-time traffic.

The following list contains the primary high-level tasks in designing the packet-switched portion of a backbone network [22]:

- Decide on the number and locations of SGSNs and GGSNs based on expected access traffic and distribution, equipment traffic handling capacity, required external interfaces, etc.
- Estimate the end-to-end aggregate traffic flow demand for each service.
- Determine the QoS requirements for each service.
- Select the appropriate transport technologies for interconnecting the nodes.
- Decide on a network topology (interconnection between traffic nodes).
- Select the appropriate network infrastructure elements (routers/switches/other traffic nodes) and configure properly to meet QoS requirements.
- Estimate the quantities of infrastructure equipment required to support the offered traffic.
- Dimension the links between the nodes.

All these tasks have to be performed for desired network reliability, requiring consideration of traffic path redundancy and thus traffic node interconnectivity, as well as cost considerations.

Two major decisions that need to be made in the design process, besides the selection of an appropriate transport technology, are a proper model for estimating the traffic demand for various services and the selection of a topology choice for interconnecting the nodes. These are discussed in the following sections.

14.9.1 Traffic Model

The most difficult information to obtain for the initial designing of a network is usually the data describing the traffic that will be using the network. Traffic data can be used to determine the placement of links, the sizing of links, and routing configurations that obtain the desired utilizations and service levels. Traffic data may be generated from actual measured traffic on a similar network or manufactured based on marketing estimates of the applications that will be using the network. In either case, a methodology for choosing the data set or sets against which to design a network must be determined. Generally, a conservative estimate such as peak-day, peak-hour traffic is assumed. When actual traffic information is unknown, it is recommended that several traffic scenarios be generated based upon whatever information is available in order to test the sensitivity of designs to changes in network traffic distributions.

Once the network is designed and operated, node configuration data can be drawn from traffic databases, configuration files, and network management systems for future capacity planning purposes. Such detailed configuration data collected from a running network can be used to accurately model routing in complex operational networks that use route-maps, route-filters, route-redistribution, etc.

The steps involve in choosing a traffic model, and using that to estimate the network link capacities, are as follows:

- Estimate the end-to-end peak hour traffic matrix based on either a similar representative network that exists or marketing estimate of the traffic the applications are expected to generate.
- Select a network topology.
- Select a traffic routing scheme.
- Size up the links connecting the nodes.

Subsequent steps for validating the initial design are:

- Fail single links and nodes, and using a network performance analysis tool, re-run the traffic through the network and make sure there are no over-utilized links. If there are, resize any over-utilized links or add new links, and re-run the traffic to make sure the bottlenecks are eliminated.
- Vary the assumed traffic flow demand within expected ranges and patterns, and see if the design can still meet it. Otherwise modify the design by resizing links or adding new ones until the sensitivity to the traffic variation is resolved.
- Fail single links and nodes, and re-run the model with the new traffic flow demand and re-iterate the design as before if needed.

14.9.2 The No Traffic Information Scenario

In this scenario, there is no information on the traffic flows in the network. An algorithm may be used to support analysis of the underlying topological design without any traffic information. Design algorithms that focus on addressing fundamental topological characteristics such as connectivity, min-hop distances, number of links, costs of links, and network hierarchy may be used to generate initial network designs. Technologies may be deployed on the model and tested for validity. Without traffic information, utilizations cannot be computed and so link sizing must be based upon user specified design criteria. This type

Table 14.1 Dimensioning factors.

GPRS core element	Key dimensioning factors
Iu/Gb interface	Average throughput per user
	Number of cells per RNC or per BSC/PCU
SGSN	BH number of attached Subscribers
	BH peak throughput
	Number of cells per SGSN
	Number of Routing Areas per SGSN
	Number of Gb E1 ports or the expected Iu throughput
Gn interface	BH number of active PDP contexts
	BH peak throughput
	Total number of Routing Areas
	Number of switch sites
GGSN	BH number of active PDP contexts
	BH peak throughput
	Number of Gi ports required
Gi interface	APN throughput requirements
	Customer security requirements

of approach is appropriate for preliminary investigations such as those conducted by pre-sales organizations or network planning groups that are exploring completely new networks designed from the ground-up. Designs based on 'traffic insensitive' approaches are appropriate. These types of design algorithms provide a method for creating low-cost solutions that are relatively insensitive to changes in the distribution or types of traffic on the network. After examining some of the basic architectural issues at this level, the best designs will generally be used as the starting point for more detailed evaluations integrating some type of traffic information.

14.9.3 Dimensioning of SGSN, GGSN, and the Interfaces

The factors that determine the dimensioning of the core network elements SGSN, GGSN, and their interfaces are given in Table 14.1.

14.9.4 Active PDP Contexts and Impact of Call Mix on Dimensioning

Subscribers who wish to access Internet services or other services offered through GPRS will need to set up an active PDP context with the gateway GGSN. The GGSN maintains knowledge of this context as long as the subscriber is attached, whether or not data is being transferred. A 95% active PDP context planning rule assumes that almost all attached subscribers have requested access to at least one APN.

For a given set of system constraints and operator planning rules, the call mix will drive the SGSN and GGSN dimensioning:

- The number of SGSNs required is driven by either *throughput* or *Attached Subscribers*, depending on the application mix on the network. That is:

- A high percentage of consumer applications will make the number of Attached Subscribers the driver. On the other hand, a large number of corporate customers with long session durations may make throughput the driver.
- Likewise the number of GGSNs required is driven by *throughput* or *attached users with active PDP contexts*:
  - A high percentage of consumer applications may make the number of active PDP Contexts the driver.
  - A large number of corporate customers may make throughput the driver.

In addition, the vendor's product constraints will also have an impact on the minimum and maximum configuration of SGSNs and GGSNs in a particular location.

14.9.5 Signaling Traffic and Link Dimensioning Guidelines

Signaling in the core network is divided here into signaling between the main core network elements, which are the SGSNs, and the GGSNs, and signaling between theses main core elements and the location databases such as the HLR and the MSC/VLR.

14.9.5.1 Signaling between SGSNs and GGSNs

Signaling between the SGSNs and GGSNs consists basically of protocol messages that are exchanged over the Gn interface for the activation and deactivation of PDP context for data transfer [8, 9].

An estimate of the SGSN-GGSN signaling traffic can be obtained by multiplying the average number of bytes involved in a PDP activation/deactivation message exchange process by the average number of expected requests for the service.

14.9.5.2 Signaling between SGSN and HLR

The signaling between the SGSN and HLR takes place through the Gr interface shown in Figure 14.5. The signaling messages are based on the SS7 MAP/TCAP protocols, which are transported over the lower layers of SS7 as shown in the figure. The signaling messages exchanged include the transfer of user profile information (i.e., subscriber data) and location

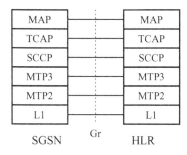

Figure 14.5 SGSN–HLR Interface.

The Core Network Technologies, Design, and Dimensioning 249

update information from SGSN to HLR in SGSN handovers, and the same from the HLR to SGSN for obtaining the necessary user information for call routing purposes.

The dimensioning of the signaling link between the SGSN and HLR is obtained by the following input:

- number of users supported by the SGSN
- busy hour service request per subscriber
- average number of signaling bytes exchanged per transaction
- average number of inter_SGSN handovers
- average number of signaling bytes exchanged per inter_SGSN handover
- busy hour number of location update and registration transactions
- average number of signaling bytes per location update and registration transactions

14.9.5.3 Signaling between SGSN and MSC/VLR

The signaling between the SGSN and the MSC/VLR takes place over the Gs interface [23], as shown in Figure 14.6.

In Figure 14.6, BSSAP+ stands for Base Station System Application Part + and is a subset of BSSAP procedures supporting signaling between the SGSN and MSC/VLR, as described in sub-clause 'Mobility Management Functionality' of Reference 23.

The signaling messages over the Gs interface are used for communicating between the VLR databases that are used for controlling circuit switched services and the SGSN data bases that are used for controlling the packet switched services. The messages exchanged over the interface include location update messages and paging messages.

The dimensioning of the signaling link between the SGSN and the MSC/VLR would be determined by the following input:

- number of subscribers served by the SGSN
- average number of location update messages per user
- busy hour number of pages received per mobile user
- average number of signaling bytes exchanged per location update
- average number of signaling messages exchanged per user paging

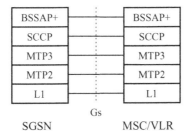

Figure 14.6 SGSN–MSC/VLR Interface.

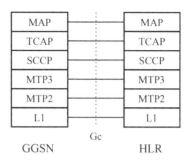

Figure 14.7 GGSN–HLR interface.

14.9.5.4 Signaling between GGSN and HLR

This optional signaling path allows a GGSN to exchange signaling information with an HLR, over the so-called Gc interface shown in Figure 14.7. There are two alternative ways to implement this signaling path:

- If a SS7 interface is installed in the GGSN, the MAP protocol can be used between the GGSN and an HLR.
- If a SS7 interface is not installed in the GGSN, any GSN with a SS7 interface installed in the same PLMN as the GGSN can be used as a GTP-to-MAP protocol converter to allow signaling between the GGSN and an HLR.

The signaling information exchanged over the Gc interface includes routing information and subscriber data, which the GGSN may request from the HLR in mobile terminated data calls. Therefore, the signaling message volume on this interface would depend on the number of mobile terminated calls received by the GGSN.

The dimensioning of the signaling link between GGSN and HLR would be based on the following input:

- number of busy hour mobile terminated service requests received by the GGSN
- number of users supported by the HLR
- average number of signaling bytes exchanged per mobile terminated service request

14.9.6 Protocol Overheads

Table 14.2 presents the overheads resulting from using various layers of protocols over the core network links.

14.10 Core Network Transport Technologies

The transport technology options, which may be used to provide the connectivity between the core network nodes (such as the SGSNs, GGSNs, etc.) at the physical and data link layers, consist of the following:

Table 14.2 Protocol overhead on core network links.

Protocol	Header overhead, bytes	SGSN–GGSN	RNC–SGSN	Node B–RNC, RNC–RNC
GTP	12	x	x	
UDP	8	x (or TCP)	X (or TCP)	
TCP	20	x (or UDP)	X (or UDP)	
IP	20	x	x	
ATM	5	option	x	x
AAL2	3	option (voice)		x
AAL5	8	option (data)	x	
PPP	8	if private lines used		
RTP	12	if packet voice used		

- Dedicated private lines
- ATM virtual circuits
- Frame relayed PVCs (permanent virtual circuits)
- Ethernet based transport

14.10.1 Dedicated Private Lines

Dedicated private lines include the T, and their European version E, carriers and their fractionals, Sonnet based optical links, and possibly line-of-site digital microwave links.

14.10.1.1 Advantages and Disadvantages of Private Lines

Advantages

- QoS and congestion not an issue (only within routers)
- Easier to engineer (size to handle peak traffic)
- Provided in more areas

Disadvantages

- Reduced reliability (usually no alternate-rerouting)
- Costs more and costs are usually distance-sensitive
- Inefficiencies involved in having to size to handle the peak traffic flow (not efficient for varying traffic flows)

14.10.1.2 Sizing Criteria for Private Lines

The bandwidth (size) of a private line between any two nodes should be based on the expected peak traffic between the two nodes. This means that any excess bandwidth not used at times of light traffic is simply wasted. On dedicated point-to-point links, the point-to-point protocol, PPP, is used to transport the core network IP formatted data.

14.10.2 ATM Virtual Circuits

ATM is a transport technology specified for the UMTS core network [24]. The ATM virtual circuits connecting the SGSN and GGSN nodes can be provided on either a permanently allocated or switched basis through an ATM network. The switched ATM virtual circuits (SVC) that can offer QoS provide the advantage of being able to offer bandwidth on demand over the permanent ATM connections. Therefore, the SVCs are most efficient when the traffic flows between nodes varies significantly. The network providing the ATM service can be either a carrier provided public network or a network of privately owned equipment serving a single wireless operator.

14.10.2.1 ATM Advantages and Disadvantages Compared to Private Lines

Advantages

- More cost effective, can provide bandwidth on demand
- Costs are not distance-sensitive in most cases
- Enhanced reliability through built in-rerouting capabilities
- Allows multiple virtual connections through a single physical port
- Dynamic bandwidth sharing
- Multi-class services to efficiently handle each traffic class and required performance

Disadvantages

- More complex engineering and configuration to handle QoS
- Involves some protocol overhead
- Can incur further packet losses and delays within the ATM switches

14.10.2.2 Sizing Parameters and Issues

ATM provides virtual circuits with different bandwidth and delay characteristics to suit different classes of services [25]. Each virtual circuit class involves parameters that need to be specified according to the bandwidth characteristics of the traffic class to be handled. ATM offers the following virtual circuits:

- **UBR (unspecified bit rate).** This circuit is for handling best effort delivery service suited mostly for handling non-real time data services, such as email, Usenet, file transfer, LAN inter-networking, etc., and does not guarantee their delivery (a higher layer protocol such TCP would be needed to ensure end-to-end delivery of the service). UBR circuits do not guarantee any amount of bandwidth, and are normally policed by the network to not exceed a specified maximum bit rate.
- **ABR (available bit rate).** This circuit is similar to UBR except that the network does guarantee a specified minimum bit rate for the service, and hence achieves better performance in terms of a more stable throughput. ABR circuits are more suited to handling web browsing and transaction services that are more delay constraint than other data services.

- **CBR (constant bit rate).** The CBR virtual circuits are used to provide circuit emulation services for handling highly delay critical traffic such as real time un-coded voice and video services. The CBR can also be used to emulate private line services through an ATM network. There is only one parameter required to specify a CBR circuit and that is the peak bit rate for the service.
- **VBR-rt (variable bit rate, real time).** This circuit is specified by three parameters consisting of the expected peak cell rate, sustained cell rate, and the maximum burst size. The network tries to minimize the end-to-end delay as well as the delay variations from cell to cell. The VBR-rt virtual circuits are suited to handling compressed voice and video services that are delay critical.
- **VBR-nrt (variable bit rate, not real time).** This circuit is specified by the same parameters as the VBR-rt, but is suited to transporting bursty data services that are less delay critical, such as banking transactions, airline reservations, process monitoring, and frame relay interworking.

14.10.3 Frame Relay

Frame relaying services are currently offered through PVCs, which are set up in advance by the network operator. All of the current frame relay providers offer fixed rate services of 56 kbps, and many offer higher rates of 128 kbps, 384 kbps, and up to 1.536 Mbps. This transport technology is not the best choice for UMTS because of its limited capacity. It is used by some operators for implementing GPRS core network links. Some of the carriers may offer a 64 kbps service in which case the pricing will be the same as the 56 kbps circuits.

PVCs can be simplex or duplex. Simplex allows only one-way traffic: Two simplex circuits can be set up with different throughput. Duplex circuits allow bidirectional transmissions and they use the same information rate in each direction.

14.10.3.1 Frame Relay Advantages and Disadvantages Compared to ATM [26]

Advantages

- Uses variable frame lengths and hence incurs less header overhead for handling certain data services that involve lengthy transmissions.
- Is often available in more regions than ATM, at the present time.
- Easier to engineer and configure in compared to ATM.
- Frame relay equipment and interface cards are less expensive compared to ATM equipment.
- Like ATM, offers distance insensitive pricing for long distance services.
- Like ATM, allows multiple virtual connections through a single physical port.

Disadvantages

- Runs at slower speeds compared to ATM, and currently is offered at a maximum speed of T1, although some vendors are developing or may already offer equipment to run at multiple T1 and up to DS3 rate.

- Basically treats all traffic the same and offers no extensive service quality features.
- Not the best in handling delay critical traffic such as conversational voice.

Frame relay's relatively low overhead and variable packet size make it an efficient protocol for transmitting data. With frame relay, the default overhead on a 128-byte frame is approximately 4% of the frame. As the frame size increases, the overhead percentage decreases dramatically. For example, with a 1600-byte frame, a size common with LAN protocols, the overhead drops to less than 0.3333%.

ATM's 53 byte fixed cell is optimized for handling multimedia traffic at high speeds. However, with ATM, approximately 10% of each full cell is overhead. In very short transactions, where cells are not full, an even greater percentage of bandwidth carries no user data.

As a general rule, the lower the speed (e.g., T1 and sub T1), the more should be the concern with overhead for data applications. At levels of T3 and above, the benefits of ATM outweigh the efficiency considerations. At relatively lower speeds (T1 and below), frame relay transports data more efficiently.

14.10.3.2 Sizing Parameters and Issues

Frame relay allows for fixed or variable rate information. The customer asks for a specific *committed information rate* (CIR), which the service provider guarantees within a certain probability. CIR is a parameter that specifies the maximum average rate at which the network may transmit without discarding data. This rate is averaged over a specified interval Tc. The user can transmit bursts of data at higher rates, but only at the risk that the extra information will be the first dropped should congestion occur in the network (by setting the Discard Eligibility (DE) bit). The parameters involved here are the Committed Burst Size (Bc) and Excess Burst size (Be). Committed Burst Size (Bc) is the largest number of consecutive bits that the network agrees to carry within a specified time interval Tc without discarding data. The Excess Burst Size parameter specifies the excess over the Committed Burst Size, which the network agrees to carry (within the time interval Tc) with a greater likelihood that some data will be discarded.

Some frame relay carriers now offer at a greatly reduced price a 0 CIR service in which no data rate is guaranteed to reach the destination – in effect all data is considered to be a burst. In a network not experiencing congestion, 0 CIR provides the same level of service provided by higher CIRs because no packets are actually dropped, and even in the face of congestion a large portion of burst packets will make it through the network. It is not yet clear if 0 CIR service will remain a practical alternative as more users begin to use frame relay and the frame relay networks become more utilized.

14.10.4 IP Transport

IP based networks such as the Internet or a private IP network may also be used to transport the IP based UMTS/GPRS applications (IP within IP as in VPNs) within the core network. However, this approach has the drawbacks of increased tunneling overheads, as well as potential performance bottlenecks. IP is a connectionless technology that has no absolute QoS mechanisms. The only QoS mechanisms offered by IP are packet prioritization and scheduling mechanisms, which help to differentiate among multiple services and provide service quality on a relative basis (no absolute guarantees of delays and throughputs). This relative QoS mechanism is not implemented throughout the Internet.

14.10.5 Transport Technology Selection for Core Network

A primary issue in the selection of a proper transport technology for the core network is how to manage QoS for packet-switched traffic. Unlike QoS objectives for circuit-switched traffic in mobile wireless networks, which are typically measured in terms of the blocking rate, QoS for packet-switched traffic is primarily measured in terms of *delay* and *throughput*. This will affect the choice of transport technology selected for the packet-switched core network, whether dedicated lines, ATM, frame relay, etc.

Dedicated private lines of course offer the best QoS because no congestion is experienced on them. However, they may not always be the best choice because of the costs involved. In particular, private lines can end up being quite inefficient when the traffic to be carried happens to be bursty, resulting in an inefficient usage of the dedicated bandwidth. Another drawback of private lines is reduced reliability. In contrast, both frame relay and ATM not only offer the opportunity for bandwidth on demand and dynamic bandwidth sharing, but also offer enhanced reliability due to the inherent redundancy built into these networks, which is shared by many users and hence is cost effective.

Since ATM is a connection-oriented packet-switching technology, which has some built-in QoS mechanisms, it offers the opportunity for integrating multi-service classes within the core network while allowing each one to meet its QoS requirements. The service quality in ATM networks is achieved by controlling delay, jitter, and loss, and dedicating bandwidth to selected traffic types. ATM switches implement proper queuing, scheduling, buffer management mechanisms, traffic shaping, and admission control techniques to provide virtual circuits with bandwidth and delay characteristics appropriate for each different traffic type. The ATM switched virtual circuits with QoS can provide bandwidth on demand between the network nodes and save on transmission costs. Frame relay on the other hand, offers very little capability for differentiating between diverse services to allow the proper sharing of the network capacity and resources and to let each service meet its delay and throughput requirements.

A more advanced form of IP over ATM-like solution is based on MPLS (multi-protocol label switched) technology. MPLS combines the intelligence of IP routing and the fast switching of ATM and thereby introduces the notion of connection-oriented forwarding to IP networks. When packets from a particular session enter the network, they are sent along the same path by giving them all the same initial label by the edge router. Each intermediate node on the path uses the packet's label as an index into a lookup table to find the next hop for the packet and a new outgoing label, hence the term label switching. In the optical domain, the label switching concept of MPLS is extended to include wavelength routed and switched light paths, and MPLS therefore takes the name multi-protocol wavelength switching (MPλS). As the name implies, in MPλS the wavelength of the light serves as its own label. In MPλS, IP packets from one or multiple sessions are mapped to a wavelength by the edge IP routers, which are then routed through a mesh of wavelength switching optical cross connect (OXC) nodes. MPλS is one technology used as the basis for optical core networks [27].

References

[1] Rahnema, M., 'Overview of the GSM System and Protocol Architecture', *IEEE Communications*, April 1993, 92–100.
[2] 3GPP TS 29.002, 'Mobile Application Part (MAP) Specifications'
[3] GSM 02.60, v 6.1.0, 'Digital Cellular Telecommunication System (Phase 2+), GPRS; Service Description; stage 12.
[4] 3GPP TS 33.102, '3G Security and Security Architecture'.
[5] 3GPP TS 23.060, 'General Packet Radio Service (GPRS), Service Description, stage 2.
[6] Kalden, R., Meirick, I. and Mever, M., 'Wireless Internet Access Based on GPRS', *IEEE Personal Communication Magazine*, April 2000, 8–18.
[7] 3GPP TS 25.413, 'UTRAN Iu Interface RANAP Signaling'
[8] 3GPP TS 24.008, 'Mobile Radio Interface Layer 3; Core Network Protocols; stage 3'.
[9] 3GPP TS 29.060, 'General Packet Radio Service (GPRS) Tunneling Protocol Across the Gn and Gp Interfaces'.
[10] RFC 2960, 'Stream Control Transmission Protocol', SCTP.
[11] 3GPP TS 24.007, 'GMM Procedures for GPRS Services'.
[12] 3GPP TS 23.121, 'Architectural Requirements for Release 1999'.
[13] 3GPP TS 23.101, 'General UMTS Architecture'.
[14] 3GPP TS 23.228, 'IP Multimedia Subsystem (IMS); Stage 2'.
[15] 3GPP TS 22.228, 'Service Requirements for the Internet Protocol (IP) Multimedia Core Network Subsystem; Stage 1'.
[16] 3GPP TS 22.800, 'IP Multimedia Subsystem (IMS) Subscription and Access Scenarios'.
[17] 3GPP TS 23.864, 'Commonality and Interoperability between IP Multimedia System (IMS) Core Networks'.
[18] 'IMS Overview and Applications', 3G America, July 2004.
[19] 3GPP TS 26.141, 'IP Multimedia System (IMS) Messaging and Presence, Media Formats and Codecs'.
[20] Perkins, C., RFC 3344, 'IP Mobility Support for IPv4', August 2002.
[21] 3GPP TS 23.228, 'All-IP Multimedia Subsystem'.
[22] Rahnema, M., 'Design Considerations for Core Networks in GPRS and UMTS, *3GIS Annual Conference*, Athens, Greece, 2–3 July, 2001.
[23] 3GPP TS 29.018, 'Gs Interface Layer 3 Specifications'.
[24] TS 23.925, 'UMTS Core Network Based on ATM transport'.
[25] Pitts, J.M. and Schorman, J.A., *Introduction to IP and ATM Design and Performance*, John Wiley & Sons, Ltd, Chichester, 2000.
[26] Rahnema, M., 'Frame Relaying and the Fast Packet Switching Concepts and Issues', *IEEE Network*, July 1991, 18–23.
[27] Rahnema, M., '3G – Via Optical Core Networking', *Electronics World*, January 2002, 26.

15

UMTS QoS Classes, Parameters, and Inter-workings

The 3G networks, of which UMTS is one example, offer a wide variety of data services besides conversational voice, with different bit rate characteristics and quality of service (QoS) requirements. This chapter first defines and discusses the various concepts and issues in QoS and with an emphasis on how the services are classified from a QoS point of view in UMTS. Then it continues with discussing the QoS parameters for each service category and the inter-working requirements and mechanisms for achieving the end-to-end QoS. The management of the end-to-end QoS requires mechanisms within the UMTS bearer components, which consist of the radio access, the Iu, and the core network [1], as well as inter-working with the IP mechanisms in the external networks that are outside its control. These are discussed in some detail and the related 3GPP and other related references are identified for further reading. A good understanding of service classes with respect to QoS, the parameters, and their inter-working across various sub-networks on the end-to-end path is critical for the optimization of user perceived service quality. The present chapter provides a necessary foundation to that end.

15.1 The QoS Concept and its Importance

Quality of service is an important commercial issue for any telecommunication service. Immature markets tend to be oblivious of poor service quality, as shown by the poor but accepted quality of early mobile networks. Once a market is mature enough for there to be serious competition, QoS often becomes as important as service pricing in winning and retaining customers. Moreover, QoS features become quite important with the network customer looking for service providers that can offer a range of services at different pricing levels to meet each application scenario.

The network QoS is ultimately concerned with the quality of the communication service as perceived by the end-user. Therefore, the significance of network QoS is on an end-to-end basis, between the communicating entities. Since the user perception of the quality of a given service strongly depends on subjective parameters such as user tolerance, social background, service circumstances, etc., one cannot always expect to be able to specify optimum values for any parameters that may be used, directly or indirectly, to quantify the quality of service. The relative weight of QoS related parameters on the influence of the user's evaluation depends on the nature of the service. The criteria will obviously be different for users of interactive applications, conversational, messaging, or retrieval services, and users of entertainment services. It is also reasonable to expect that users may accept a different level of response time tolerances such as in the interactive query/response time under mobile and fixed conditions. For instance, a mobile user trapped in the middle of nowhere would be quite happy with a few seconds or more of response time to a query on the nearest tow track for fixing a car that's broken down on the road. Whereas a user sitting at his or her desk and searching the web for various types of scientific/technical information on a routine basis could find a response time of more than one or two seconds unsatisfactory.

The user's perception of the service quality can be related to a set of network performance parameters (NP) that are measurable. It is then the operator's most important task to plan and optimize the network in such a way that the combined effect of the network performance parameters on the user's perception is satisfactory and in accord with acceptable costs and under the circumstances in which the service is used.

The relationship between QoS and NP is usually not a simple one and it is difficult to determine the range of each NP parameter producing any particular desired QoS level. Most often, several sets of NP parameters lead to an acceptable QoS. The economical factor intervenes here to determine the most appropriate solution. Some applications such as voice have stringent delay requirements and can tolerate minimal packet loss, whereas others such as FTP cannot tolerate packet loss but do not have tight delay requirements.

In a mobile and wireless network, the nature of the medium itself introduces problems owing to limited bandwidth and high bit error rate. The signal strength determines how well traffic is received; in addition, multipath effects have to be dealt with and the mobility of users requires procedures such as handovers to be executed, which has a direct impact on quality of service. Handover is a process whereby a connection may be lost temporarily (in the case of hard handoffs such as in inter-system handovers), during which time traffic may get lost. The resulting delays and loss of traffic have a direct bearing on the deterioration of quality of service.

15.2 QoS Fundamental Concepts

The fundamental concepts for QoS are as follows [2, 3]:

- **QoS requirements and policies**, which define/determine the users' perception of quality and service satisfaction.
- **QoS categories**, such as voice and video (time-critical applications) and data services (Internet-browsing, email, file transfer, etc.).
- **QoS characteristics**, which describe the fundamental aspects of QoS to be managed.
- **QoS mechanisms**, which are used to manage the QoS characteristics and satisfy the QoS requirements.

QoS requirements can be characterized by call blocking probability, call set up delays, and call clarity for voice calls, for instance. In packet mode communication, bandwidth (bits per second), bounds on end-to-end delays, and bounds on packet losses and delay jitter, specify some of the QoS requirements as required by the application.

QoS categories include real time services such as voice and video, interactive data such as transaction oriented services, streaming, and delay non-critical applications such as email and file transfer.

QoS characteristics includes network and non-network related issues. Examples of network related QoS characteristics are call blocking probability, packet loss rate, inter-packet arrival variation (delay jitter), end-to-end latency, throughput (average bits/s), and generally quantifiable parameters, whose allowed value ranges with regards to satisfactory service define the QoS requirements. Examples of non-network related QoS characteristics are the quality of the service management such as service provisioning, modification or cessation done correctly, quickly, and at the agreed time, and the quality of the billing with required accuracy and timeliness.

There are certain QoS characteristics for each application or QoS category that are critical to its satisfactory performance. For instance, reducing packet loss is more important to voice and video because the time required to detect errors and retransmit lost packets is too long for their real time requirements. Delay jitter is also detrimental to voice and video: It can complicate the playback of voice and video packets within the receiver.

QoS mechanisms are implemented by the network to achieve cost effective predictable QoS performance with limited resources and statistical sharing.

The QoS mechanisms include:

- Packet classifying and priority queuing.
- Queue scheduling to manage the priority of flows.
- Resource reservation and call admission control.
- Capacity planning and upgrade.
- Flow shaping to regulate the flow according to a statistical representation of the required bandwidth.
- Flow policing to ensure that the QoS contracted by the provider and the user is being maintained.
- Flow control to adapt the flow to the fluctuations in the available network resources.
- Flow synchronization to control event ordering of delay sensitive sub-flows, such as audio and video (normally implemented by multimedia protocols).

15.3 QoS Monitoring Process

In order to monitor the QoS parameters, and make sure that the QoS guarantees are met, the following steps are proposed:

- Identify the system observation points for QoS parameters.
- Decide on the frequency of the measurement and the length of the interval over which the QoS related data are measured.
- Define procedures to compute QoS parameters from observed values.

15.4 QoS Categories in UMTS

The QoS mechanisms provided in the cellular network have to be robust and capable of providing reasonable QoS resolution. The UMTS Specifications [2, 3] have defined four different QoS categories (or traffic classes), which may be used as a guide to QoS resolution within the UMTS networks:

- Conversational class
- Streaming class
- Interactive class
- Background class

The fundamental characteristics and examples for each of these traffic classes are given in Table 15.1.

The main distinguishing factor between these classes is how delay sensitive the traffic is. Conversational class is meant for traffic that is very delay sensitive whereas Background class is the most delay insensitive traffic class. Conversational and Streaming classes are intended mainly to carry real time traffic flows. The main divider between them is how delay sensitive the traffic is. Conversational real time services, like video telephony, are the most delay sensitive applications and those data streams should be carried in Conversational class.

The Interactive and Background classes are meant mainly to be used by traditional Internet applications such as WWW, email, Telnet, FTP, and News. Due to looser delay requirements compared to the Conversational and Streaming classes, both provide better error rate by means of channel coding and retransmission. The main difference between the Interactive and Background classes is that Interactive is mainly used by interactive applications, for example, interactive email or interactive web browsing, whereas

Table 15.1 UMTS QoS classes.

	Traffic classes			
	Conversational conversational RT	Streaming streaming RT	Interactive Interactive best effort	Background Background best effort
Fundamental characteristics	Preserve time relation (variation) between information entities of the stream.	Preserve time relation (variation) between information entities of the stream.	Request response pattern. Preserve payload content	Destination is not expecting the data within a certain time.
	Conversational pattern (stringent and low delay).			Preserve payload content.
Example of the application	Voice	Streaming video	Web browsing	Background download of emails

Background is meant for background traffic, for example, background download of emails or background file downloading. Separating the interactive and background applications ensues responsiveness of the interactive applications. Traffic in the Interactive class has a higher priority in scheduling than Background class traffic, and so background applications use transmission resources only when interactive applications do not need them. This is very important in wireless environments where the bandwidth is low compared to fixed networks.

The characteristics for each of these classes are discussed in more detail in the following subsections.

15.4.1 Conversational Traffic

Examples of conversational traffic include interactive voice communication. For interactive voice applications, experiments have determined that the one-way end-to-end delay must be in the range of 100–150 ms; because, coding, interleaving, decoding, and de-interleaving can consume a substantial portion of this delay budget, an additional delay of 20 ms is on the boundary of acceptable performance. In packet mode voice communication, the inter-packet arrival time variation known as delay jitter should be made negligible to avoid discontinuities in the voice syllables. However, a small amount of packet delay variation can be tolerated because of buffering and jitter compensation within the voice decoders.

15.4.2 Streaming Traffic

When the user is looking at (listening to) real time video (audio) the scheme of real time streams applies. The real time data flow is always aiming at a live (human) destination. It is a one-way transport.

This scheme is one of the newcomers in data communication, raising a number of new requirements in both telecommunication and data communication systems. It is characterized by the fact that the time relations (variation) between information entities (i.e., samples, packets) within a flow must be preserved, although it does not have any requirements on low transfer delay.

The delay variation of the end-to-end flow must be limited, to preserve the time relation (variation) between information entities of the stream. But as the stream is normally time aligned at the receiving end (in the user equipment), the highest acceptable delay variation over the transmission media is given by the capability of the time alignment function of the application. Acceptable delay variation is thus much greater than the delay variation given by the limits of human perception.

15.4.2.1 Streaming Packet Switched QoS

Streaming Packet Switched Quality of Service enables streaming class Radio Bearers in the RAN for packet switched connections. The Streaming class RAB (Radio Access Bearer) is called real time (RT) RAB. The basic difference between RT RAB and NRT RAB (non-real time RAB) is the guaranteed bit rate and the required transfer delay, which the RAN must be able to fulfill. The radio network controller (RNC) [4] allocates a permanent DCH to the Streaming RAB, unlike to the NRT RAB, to fulfill real time requirements. A permanent RAB means that the DCH of certain constant bit rate is allocated for the whole time of the RAB's existence in the RAN.

15.4.3 Interactive Traffic

When the end-user, either a machine or a human, is online requesting data from remote equipment (e.g., a server), the interactive traffic scheme applies. Examples of human interaction with the remote equipment are: web browsing, data base retrieval, and server access. Examples of machine interaction with remote equipment are: polling for measurement records and automatic database enquiries (tele-machines).

Interactive traffic is the other classical data communication scheme that on an overall level is characterized by the request response pattern of the end-user. At the message destination there is an entity expecting the message (response) within a certain time. Round trip delay time is therefore one of the key attributes. Another characteristic is that the content of the packets must be transparently transferred (with low bit error rate).

15.4.4 Background Traffic

When the end-user, typically a computer, sends and receives data-files in the background, the background traffic scheme applies. Examples are background delivery of e-mails, SMS, download of databases, and reception of measurement records.

Background traffic is one of the classical data communication schemes that on an overall level is characterized by the fact that the destination is not expecting the data within a certain time. The scheme is thus more or less delivery time insensitive. Another characteristic is that the content of the packets must be transparently transferred (with low bit error rate).

15.5 Instant Messaging

Instant messaging is an Internet-based application that involves the transmission of short messages in real time, without storage. Instant messaging applications include chatting, stock quotes and alerts, traffic and weather alerts, etc.

All interactions with the business instant messaging should occur in a time frame expected by the user from past experience using instant messaging and web software. All transactions should seem virtually instantaneous to the user or at best relatively fast. Therefore, we may classify instant messaging in the same QoS performance category as the interactive services.

15.6 UMTS Bearer Service Attributes

Table 15.2 defines the UMTS bearer attributes and their relevancy for each bearer class [2]. Observe that traffic class is an attribute itself.

Transfer Delay (ms)
The transfer delay attribute indicates maximum delay for 95th percentile of the distribution of delay for all delivered SDUs during the lifetime of a bearer service, where delay for an SDU is defined as the time from a request to transfer an SDU at one Service Access Point (SAP) to its delivery at the other SAP.

Traffic Handling Priority
The traffic handling priority specifies the relative importance for handling of all SDUs belonging to the radio access bearer compared to the SDUs of other bearers. Within the

UMTS QoS Classes, Parameters, and Inter-workings 263

Table 15.2 UMTS bearer attributes defined for each bearer class.

Traffic class	Conversational class	Streaming class	Interactive class	Background class
Maximum bit rate	X	X	X	X
Delivery order	X	X	X	X
Maximum SDU size	X	X	X	X
SDU format information	X	X		
SDU error ratio	X	X	X	X
Residual bit error ratio	X	X	X	X
Delivery of erroneous SDUs	X	X	X	X
Transfer delay	X	X		
Guaranteed bit rate	X	X		
Traffic handling priority			X	
Allocation/retention priority	X	X	X	X

interactive class, there is a need to differentiate between bearer qualities. This is handled by using the traffic handling priority attribute, to allow RAN to schedule traffic accordingly. By definition, priority is an alternative to absolute guarantees, and thus these two attribute types cannot be used together for a single bearer.

Allocation/Retention Priority
The allocation/retention priority specifies the relative importance compared to other radio access bearers for allocation and retention of the radio access bearer. For PS services, the allocation/retention priority attribute of the radio access bearer is derived from the UMTS bearer service attributes allocation/retention priority. Other attributes may be used in addition. For CS services, the allocation/retention priority attribute of the radio access bearer is derived from the eMLPP priority level attribute (TS 23.067) and/or the CS allocation/retention priority attribute (3GPP TS 23.008) and/or the mobile station category attribute (3GPP TS 23.008) (which is only available for subscribers in their home PLMN)[2].

Delivery of Erroneous SDUs
This attribute determines whether error detection shall be used and, if so, whether SDUs with error in a certain sub-flow are delivered or not.

Residual Bit Error Ratio
The residual bit error ratio specifies the bit error ratio for undetected delivered bits.

15.6.1 Ranges of UMTS Bearer Service Attributes

Table 15.3 lists the value ranges of the UMTS bearer service attributes: The value ranges reflect the capability of UMTS network.

For inter-working with GPRS Release 97/98, the mapping with the above UMTS (Release 99) is performed as in Reference 2.

Table 15.3 Value ranges specified for the UMTS bearer service attributes.

Traffic class	Conversational class	Streaming class	Interactive class	Background class
Maximum bit rate (kbps)	< 2000	< 2000	< 2000 − overhead	< 2000 − overhead
Delivery order	Yes/No	Yes/No	Yes/No	Yes/No
Maximum SDU size (octets)	< 1500	< 1500	< 1500	< 1500
SDU format information				
Delivery of erroneous SDUs	Yes/No	Yes/No	Yes/No	Yes/No
Residual BER	$5*10^{-2}, 10^{-2},$ $10^{-3}, 10^{-4}$	$5*10^{-2}, 10^{-2},$ $10^{-3}, 10^{-4},$ $10^{-5}, 10^{-6}$	$4*10^{-3}, 10^{-5},$ $6*10^{-8}$	$4*10^{-3}, 10^{-5},$ $6*10^{-8}$
SDU error ratio	$10^{-2}, 10^{-3},$ $10^{-4}, 10^{-5}$	$10^{-2}, 10^{-3},$ $10^{-4}, 10^{-5}$	$10^{-3}, 10^{-4}, 10^{-6}$	$10^{-3}, 10^{-4}, 10^{-6}$
Transfer delay (ms)	100 − maximum value	500 − maximum value		
Guaranteed bit rate (kbps)	< 2000	< 2000		
Traffic handling priority			1,2,3	
Allocation/retention priority	1,2,3	1,2,3	1,2,3	1,2,3

The *SDU format information* attribute defines the exact sub-flow format of SDU payload.

15.6.2 Ranges of Radio Access Bearer Service Attributes

Table 15.4 lists the value ranges of the radio access bearer service attributes: The value ranges reflect the capability of UTRAN.

15.7 UMTS QoS Mechanisms

To achieve the end-to-end QoS, UMTS has to perform its share over the various UMTS bearer components. These consist of the radio bearer, which is made up of the radio bearer, the Iu (the link between the RNC and the core network edge nodes), and the core network bearer. In addition where the UMTS network is using external IP networks such as the Internet as part of the bearer service, it will be necessary to inter-work the QoS requirements with an external IP resource manager. In this case, the inter-working function will involve the translation and mapping between mechanisms and parameters used within the UMTS and the external IP bearer service (see Section 15.9).

The radio bearer refers to the radio link and requires the assignment of adequate radio resources (codes, transmit power, channels, etc.) over the air [5, 6]. In the UTRAN, it is the

Table 15.4 Value ranges of radio access bearer service attributes.

Traffic class	Conversational class	Streaming class	Interactive class	Background class
Maximum bit rate (kbps)	< 2000	< 2000	< 2000 – overhead	< 2000 – overhead
Delivery order	Yes/No	Yes/No	Yes/No	Yes/No
Maximum SDU size (octets)	< 1500	< 1500	< 1500	< 1500
SDU format information				
Delivery of erroneous SDUs	Yes/No/–	Yes/No/–	Yes/No/–	Yes/No/–
Residual BER	$5*10^{-2}, 10^{-2}, 10^{-3}, 10^{-4}$	$5*10^{-2}, 10^{-2}, 10^{-3}, 10^{-4}, 10^{-5}, 10^{-6}$	$4*10^{-3}, 10^{-5}, 6*10^{-8}$	$4*10^{-3}, 10^{-5}, 6*10^{-8}$
SDU error ratio	$10^{-2}, 10^{-3}, 10^{-4}, 10^{-5}$	$10^{-2}, 10^{-3}, 10^{-4}, 10^{-5}$	$10^{-3}, 10^{-4}, 10^{-6}$	$10^{-3}, 10^{-4}, 10^{-6}$
Transfer delay (ms)	80 – maximum value	500 – maximum value		
Guaranteed bit rate (kbps)	< 2000	< 2000		
Traffic handling priority			1,2,3	
Allocation/retention priority	1,2,3	1,2,3	1,2,3	1,2,3
Source statistic descriptor	Speech/unknown	Speech/unknown	Speech/unknown	Speech/unknown

RNC (radio network controller) that manages and assigns the radio resources based on examining the QoS parameters conveyed by the core network [1]. The RNC and the core network map the end-to-end QoS requirements over the Iu interface and the RNC handles the QoS requirements over the radio path. The Iu bearer service requires QoS management for IP such as scheduling priorities at the interface, and ATM transport [7] as was discussed in Chapter 14.

The provisioning of QoS in the core network bearer is based on IP QoS mechanisms, which include 'differentiated Service' (DiffServ) [8] and 'IntServ' [5]. The exact mechanisms used for realization of the QoS attributes are not a subject of the specifications.

15.8 UMTS QoS Signaling

The QoS signaling is concerned with how QoS parameters are exchanged and interpreted between different network bearer components to provide the end-to-end QoS [9]. The UMTS bearer QoS requires signaling among the UE, RNC, and the core network nodes SGSN and GGSN. UMTS manages the QoS signaling during the session setup and in the traffic exchange process. During the session setup, QoS parameters are exchanged in signaling messages. The format of the signaling messages and the parameters vary between different

Table 15.5 RNC mapping of QoS parameters.

QoS parameter	Value	Radio resource requirement
Maximum bit rate	128 kbps	64 chip spreading code
Maximum SDU size	1500	Mapped to transport formats
Residual BER	10^{-6}	1/3 Turbo Encoder
Transfer delay	specified	Interleaver = 40 or 80 ms
Guaranteed bit rate	64 kbps	Smallest channel rate
Delivery order	Yes	Use acknowledged mode RLC
SDU error ratio	1%	
Delivery of erroneous SDU	No	Use acknowledged mode RLC

network domains. The application within the UE requests the UMTS bearer QoS that is translated by the UE into appropriate QoS parameters and sends them through the Packet Data Protocol (PDP) context activation requests to the SGSN. The SGSN performs admission control, which involves first verifying the subscriber's QoS profile with the Home Location Register (HLR), and then checking the required resource availability within the core network. If the resources are available, it requests for the radio resources from the RAN.

The SGSN uses the Radio Access Network Application Protocol (RANAP) part [10] to translate the QoS requirements into a set of parameters for the RAN. The radio Network Controller (RNC) uses the radio resource control protocol [11] to map the RANAP indicated QoS parameters to the required radio resource parameters, which are then checked against the available radio resources in the cell, the impact on existing connections, the user priority, admission policies, etc. [9] (as discussed in Chapter 9) before allocating the necessary resources. The typical RNC mapping of the QoS parameters is illustrated in Table 15.5.

The RNC subsequently directs the Node B to configure the allocated resources using the Node B Application Part (NBAP) protocol [12]. When the radio resources are configured, the RNC informs the UE of the allocated resources using the RRC messages [11], and also sends the RAB Assignment Response to the SGSN.

The SGSN then requests that a PDP context be established with the GGSN using the GPRS Tunneling Protocol (GTP) [13]. The GGSN in turn allocates the appropriate resources for the session according to the QoS profile conveyed from the SGSN, and the UMTS bearer is set up. The SGSN translates the application specified QoS parameters into appropriate parameters of the backbone QoS mechanisms such as DiffServ [14]. The DiffServ is used to implement QoS mechanisms over the Iu and within the core and external network. In the DiffServ approach, no resources are reserved from individual flows or users. Instead, each packet belonging to a particular flow (voice, video, or data) is marked with a class of service priority, which is used by the network routers and interfaces to determine the transmission priority of the packet at a queue. By placing packets from different classes into different priority levels, DiffServ helps to expedite the transmission of higher priority traffic but does not guarantee a quantifiable QoS (delay) for any specific session. All routers within the core network are required to support defined class of services and QoS mechanisms.

Likewise, when RNC receives a packet from the UE, it marks the packet with the appropriate class of service before forwarding it to the SGSN. The SGSN then marks the class of service and sends it through the core network to the GGSN. Similarly GGSN marks the packet with the appropriate QoS class before forwarding it to the external network.

The RSVP mechanism is based on bandwidth and resource reservation within the links and nodes of the core network for the session. Differentiated services would require that there is either one QoS profile for each traffic type or alternatively the priority and traffic type information is included in the data packets. The priority information is then used for scheduling the packets from the session within the network routers and link interfaces in the traffic exchange process.

15.9 UMTS–Internet QoS Inter-working/Mapping

In the case of Internet applications, the selection of the class and appropriate traffic attribute values is made according to the Internet QoS parameters. Internet applications do not directly use the services of UMTS but they do use Internet QoS definitions and attributes, which are mapped to UMTS QoS attributes at API (application program interface). Currently there are two major Internet (IP) QoS concepts, namely IntServ and DiffServ. IP-based QoS models must be supported for PDP contexts, meaning both IntServ based on the resource reservation protocol (RSVP) [15] and DiffServ (which uses a 6-bit QoS parameter on each IP packet to specify its class of service or priority). Both mechanisms are controlled by applications residing in the user equipment allowing different application specific QoS levels for the same PDP context. Application level IP based QoS must be mapped to the UMTS packet core QoS by an IP bearer service manager in a network element at the border of the network, such as the GGSN node. RSVP support would require flow establishment and the reservation of an appropriate amount of bandwidth and resources within all the routers and links in the backbone network. IntServ based on RSVP is not mandated for UMTS bearer QoS. However, it may be used between the UE and the end node to reserve resources for the external part of the bearer. RSVP requires that the source, destination, and all intermediate nodes support the RSVP protocol to guarantee QoS end-to-end.

UMTS defines inter-working with IP QoS mechanisms in the UMTS External bearer QoS architecture [3]. The external bearer architecture is designed to support multimedia services in UMTS Release 5. UMTS defines a logical component known as the IP Bearer Service Manager in the UE and GGSN to mange IP bearer QoS. The IP bearer service manager uses DiffServ or IntServ to set up an IP bearer QoS, where the QoS parameters for the IP bearer are derived from the UMTS bearer parameters, and vice versa. UMTS Release 5 uses the session initiation protocol (SIP) [8] to set up an IP bearer for a voice over IP or multimedia service.

The UMTS external bearer QoS architecture also includes the IP Policy Decision Point (IP PDP) that implements IP bearer policies. The IP PDP makes decisions on QoS policies for each IP bearer service, which is obtained from the application servers.

15.10 End-to-End QoS Delay Analysis

One-way, end-to-end delay seen by a stream is the accumulated delay through the entire data flow pipeline including sender processing (such as packetization and coding), access protocol, network transmission, and reception. Some delays such as coding and decoding are of fixed duration whereas others are non-deterministic due to highly dynamic network or process scheduling conditions. The network transmission delay is contributed by the core access, the core network, and the external network such as the Internet, PDNs, and PSTN.

The delay components on the radio access network include the following:

- random access
- access grant queuing
- access grant transmission
- switching to a traffic channel
- traffic channel queuing

These parameters can be estimated from the radio frame structure, access protocol, and traffic request volume on the air interface.

The delay components on the core access include:

- processing time within the BSC or the RNC (dependent on load and CPU capacity)
- queuing within the PCU (packet control unit)
- packet insertion time on the BSC/RNC–SGSN link (= packet length/propagation speed).

(A good general rule for propagation speed on wired mediums is 10ms/1000 miles.)

The delay components in the core network include:

- packet insertion time (the packet length/packet node port speed)
- propagation delay (distance/speed of propagation)
- queuing delays (depends on network load status and queuing disciplines of the nodes)
- processing delays within the nodes (depends on CPU capacity within the nodes and load status)

Delays and delay variation contributed by the core network may be measured by examining time stamped control packets that are used in loopback mode (an example would be ICMP Ping packets in IP networks). It is also possible to get an indication of the number of packet losses by examining the number of TCP re-transmissions reported by the network element managers. The number of RLP NACKs (negative acknowledgements) received on the radio link can be used as an indication of losses on the radio access channel.

The delay component within the external network such as the Internet is beyond the control of the operator in most cases. However, different ISPs may optimize their networks to provide different levels of quality of service. The cellular operator may choose the ISP that provides better connectivity and capacity for reaching the various information resources on the Internet as possibly specified by service level agreements (SLAs) between the operator and the ISP. The cellular operator can also implement caching services within the core network to minimize the delays and eliminate the need to have to cross the Internet each time a user tries to get information from the same remote source (on the Internet).

15.11 ATM QoS Classes

The connection oriented nature of ATM allows it to allocate resources to a virtual circuit during connection setup, which remains dedicated during the connection. Calls for which the required resources are not available are simply rejected by the network call admission control mechanism. In this way ATM is able to satisfy the real time delivery requirement of interactive voice and video services. ATM provides virtual circuits with different bandwidth and delay characteristics to suit different classes of services. Each virtual circuit class involves parameters that need to be specified according to the bandwidth characteristics of the traffic class to be handled [16]. ATM offers the following virtual circuits (or service classes):

- **UBR (unspecified bit rate).** This is for handling best effort delivery service suited mostly for handling non-real time data services such as email, Usenet, file transfer, LAN inter-networking, etc., and does not guarantee their delivery (a higher layer protocol such TCP would be needed to ensure end-to-end delivery of the service). UBR circuits do not guarantee any amount of bandwidth, and are normally policed by the network to not exceed a specified maximum bit rate.
- **ABR (available bit rate).** This is similar to UBR except that the network does guarantee a specified minimum bit rate for the service, and hence achieves better performance in terms of a more stable throughput. ABR circuits are more suited to handling web browsing and transaction services, which are more delay constrained than other data services.
- **CBR (constant bit rate).** The CBR virtual circuits are used to provide circuit emulation services for handling highly delay critical traffic such as real time un-coded voice and video services. The CBR can also be used to emulate private line services through an ATM network. There is only one parameter required to specify a CBR circuit and that is the peak bit rate for the service.
- **VBR-rt (variable bit rate, real time).** The VBR-rt is specified by three parameters: expected peak cell rate, sustained cell rate, and maximum burst size. The network tries to minimize the end-to-end delay as well as the delay variations from cell to cell. The VBR-rt virtual circuits are suited to handling compressed voice and video services, which are delay critical.
- **VBR-nrt (variable bit rate, not real time).** The VBR-nrt is specified by the same parameters as the VBR-rt, but is suited to transporting bursty data services that are less delay critical such as banking transactions, airline reservations, process monitoring, and frame relay inter-working.

The ATM Operations, Administration, and Maintenance (OAM) protocol offers standardized, in-service tools for network monitoring, including real time measurement of the quality of ATM connections (block-level QoS parameters). The ATM QoS measurement parameters normally reported are the following:

- CER – Cell Error Ratio
- SCEBR – Severely Errored Cell Block Ratio
- CLR – Cell Loss Ratio
- CTD – Cell Transfer Delay
- CDV – Cell Delay Variation
- CMR – Cell Misinsertion Rate

15.12 More on QoS Mechanisms in IP Networks

Currently, the public Internet provides virtually no QoS. However, there are QoS mechanisms that have recently been implemented by the routing devices, which can be taken to advantage in providing certain measures of QoS in private IP networks.

There are two approaches to achieving QoS in IP networks. The first is referred to as the IETF's differentiated services (DiffServ). This model combines flows with the same service requirements into a single aggregated flow, which then receives levels of service relative to other traffic flows. The DiffServ uses the Type of Service (TOS) field of IPv4 or the Traffic Class field of IPv6 to prioritize packets with different QoS requirements. This approach is also classified as a packet-oriented classification mechanism for different QoS classes. This solution helps to decrease the end-to-end delay and jitter due to offering a 'better' *best effort* service. However, since packets are classified individually based on their priorities, there is no chance to negotiate end-to-end QoS. Currently, three significant bits within the TOS field define eight IP Precedence states. The Internet Engineering Task Force (IETF) is redefining the (TOS) byte used in IPv4. The IETF DiffServ working group has proposed expanding the number of bits to six, bringing the available states to 64. These six bits are called the Differentiated Service Code Points (DSCP). It has also been proposed that the last two bits in the TOS byte be used for flow control.

The mapping from UMTS QoS classes to Diffserv codepoints will be controlled by the operator. The mapping depends on bandwidth and provisioning of resources among the different Diffserv classes, which the operators control to satisfy their cost and performance requirements. Interoperability between operators will be based on the use of service level agreements (SLAs), which are an integral part of the Diffserv Architecture.

The second approach, usually referred to as the IntServ model, tries to propagate the QoS requirements of a media stream (flow) to the network by means of a resource reservation protocol such as the Resource Reservation Protocol (RSVP). RSVP, a proposed IETF standard, allows Internet applications to request reservation of resources along the path of a data flow so that applications can obtain predictable quality of service on end-to-end basis. Intervening nodes put the necessary resources (i.e., bandwidth, processing time) aside to guarantee packets of flow with the requested QoS. This approach can be classified as a *flow-based resource reservation* mechanism where packets are classified and scheduled according to their flow affiliation. Resource reservation has the potential for end-to-end QoS due to the concept of flows: a flow originating from a defined endpoint and destined for another. Thus, this approach is preferable to the former solution to real time streaming due to its awareness of end-to-end service characteristics.

The two solutions are not interchangeable because of the IntServ's serious scalability problems (requires per flow state maintenance within the network nodes), but on the other hand, DiffServ has administration problems in the allocation and accounting of the resources needed by applications.

15.13 IP Precedence to ATM Class of Service Mapping

For hybrid IP+ATM network designs that involve an ATM core or traverse an ATM service, it may be desirable to map classified IP precedence or DSCP values to an ATM COS. For example, to ensure end-to-end QoS from an Ethernet LAN through an ATM core, you may want to map as follows: map all traffic with IP precedence level four to a Virtual Circuit Connection (VCC) provisioned as Unspecified Bit Rate (UBR); map packets with IP precedence level six to a VCC provisioned as Available Bit Rate (ABR); and map packets with IP precedence level seven to Constant Bit Rate (CBR).

15.14 Web Traffic Classification for QoS

Emerging multimedia and web-based applications require deeper packet inspection and cannot be identified by a static layer 4 port number. Multimedia traffic may require unique classification because it may have stringent delay requirements or use large amounts of bandwidth, which could impact mission-critical application performance. Yet web-based applications may include mission-critical applications, such as web-based network management, casual browsing, and non-mission-critical traffic.

Web-based applications are ever more important and ubiquitous. Because most web-based applications use the well-known HTTP port 80, classification as a function of the URL address is required. Cisco routers, for instance, have new capabilities that enable them to classify web traffic by URL address.

15.15 QoS Levels of Agreement

Different levels of agreement can be imposed or negotiated for any QoS requirement, defined as follows:

- Best efforts
- Compulsory
- Guaranteed

The weakest agreement is that all parties use their *Best Efforts* to meet the users' requirements, but understand that there is no assurance that the QoS will in fact be provided and no undertaking is given to monitor the QoS achieved or to take any remedial action should the desired QoS not be achieved in practice.

In the *Compulsory* level of agreement, the achieved QoS must be monitored by the provider and the service aborted if it degrades below the compulsory level; however, the desired QoS is not guaranteed, and indeed it may be deliberately degraded and the service aborted, for example to enable a higher-precedence demand for service to be satisfied.

In the *Guaranteed* level of agreement, the desired QoS must be guaranteed so that the requested level will be met, barring 'rare' events such as equipment failure. This implies that the service will not be initiated unless it can be maintained within the specified limits.

References

[1] 3GPP TS 24.008, 'Mobile Radio Interface Layer 3 Specification; Core Network Protocols'.
[2] 3GPP TS 23.107 version 640, 'Quality of Service, Concept and Architecture'.
[3] 3GPP TS 23.207, 'End-to-end Quality of Service Concept and Architecture'.
[4] 3GPP TS 23.101, 'General UMTS Architecture'.
[5] Dixit, S., Guo, Y. and Antoniou, Z., 'Resource Management and Quality of Service in Third Generation Wireless Networks', *IEEE Communications*, February 2001, **39**(2), 125–133.
[6] Jorguseski, L., Fledderus, E., Farserotu, J. and Prasad, R., 'Radio Resource Allocation in the Third-generation Mobile Communication Systems', *IEEE Communications*, February 2001, 117–123.
[7] 'Traffic Management Specifications, version 4.9', ATM Forum, February 1996.
[8] RFC 2543, 'Session Initiation Protocol (SIP)', IETF, March 1999.
[9] 3GPP TS 29.208, 'End-to-end Quality of Service (QOS) Signaling Flows'.
[10] 3GPP TS 25.413, 'UTRAN Iu Interface RANAP Signaling'.
[11] 3GPP TS 25.331, 'Radio Resource Control (RRC) Protocol Specifications'.

[12] 3GPP TS 24.433, 'UTRAN Iub Interface NBAP Signaling'.
[13] 3GPP TS 29.060, 'GPRS Tunneling Protocol (GTP) across the Gn and Gp Interface'.
[14] RFC 2475, 'An Architecture for Differentiated Services', IETF, December 1998.
[15] RFC 2205, 'Resource Reservation Protocol (RSVP)', IETF, September 1997.
[16] Ibe, O.C., *Essential of ATM Networks and Services*, Addison-Wesley, Reading MA, 1997.

16

The TCP Protocols, Issues, and Performance Tuning over Wireless Links

With a main feature of 3G networks being the provisioning of high-speed non-real time services such as web services and general information transfer, the existence of well tuned protocols for the efficient reliable end-to-end transfer of data becomes very critical. Transmission control protocol (TCP) is an end-to-end transport protocol, which is commonly used by network applications that require reliable guaranteed delivery. The most typical network applications that use TCP are File Transfer Protocol (FTP), web browsing, and TELNET. TCP is a sliding window protocol with time out and retransmits. Outgoing data must be acknowledged by the far-end TCP. Acknowledgements can be piggybacked on data. Both receiving ends can flow control the far end, thus preventing a buffer overrun. As is the case with all sliding window protocols, TCP has a window size. The window size determines the amount of data that can be transmitted before an acknowledgement is required. For TCP, this amount is not a number of TCP segments but a number of bytes. However, the reliable TCP protocols such as the TCP-Reno type are tuned to perform well in traditional fixed networks where transmission errors are relatively very low and packet losses are assumed to occur mostly because of congestion. TCP responds to all packet losses by invoking congestion control and avoidance. Therefore, it results in degraded network throughput performance and delays in networks and on links where packet losses mainly occur due to bit errors, signal fades, and handovers rather than congestion as in the mobile communication environment.

A number of different strategies and protocol variations have been proposed to provide efficient reliable transport protocol functionality on wireless networks. Some of these schemes rely on specialized link layer protocols, specified by the 3GPP standards, which enhance the reliability of the wireless link. These are called link layer solutions, which when combined with certain options or versions of available TCP implementations, help greatly to address and resolve the wireless problems, as we will discuss in detail. Other schemes require special measures to be taken such as splitting the TCP connection in the base station, providing sniffing proxies, or implementing non-standard modifications to the TCP stack within the end systems (both the fixed and the mobile hosts). We will also discuss the latter category of solutions and their merits in order to provide the reader with a broad knowledge base and insight for tackling specific problems and issues in this particular field as they arise.

16.1 The TCP Fundamentals

The TCP was adopted in 1983 as the official Internet transport level protocol for the reliable end-to-end transfer of non-real time information, including web browsing, FTP, and email. This protocol has been originally developed to optimize its performance over fast and fairly reliable wired networks. The TCP provides a byte streamed transmission. The TCP sender divides the data into segments, which are called packets at the IP layer. The segments are limited to a maximum segment size (MSS) (excluding the TCP Header and options field), which can be managed by the TCP sender and receivers, and are communicated in the SYN messages at connection set-up. The segments are sequentially numbered and transmitted. The MSS value must fit into the IP MTU (maximum transmission unit size) in order to avoid fragmentation at the TCP layer within network routers. Therefore, the choice of an appropriate MSS depends on the MTU used at the IP layer. In practice, to find the optimum MSS the MTU Discovery procedure (RFC 1191), which allows the sender to check the maximum routable packet size in the network, can be used. Then the MSS is set as a parameter. However, the MSS may interact with some TCP mechanisms. For example, since the congestion window is counted in units of segments, large MSS values allow the TCP congestion window to increase faster.

TCP operates in acknowledged mode; that is, the segments that are received error free are acknowledged by the receiver. The TCP receiving end uses cumulative acknowledgement messages in the sense that one single ACK message may acknowledge the receipt of more than one segment, indicated by the sequence number of the last in-sequence error free received packet. However, the receiver doesn't have to wait for its receiver buffer to fill before sending an ACK. In fact, many TCP implementations send an ACK for every two segments that are received. This is because when a segment is received, the TCP starts a delayed ACK, waiting for a segment to send with which the ACK can be piggybacked. If another segment arrives before the 200 ms tick (TCP has a timer that fires every 200 ms, starting with the time the host TCP was up and running), the sender will send the second ACK. If the tick occurs before another segment is received or the receiver has no data of its own to send, the ACK is sent. Packets that are not acknowledged in due time are retransmitted until they are properly received.

TCP is a connection oriented protocol meaning that the two applications using TCP must establish a TCP connection with each other before they can exchange data. It is a full duplex protocol, in that each TCP connection supports a pair of byte streams, one flowing in each direction. The acknowledgment of TCP segments for one direction is piggybacked to transmission in the reverse direction or sent alone when there are no reverse data transmissions. TCP includes a flow-control mechanism for each of these byte streams, which allow the receiver to limit how much data the sender can transmit. TCP also implements a congestion-control mechanism to avoid congestion within the network nodes.

16.1.1 TCP Connection Set Up and Termination

The TCP sender and receiver must establish a connection before exchanging data segments as already mentioned. This involves a three-way hand shaking process with the following steps:

1. The client end system sends TCP SYN control segments to the server and specifies the initial sequence number.
2. The server end system receives the SYN, and replies with the SYN ACK control segment.
   1. In the meantime it allocates a buffer, specifies a server, and receives the initial sequence number.
3. The client also allocates buffers and variables to the connection. The server sends another SYN segment, which signals that the connection is established.

Connection Termination Phase

1. The client end system sends TCP FIN control segment to server.
2. The server receives FIN, and replies with ACK. If it has no data to transmit, it sends a FIN segment back with the ACK. Otherwise, it keeps sending data, which is then acknowledged by the receiving end (the client end). Then after it finishes, it sends the FIN segment.
3. The client receives FIN, and replies with ACK. It then enters time out 'wait' before it closes the connection from its end.
4. The server receives the ACK and closes the connection.

16.1.2 Congestion and Flow Control

Congestion and flow control are performed by the use of a variable size congestion window (CWND) and a receiver advertised window, respectively [1]. The receiver advertised window is dependent on the amount of buffer space that the receiver can or decides to allocate for the connection. The maximum number of data segments that the TCP sender can at any time inject into the network with outstanding acknowledgements is determined by the minimum of the receiver advertised window and the sender's congestion control window at the time. The congestion control window size is dynamically changed by the TCP sender to limit the transmission rate into the network based on indirect detection of congestion within the network. The TCP sender uses two mechanisms for the indirect detection of congestion within the network: the time outs for ACKs from the receiving end based on a retransmission time out timer (RTO); and the receipt of duplicate ACKs (DUACKs) from the receiving end. The receiver transmits duplicate ACKs for every segment that is received out of order. The DUACKS are not delayed and are transmitted immediately. The sequence number within a duplicate ACK refers to the sequence number of the last in-sequence error free received packet. The RTO is based on a jitter compensated smoothed version of the round trip time (RTT) measurement samples. A RTT sample is the time from when the sender transmits a segment to the time when it receives the ACK for the segment.

The TCP congestion control starts with probing for usable bandwidth when the data transmission starts following connection set up. This is called the slow start congestion control phase, in which the congestion window (CNWD) starts from a small initial value and grows rapidly with an exponential rate. Then, when congestion is detected, the CWND is restarted from a lower value and is incremented only linearly. This phase is called congestion avoidance. The boundary between the initial low start and the congestion avoidance phases are controlled by the two important variables of CNWD and a slow start threshold (SS-Thresh). The TCP sender thus uses the slow start and congestion avoidance procedures to control the amount of outstanding data that are being injected into the network. The two phases of slow start and congestion avoidance are discussed in the following two sections.

16.1.2.1 Slow Start Congestion Control Phase

The slow start phase is also known as a probing phase in which the TCP sender tries to probe for the available network bandwidth at initial start up or after a loss recovery phase. The TCP sender starts by transmitting one segment and waiting for its ACK. When that ACK is received, the congestion window is incremented from one to two, and two segments can be sent. When each of those two segments is acknowledged, the congestion window is increased to four. And then up to the minimum of the updated congestion window and the receiver advertised window can be sent, and so on. This provides an exponential growth in the transmission rate, although it is not exactly exponential because the receiver may delay its ACKs, typically sending one ACK for every two segments that it receives.

This process continues until either the CWND window reaches the initial value set in the slow start threshold parameter (SS_Thresh) or when congestion is observed. TCP interprets packet losses as a sign of congestion. The SS-Thresh is initially set to an arbitrarily high value (equivalent to say 65535 bytes) as it will adjust itself in response to congestion as will be discussed in the next section. Congestion causes packet losses and occurs when the data packets arrive through a large pipe and are queued for transmission over a smaller pipe, or when multiple input streams arrive at a router whose output capacity is less than the sum of the inputs.

Generally, all TCP versions use either retransmission time outs or the receipt of DUACKs, indicating the receipt of an out-of-sequence received packet, as a sign of packet loss or delay due to congestion. The exact way in which time out events or DUACKs are used for this purpose depends on the particular TCP version implemented, as will be discussed later in the chapter. However, once the TCP sender determines that packets are getting lost, it starts to perform congestion avoidance measures.

16.1.2.2 Congestion Avoidance Phase

When congestion occurs, TCP must slow down its transmission rate of packets into the network. The TCP sender slows down its transmission rate by entering either the congestion avoidance phase or restarting the slow start phase. If congestion was detected by the time out event, it restarts the slow start phase to begin probing the network bandwidth again. The CWND is reset to one segment, but this time the SS-Thresh is set to one half of the minimum of the CWND (the receiver advertised window. This continues until either the CWND reaches the newly updated SS_Thresh, or another time out event occurs. The important variables that define the boundary between the slow start phase and the congestion avoidance phases are the SS_Thresh and the CWND.

The congestion avoidance phase is entered if either the CWND exceeds the SS-Thresh or a certain number of DUACKs, depending on the TCP version implemented, are received. In the congestion avoidance phase, the TCP sender increments the CWND by one segment for each ACK received, resulting in a linear growth of the transmission window. Note that the ACKs in TCP can be cumulative in the sense that one ACK from the receiver may acknowledge the proper receipt of more than one data segment. In the congestion avoidance phase, a cumulative ACK causes the CWND window to increment by only one regardless of how many data segments are acknowledged by that ACK. This is in contrast to the slow start phase in which a cumulative ACK will cause the CWND to increment by the number of the data segments acknowledged by the single ACK message.

16.1.2.3 TCP Congestion Algorithm Bottlenecks in Wireless Networks

Since the TCP protocol was originally designed and developed for the fixed networks with very small error rates (packet losses much below 1%), all packets that are not acknowledged by the receiving end in due time are interpreted by the sending end to be due to congestion and drop outs within the network, as opposed to packet corruptions due to bit errors. Therefore the original TCP runs into a number of performance problems when it is implemented on networks that involve wireless links. The wireless networks incur considerably higher packet corruptions (which result in packet losses) due to burst errors caused by channel fades and packet losses that occur in the mobility handling processes such as cell reselections and hard handovers. For that reason, a number of different measures and mechanisms have been developed for adopting the use of TCP in wireless networks, which we will discuss in detail later in this chapter (Section 16.4).

16.1.3 TCP RTO Estimation

The accurate estimation of the retransmission time out parameter (RTO) is very important because it helps to avoid spurious retransmissions and the activation of congestion avoidance mechanisms. The RTO estimation is based on processing and filtering the network round trip time (RTT) samples obtained by the TCP sender. The RTT is the time elapsed from the instant a segment is transmitted to the reception of an ACK for that segment. The TCP sender measures the RTT samples by the time stamping technique: The sender places a timestamp within the data segment. Upon receiving the segment, the receiver puts the timestamp on a timestamp echo field within the ACK message. Using these two values, the sender is able to calculate the RTT properly. This requires that the two communicating ends be time synchronized or be aware of each other's time.

The RTT can change from time to time even for a given connection due to the changing traffic and condition in the network. Thus, the RTO estimation should be adaptive. Various methods have been developed and evolved for the optimum estimation of the RTO [2, 3]. However, all methods are based on some filtering and jitter compensated version of the sample RTTs measured by the TCP sender. However, the RTO estimation procedures are now standardized [4], with the current procedure in use being due to a method proposed by Jacobson [5]. This procedure uses the sample RTT measurements obtained by the TCP sender to calculate an exponentially smoothed average of the sample RTTs (SRTT) and adjust the final estimation by a deviation correction to account for rapid jitters in the RTT. The algorithm is given below:

$$\text{SRTT} = (1 - g) \times \text{SRTT} + g \times \text{sample RTT} \tag{1}$$

mean Variance:

$$\text{RTT}_{VAR} = (1 - h) \times \text{RTT}_{VAR} + h \times |\text{sample RTT} - \text{SRTT}| \tag{2}$$
$$\text{RTO} = \text{SRTT} + 4\text{RTT}_{VAR} \tag{3}$$

Recommended values for the coefficients g and h are:

$$g = 1/8 = 0.125, h = \tfrac{1}{4} = 0.25$$

with the initial RTO value set to 3 seconds.

The time out values estimated from Equation (3) are converted into multiples of a timer granularity, dependent on the implementation. In typical implementations, the timer granularity is from 200 to 500 ms.

When a segment is retransmitted upon the timer expiration (upon RTO expiration), the new RTO is initialized to double its value at the time. This results in an exponential timer backoff, and is bounded by a maximum value of at least 60 seconds (RFC 2988).

The RTT samples are not made on segments that were retransmitted (and thus for which it is ambiguous whether the reply is for the first instance of the packet or a later instance). The problem with this solution is that if the RTT suddenly increases, all packets will be retransmitted before the ACKs arrive, as a result of using a lower outdated RTO.

The algorithm suggested to solve the retransmission caused ambiguity problem in the RTT estimation is known as Karn's Algorithm [6]. When an ACK arrives for a packet that has been retransmitted at least once, any RTT measurement based on this packet is ignored. Additionally, the backed-off RTO for this packet is kept for the next packet. Only when it (or some other future packet) is ACKed without retransmission is the RTO recalculated from the SRTT. The convergence of the SRTT to new RTT values depends on the exponential backoff mechanism and the SRTT filtering used, but it has been shown that it typically converges quite fast [6].

16.1.4 Bandwidth-Delay Product

The bandwidth delay product (BDP) obtained by the product of the network round trip time and the network transmission bit rate is a measure of the network 'data holding in transit capacity'. This means that without any additional buffering within the network nodes, the link between the transmitter and receiver will not be able to hold more than the bandwidth-delay product, regardless of the CWND and the advertised window. Therefore, the bandwidth-delay product sets a limit to the TCPs maximum window size (for both the congestion and the receiver windows). When the window size is set to the bandwidth-delay product, there will be a constant flow of data from the sender to the receiver, and the link is utilized effectively. However, since the TCP protocol is not aware of the BDP, appropriate measures such as providing extra buffers within the network (beyond what is provided by the

BDP) should be taken to avoid packet losses and the consequences when the receiver window is larger than the BDP.

In networks with a very high bandwidth-delay product, also known as LFN (Long Fat Networks), such as the geo-stationary satellite links, care must be taken in the calculation of the retransmission time out (RTO). This is because in such networks, the window size is very large and the standard RTT measurements don't measure the RTT of every segment, just a window at a time [7].

On the other hand, in networks with a very small bandwidth-delay product such as 1, the effect of the slow start phase is null because the BDP is fulfilled by the initial value of the congestion control window, which is set to 1.

16.2 TCP Enhanced Lost Recovery Options

TCP implements a few options that can be used in the speedy recovery from lost packets. There are the fast retransmit, fast recovery, and selective acknowledgements, all of which were developed for improving performance on fixed networks. However, some of these (as we will see later) happen to be useful features for handling some of the packet loss characteristics of wireless links.

16.2.1 Fast Retransmit

In the fast retransmit mechanism, three duplicate acknowledgements carrying the same sequence number trigger a retransmission without waiting for the associated time out event to occur. The window adjustment strategy for this early time out is the same as for the regular time out and Slow Start is applied.

When an out-of-sequence data segment is received, the TCP receiver sends an immediate DUACK acknowledging the last received error free in-sequence packet. The purpose of this immediate DUACK is to let the other end know that a segment was received out of order, and to tell it what sequence number is expected next. If three or more DUACKs are received in a row for the same segment sequence number, it is taken as a strong indication that a segment has been lost. Then TCP performs a retransmission of the missing segment, without waiting for the retransmission timer (RTO) to time out. If only one or two DUACKs are received in a row, it is interpreted as an indication that segments may simply be reordered in the transmission process, and TCP waits before retransmitting the indicated segment expected.

16.2.2 Fast Recovery

Fast Recovery pertains to starting the congestion avoidance phase as opposed to re-starting the slow start phase following a fast retransmit. The reason for not performing slow start in this case is that the receipt of three DUACKs are taken to mean that, although a segment is missing, data is flowing. The receiver can only generate the DUACKs when another segment is received, implying that there is still data flowing between the two ends and no need to reduce the flow abruptly by re-starting the slow start phase. Then, when a new data segment is ACKed, the congestion window is halved and transmission continues in the congestion avoidance phase. The fast retransmit and fast recovery options are usually implemented together [8], which helps to achieve better utilization of the connection available bandwidth.

16.2.3 Selective Acknowledgement (SACK)

The TCP Selective Acknowledgement (SACK) is a TCP option that allows the receiving end to specify precisely which data segments are received and which are not, in the presence of multiple non-contiguous data segment losses. It uses a bit vector that specifies lost packets. TCP SACK is an Internet Engineering Task Force (IETF) proposed standard (RFC 2018, Oct 1996), which is implemented for most major operating systems. SACK allows the TCP sender to recover from multiple-packet losses faster and more efficiently. This option enables quicker recovery from multiple packet losses particularly when the load is not light, and the end-to-end delays are large such as on satellite links.

16.2.4 The Timestamp Option

Traditionally the TCP implementations time only one segment per window resulting in one RTT measurement at a time (per round trip time). This yields an adequate approximation to the RTT for small windows (e.g., a 4–8 segment window), but it results in an unacceptably poor RTT estimate for long delay links (LFNs) (e.g., 100 segment windows or more).

When using the timestamp option, each data segment is time stamped and used in more accurately estimating the RTO. RFC 1323 suggests that TCP connections utilizing large congestion windows should take many RTT samples per window of data to avoid aliasing effects in the estimated RTT.

16.3 TCP Variations as used on Fixed Networks

There are a variety of versions of TCP implementation with respect to congestion control mechanisms, their triggering, and recovery from packet losses. Though these different versions were all developed and evolved for improving performance on fixed wired networks, some have features that make them more suitable for implementation on wireless networks. Some can be complemented with other measures to make them more suitable for wireless networks. These are discussed in the following sections first before we focus on the peculiarities and the requirements of the wireless networks, and particularly UMTS. A comparative discussion of the different versions of TCP are provided in Reference 9.

16.3.1 TCP Tahoe

TCP Tahoe is the oldest and most famous version of TCP. In the literature it is sometimes just called TCP. The TCP Tahoe congestion algorithm includes Slow Start and Congestion Avoidance. TCP Tahoe uses three retransmission time out events to determine that a packet has been lost, in which case it performs the congestion avoidance measures. This is its main drawback [10], because when a segment is lost it would take a long time for the sender side to decide so, and take the appropriate measures.

16.3.2 TCP Reno

TCP Reno includes the Slow Start and Congestion Avoidance algorithms, and also implements Fast Retransmit. In the Fast Retransmit mechanism, three duplicate acknowledgements carrying the same sequence number trigger a retransmission without waiting for the associated time out event to occur. TCP Reno uses the same window adjustment strategy for this early time out as was discussed earlier for the regular time out and Slow Start phases.

16.3.3 TCP New Reno

TCP New Reno introduces Fast Recovery in conjunction with Fast Retransmit as discussed in Section 16.2.2 [1].

16.3.4 TCP SACK

TCP SACK implements the Selective Acknowledgement option discussed in Section 16.2.3.

16.4 Characteristics of Wireless Networks and Particularly 3G

The inherently different characteristics of the wireless networks cause them to experience different performance impacts by the TCP parameters and mechanisms. The TCP retransmission mechanisms, for instance, are either due to a time out or triple-duplicate ACKs, which are caused mainly by packet loss in a wired environment. Adding a wireless mobile portion to this setting with its significant error figures will inevitably trigger TCP congestion control and retransmission backoff mechanisms against error-caused packet losses, and not the congestion caused.

This section discusses and highlights those characteristics of the wireless networks that are considered highly relevant in impacting the TCP performance. These characteristics are then used in subsequent sections as the framework to discuss the complementary measures, as well as the modifications to the TCP protocol that have been proposed for implementing the protocol on wireless networks.

16.4.1 BLER, Delays, and Delay Variations

High block error rates, compared to what TCP was designed for, occur on the radio interface side of wireless networks due to channel fades, and mobility such as hard handovers and cell reselections. The radio interface uses powerful error correction codes and interleaving to combat the burst errors and reduce the error rates. But uncorrectable errors still occur, which can result in significant frame errors over the link. The uncorrectable frame errors are detected by errors detection codes, and that results on average in a BLER (Block Error Rate). Then if the frame losses due to BLER are not locally corrected, they are interpreted by the end-to-end TCP protocol as a sign of congestion in the network, and trigger congestion control procedures that will reduce the CWND and reset retransmission timers. This will slow down the packet flow rate into the network and result in high delay variations and underutilization of the allocated bandwidth. The reduction in the congestion window is a necessity to avoid a collapse from congestion when the network is experiencing congestion, but it is not required if packet losses occur due to corruption of data on the wireless link. This unnecessary reduction in the congestion window for corruption losses is the main reason that poor TCP performance results on wireless links.

The link level retransmission protocols known as RLC (Radio Link Control) are implemented to handle the corrupted frames through local retransmissions over the radio channel. The link level retransmissions hide some of the data losses from the TCP, but can result in increased transmission delays with significant delay variances. The latter occurs as a result of the varying degree of retransmissions required due to the dynamically changing radio channel. The retransmissions further add to the delays involved in the complex error coding and interleaving performed to combat errors. In Reference 11, measurements show latencies of the order of 179 ms to over 1 s for code-division multiple-access (CDMA)-based 3G1X networks. However, the RTT times of 3G networks are considerably shorter than in GPRS. For instance, measurements reported [12, 13] indicate that only 10% of the RTT measurements collected in UMTS exceeded 1 s, whereas in the case of GPRS 70% of the RTT measurements exceeded 1 s.

Packets can also be lost in the process of hard handovers and cell reselections such as in GPRS and UMTS. In cell reselections, packets buffered at the base station for a given cell may be completely flushed out before they are transmitted over the radio link once a cell reselection or hard handover to a cell under a different base station takes place. This further adds to packet loss rates and the end-to-end transmission delays, which are perceived by the TCP as being due to congestion. In the hard handover or cell reselection processes, the connection is paused for a certain time, allowing several retransmissions of the same segment to get lost during the pause. Due to the exponential backoff mechanism of TCP, the time interval between retransmissions is doubled each time until it reaches a significant limit, which can drastically degrade the performance when the mobile resumes its connection.

16.4.2 Delay Spikes

A delay spike is a sudden increase in the latency of the link. Delay spikes in 3G networks can occur due to a number of mechanisms including the following: momentary link outages due to loss of radio coverage such as caused by blockage or passage through a tunnel; hard inter-system and inter-carrier handovers; blockage by higher priority traffic in the packet scheduling processes such as on the HS-DSCH; and channel switching in the RRC (radio resource control) state transitions for providing bandwidth on demand to the users in real time, which may take up to 500 ms. Delay spikes also result from cell reselections in UMTS and GPRS, which can result in degraded performance even when no packets are lost (no flush-outs). Voice call preemptions in GPRS is another source of delay spikes. The voice calls can take time slots away from an existing data call thus reducing the available bandwidth to the data call. The RTO estimation algorithms do not capture the sudden sharp increase in RTT delays caused by delay spikes. This causes the TCP sender to time out and perform spurious (unnecessary) retransmissions and multiplicative reductions of the transmission window, resulting in reduced utilization of the allocated bandwidth [12, 14].

16.4.3 Dynamic Variable Bit Rate

The data bit rates experienced by users in 3G networks are highly dynamic and not steady, as would more or less be the case in fixed networks. The user mobility and changing locations within the cell and between cells can result in different bit rates experienced by the user. So would the changing traffic and interference within the cell due to the dynamics of other users impacting the bit rates experienced by a given user over the data session. The real time changing data rates can result in large delay variations and delay spikes. The delay spikes then cause spurious time outs in TCP where the TCP sender incorrectly assumes that a packet is lost while the packet is only delayed. This can then unnecessarily force the TCP into slow start, adversely impacting the protocol performance.

16.4.4 Asymmetry

Asymmetry with respect to TCP performance occurs if the achievable throughput is not solely a function of the link and traffic characteristics of the forward direction, but also significantly depends on those of the reverse direction [15]. The asymmetry can affect the performance of TCP because TCP relies on feedback through ACKs messages. Asymmetry in the network bandwidth can result in variability in the ACK feedback returning to the sender. For ACKs on the reverse path, a very simplified example is given to show the effect of asymmetry on TCP performance. Assume that we are downloading a file using a link with 384 kbps in the downlink and 8 kbps in the uplink. Furthermore, assume that 1500 byte segments are used in the downlink while the only traffic in the uplink is 40 byte ACKs. The time required for transmission of one data packet is $1500 \times 8/384$ ms; that is 31.25 ms, whereas the time required for one ACK is $40 \times 8/8 = 40$ ms. Since TCP is ACK-clocked, that means at steady state it cannot send more ACKs than it can receive. Therefore, the effective data rate of the link is reduced from 384 kbps to 300 kbps ($384 \times 31.25/40$) [14].

Generally, a normalized bandwidth ratio defined as the ratio of the transmission time required for ACK packets to that of data packets [15], can be used to quantify the asymmetry measure. Then, when this ratio is greater one, the asymmetry will reduce the effective bandwidth available to TCP. A compromise can be made by using delayed ACKs, or another cumulative ACK scheme, in order to stop the reverse link from being the bottleneck. But this will slow down the TCP's slow start phase further.

16.5 TCP Solutions Proposed for Wireless Networks

Several proposals have been made by researchers to improve the performance of TCP over wireless networks, some of which are already in use on networks [16]. Some of these solutions include features that originated from the attempt to further improve the TCP performance on wired networks, as stated in earlier sections. Later, it turned out that some of these provided useful features for implementation on wireless networks. We will discuss here the solutions and their optimization that are specifically targeted for wireless networks. We will also discuss some of their reported performances in GPRS and UMTS. These solutions are based on optimizing parameters and implementing features in lower layer protocols that impact TCP performance as well as making intelligent use of features provided in various TCP versions and optimizing the relevant parameters (link layer solutions), inclusion of intelligent performance enhancement elements such as proxies inside the network nodes (proxies), split connection solutions, and making modifications to TCP end systems (TCP end-to-end solutions). We will discuss each of these categories, present examples, and reported performance results.

16.5.1 Link Layer Solutions

The link layer solutions help to hide the lossy characteristics of the radio link between the mobile and the network from the TCP protocol. They achieve this basically through local retransmission of packets that experience uncorrectable errors over the radio links. A variety of link layer solutions have been proposed in the literature for use on wireless links. However, the solutions implemented by GPRS/EDGE and UMTS are based on specifications of RLC protocols that attempt to make the radio link more reliable for TCP data and signaling transmissions. The performance achieved with the RLC will depend on the application and how it is configured and used. We will focus on discussing the UMTS RLC, the related parameters, and issues, and how its configuration can impact the performance achieved. Then we will discuss how certain critical TCP parameters and options can be tuned and selected to provide the best performance in conjunction with the functionality provided by the link layer solutions used in GPRS/EDGE, and UMTS.

16.5.1.1 The RLC Solution for TCP Connections

Both UMTS and GPRS/EDGE implement local retransmission of packets that are corrupted over the radio channel. This helps to mitigate or to reduce the TCP end-to-end retransmissions. Reducing the TCP retransmissions triggered by packet losses helps to prevent TCP congestion control and retransmission timer backoff mechanisms that result in degraded throughput and increased delays. The local retransmissions of corrupted packets are performed by the RLC protocol in acknowledged mode [17]. RLC implements powerful forward error correction coding (FEC) using convolutional and Turbo codes (in UMTS), along with interleaving to randomize and correct the burst errors over the radio channel. The packets left with uncorrectable errors are detected through error correction detection codes such as CRC (Cyclic Redundancy Check) bits and are locally retransmitted over the link between the mobile and the base station when the acknowledged mode (AM) of the RLC protocol is used. With RLC retransmissions, it is possible to reduce the residual packet error rates down to 1% on average [17], thus minimizing the impact of loss to TCP. When the packet loss rates over the link stay below 10%, TCP throughput suffers about 10% (TCP Reno or Vegas). The degradation in the TCP throughput becomes extremely high when the packet losses exceed 10% [18].

RLC uses the ARQ (Automatic Retransmission Request) mechanisms to request the retransmission of corrupted packets due to uncorrectable errors. Since the link layer solutions, such as provided by RLC, are TCP unaware, it is important to configure the RLC to provide as much of the reliability required for TCP operation, and prevent inefficient interactions between the two protocols. The optimal performance is achieved when the bandwidth provided by the RLC, which can vary over time due to the RLC retransmissions, is fully utilized by TCP. No optimal utilization results when the TCP sender leaves the link idle or retransmits the same data that RLC is already retransmitting, RLC tuning for TCP performance optimization requires consideration of a number of parameters and measures, which are discussed in the following sections.

The Interleaving Depth
In UMTS specifications, different configurations can be chosen for a given required throughput so that an RLC PDU (packet data unit) can fit in 1 to up to 8 frames of 10 ms. Likewise, a given bit rate can be obtained through one or several channelization codes with multi-code transmissions. Depending on what configuration can be chosen within the vendor's implementation, various interleaving depths can be achieved. The interleaving

depth has a proportional impact on the data transfer delays. Larger interleaving depths result in larger delays but in more randomization of burst errors caused by radio channel fades. More randomization of burst errors helps more of the errors to get corrected by the error correcting codes. However, for the same average packet error rates (errors left at the decoder output), TCP favors the bursty over the uniform errors. A smaller interleaving depth in the case of the more bursty residue errors results in reduced end-to-end data transmission delays, as confirmed by simulations [10], and these reduced end-to-end delays can help to reduce the number of TCP packet retransmission time outs, which would trigger unnecessary congestion control and retransmission timer backoffs. Therefore, choosing the optimum interleaving configuration for a given throughput and the achievable error rates is one area of experimentation within the vendor's implementation flexibility. The tradeoffs are the reduced delays achieved with shorter interleaving depths and the achievable packet error rates in each scenario.

RLC Retransmission Timer
The retransmission timer controls the ARQ retransmission time out for retransmitting of lost RLC blocks in acknowledged mode. This timer determines the number of radio frames waited before a dropped block is retransmitted, and impacts the retransmission delay. The retransmission delays at the RLC level in turn are reflected in the end-to-end transfer delays seen by the TCP, which can result in the triggering of congestion avoidance measures and loss of throughput. Therefore, it is important to pay attention to the retransmission time outs values that are configured for the RLC, and make sure that it is configured to reflect properly the expected delay on the radio link under normal conditions.

Number of RLC Retransmissions
In theory, with an infinite number of retransmissions configured, the RLC protocol will provide an error free packet service to the upper layer. However, in practice the maximum number of RLC retransmissions is configurable up to a maximum of 40, with a value of 5 usually used. The more transmissions that are permitted for a single packet, the more reliable the channel will look to the TCP. The retransmissions end up consuming part of the link bandwidth. Therefore TCP will see the wireless link as a reliable link with a smaller bandwidth and a higher delay (the delays caused by the RLC retransmissions). The effective bandwidth seen by the TCP will reduce with increased packet error rates and hence the link level retransmissions. However, the RLC local retransmission of corrupted packets results in huge gains in the TCP Goodput, defined as the ratio of TCP data segments (excluding TCP Headers, which are not useful data) correctly received to the bandwidth of the wireless link. This has been verified by the simulation results shown in Figure 16.1 [10]. The results are shown for different values of the retransmission parameter, MAXDAT, in the RLC. The case in which MAXDAT $= 0$ corresponds to the pure TCP end-to-end retransmission.

The huge gains achieved in the throughput with RLC retransmissions is due to the fact that retransmitting packet at the TCP level has much more long-term harmful consequences for the throughput than for the retransmission itself. The missing packets arriving at the TCP level will trigger the congestion control mechanisms, which result in reducing the throughput. On the other hand, a retransmission at the RLC level results in just an additional delay other than the retransmission itself. The RLC in the AM performs the in-sequence delivery of the packets at the TCP layer, hence avoiding spurious duplicate

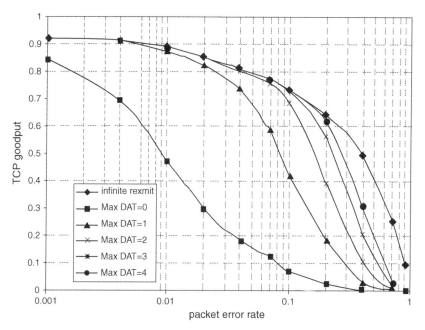

Figure 16.1 The positive influence of RLC retransmissions on TCP Goodput [10] [Reproduced from Lefevre, F. and Vivier, G., 'Optimizing UMTS Link Layer Parameters for a TCP Connection', *Proceedings IEEE Conference on Vehicular Technology* (VTC 2001), volume 4, p. 2319, © 2001 IEEE].

acknowledgements that could trigger the fast recovery procedures. The RLC retransmissions are particularly beneficial for TCP-based applications when the delays over the wireless link are small compared to the delays incurred within the fixed core network. Therefore, the number of maximum allowed link level retransmissions should be set to the highest possible value. Also, for effective retransmissions at the link layer, it is important to have the granularity of the link layer timers much finer compared to the TCP timers. Otherwise, contention between the two timers can reduce the throughput.

However, this can potentially become problematic if the wireless delays are predominant. In this case, a high value for the allowed number of retransmissions can result in high RTT and RTT variations as seen by the TCP. The high RTT variations are caused by the changing radio channel conditions, which result in different number of retransmissions until a packet is successfully transmitted over the radio link. The high variations in RTT can result in unstable RTO (time outs) and trigger TCP retransmission time outs that result in competing TCP recovery. That is, TCP starts the congestion control mechanisms and retransmission of packets that are already under retransmission by RLC. However, for low residue error rates on the radio channel, the RTT jitter due to retransmissions will be small, and hence there will be less likelihood of high competing interactions between TCP and RLC. In fact, the simulation results presented in Reference 10 confirm a high degree of robustness of the TCP time out mechanisms and a non-existence of competing retransmissions even on wireless links with high delays. This is particularly the case when the RTO estimation algorithm considers several consecutive RTT samples, as well as their variances.

The problem of competing error recovery between TCP and other reliable link layer protocols resulting from spurious time outs at the TCP sender is also investigated in Reference 19. The same conclusion is reached as in Reference 20, which performed measurements during bulk data transfers with very few time outs due to link layer delays. The measurements performed in Reference 20 indicate that most TCP time outs are due to the radio link protocol resets resulting from unsuccessful transmission of data packets under weak signal strengths; TCP was seen to provide a very conservative retransmission timer allowing for extra delays due to link layer error recovery beyond 1200 ms. This is the upper range of typical RTT measurements observed in some of the existing UMTS networks. Measurements performed on a major provider's UMTS network [21] over different times of the day have indicated that only 10% of the measured RTTs exceeded 1 s. Therefore, the number of RLC retransmissions can be set to a high value to make the link more reliable, without causing excessive link delays that would be seen by TCP.

Setting the number of RLC retransmissions to an adequately large number will also help to prevent link resets, which can have a major impact on TCP performance when TCP header compression is implemented to reduce the per segment overhead. Header compression uses differential encoding schemes, which rely on the assumption that the encoded 'deltas' are not lost or reordered on the link between the compressor and the de-compressor. Lost 'deltas' lead to generation of TCP segments with false headers at the de-compressor. Thus, once a 'delta' is lost, an entire window worth of data is lost and has to be retransmitted. This effect is made worse because as the TCP receiver will not provide feedback for erroneous TCP segments and forces the sender into time out. The RLC link resets result in a failure of the TCP/IP header decompression that in turn causes the loss of an entire window of TCP data segments. Therefore, making the link layer retransmissions persistent has multiple benefits when transmitting TCP flows. Finding a reasonable value for the link layer retransmissions is a challenge of TCP optimization over wireless links.

RLC Buffer Influence

The RLC transmission buffer stores all the packets arriving from the upper layer that have not yet been sent by the RLC. In theory, the TCP connection bandwidth is fully utilized when the transmission window is set to the bandwidth-delay product (BDP). The BDP is the minimum amount of buffer that is needed for optimum performance when there is a constant flow of data over the link. However, additional buffer space is required at the RLC link, which is the slower link in the network, to absorb the delays that result from the RLC retransmissions as well as the delay variations caused by contention and queuing within the core network nodes. The latter results in the irregular arrival of data at the RNC. The RLC buffers help to provide enough packets in store for continuous transmission and utilization of the link in times when packet arrivals are delayed from the core network. In this way, the link is not left idle and full utilization can be achieved. It is important to size the RLC buffers and manage the queuing properly to prevent overflow and packet drop outs. Packet drop outs occur when packets arrive and the buffer is full. An overflow of the RLC buffer can result in multiple packet drops in a row with disastrous impact on the TCP Goodput [10].

The RLC transmission buffers can quickly fill up when the receiver advertised window is considerably larger than the connection BDP. The TCP sends data up to the limit of the minimum between the receiver advertised window (Rxwnd) and the congestion control window (CWND) per RTT. But the congestion window is periodically increased until the

TCP receiver's advertised window is reached. The latter figure usually corresponds to the default socket buffer size (commonly 8 or 16 kbytes). When the latter overtakes the BDP, TCP injects more packets into the network per RTT than can be handled by the wireless link, since TCP is not aware of the BDP limitation of the connection.

The over-buffered RLC links result in TCP sending too many packets beyond what can be handled by the effective bandwidth of the wireless link. This excess queuing results in inflated end-to-end delays, which can cause a number of problems including unnecessary long time outs in cases of unsuccessful RLC transmission of a data packet. The in-advance accumulation of large chunks of data within the RLC buffers also has certain other side effects, for instance, when a user aborts the download of a web page in favor of another. The stale data must then first drain from the queue, which in the case of a narrow bandwidth link can take a significant time. Solutions around this problem are either to adjust the allocated RLC buffer to be consistent with the connection's bit rate or implement advanced queue management with explicit congestion notification (ECN) within the base station [22, 23]. The objective is to adapt the buffer size allocated to the connection bit rate and the worse-case RTT.

16.5.2 TCP Parameter Tuning

There are certain critical parameters in TCP that can be experimented with and tuned to provide the best performance when used on wireless links. These measures result in more effective utilization of the benefits achievable by the link layer solution based on RLC, and provide complementary benefits. They are therefore discussed here as a continuation of the link layer solution section. The TCP parameters that can be tuned are the receiver advertised window and the maximum TCP data segment size (MSS).

16.5.2.1 TCP Rxwnd tuning

The maximum number of data segments that TCP may have outstanding (unacknowledged) in the network at any time is ruled by the minimum between the receiver advertised window (Rxwnd) and the congestion control window at the time CWND). TCP sends data up to the limit of the minimum between the Rxwnd and the CWND per RTT. Therefore, it is important to configure a suitable value for the Rxwnd to guarantee that the connection bandwidth is fully utilized. The connection bandwidth is fully utilized when the transmission window is at least the size of the BDP. Therefore, a lower limit for the Rxwnd is set by the following relation:

$$\text{Rxwnd} > \text{BDP} \quad (4)$$

On the other upper side, any packets transmitted over the BDP limit in one RTT are buffered within the network nodes, and particularly at the RLC buffer. Since the wireless link is the bottleneck link in the network, buffering within the fixed part of the network is assumed to be negligible. This implies that all packets transmitted over the connection capacity (BDP) in one RTT are buffered within the RLC buffer. Therefore, to prevent overflow of the network buffering capacity, we should ensure that the receiver window is upper limited by the following relation:

$$\text{Rxwnd} < \text{BDP} + \text{B} \quad (5)$$

where B is the size of the RLC buffer.

Combining Equations (4) and (5) in one relation results in the following guideline for sizing the receiver advertised buffer:

$$BDP + B > Rxwnd > BDP \qquad (6)$$

in which the value of BDP can be estimated based on typical measured RTTs in the network.

The above equation is based on a static setting of the TCP receiver window. But in the radio environment the effective BDP and packet queuing incurred within the network can change over the course of a connection due to changing radio channel conditions, mobility, and delays caused by rate variations and channel scheduling in 3G. Then, for small RLC buffer sizes, setting the receiver window statically to the buffer size results in significant under utilization of the link because the full buffer of packets is not able to absorb the variations over the wireless link. The Next Generation TCP/IP stack supports Receive Window Auto-Tuning, which continually determines the optimal receive window size by measuring the bandwidth-delay product and the application retrieve rate, and adjusts the maximum receive window size based on changing network conditions.

A dynamic window regulation approach is presented in Reference 24 for improving the end-to-end TCP performance over CDMA 2000-1X based 3G wireless links. Since TCP inserts data into the network up the minimum between the TCP transmission window (i.e., congestion control window) and the receiver window, it is possible to regulate the flow to the maximum link utilization while preventing excess network buffering by dynamically adjusting the TCP window in real time. TCP does adjust its transmission window in response to the ACK messages received from the receiver, but basically regardless of the BDP; TCP is not aware of the connection bandwidth-delay product. The dynamic window regulator algorithm proposed in Reference 24 uses a strategy based on regulating the release of the ACK messages – in response to packet buffering status for instance in the base station – to adjust adaptively the TCP transmit window to the optimum range given by Equation (6). The algorithm provides the advantage of dynamically adjusting the window size to the connection's varying BDP. The algorithm is shown to result in the highest TCP Goodput and performance for even small amounts of buffer per TCP flow (at the RLC link), and to minimize packet losses from buffer overflows.

16.5.2.2 TCP Maximum Segment Sizing (MSS)

The maximum data segment size (MSS) is also a parameter that impacts the TCP performance. The optimum data segment size used by a TCP connection is a tradeoff between avoiding segment fragmentation within the IP routers and rapidly achieving the maximum link throughput as set by the maximum transmission window size. Fragmentation at the IP level is not recommended because it increases processing time. Furthermore, the loss of a single fragment implies the retransmission of the entire original datagram. On the other hand, using too small MSS can hurt the transmission of small files that take short TCP flows, because the full link capacity is not used during an important part of the transmission time due to the TCP slow-start phase. The effect is more pronounced when a high bandwidth-delay product is involved. For example, an MSS of 1460 bytes can increase the application throughput by more than 30% compared to an MSS of 536 bytes in EGPRS [25]. When the maximum transmission unit is not known within the fixed network routers, the MTU discovery feature provided within the TCP protocol stack can be used to obtain the optimum value for the MSS.

16.5.2.3 Initial Transmission Window

When a TCP connection is started, there is an initial window that by default has the size of two segments. It is possible to increase the size to speed up the slow start phase of a TCP connection particularly when most connections involve shorts TCP flows. As a guide, the initial CWND may be set to the following:

$$\text{Initial CWND} = \text{Minimum}(4\,\text{MSS}, \max(2\,\text{MSS}, 4380\,\text{bytes}))$$

as recommended in RFC 2414.

16.5.3 Selecting the Proper TCP Options

A few options and features have been added to various available TCP implementations for improving the protocol's performance in various scenarios. Some of these features and options offer merits that are well suited to certain applications and cases in wireless networks. The available TCP options relevant to improved performance in wireless networks include the TCP Selective Acknowledgement (SACK), the TCP timestamp option, and the fast transmit and recovery mechanisms. We will discuss the benefits that these options provide with some of the reported results on their trials or simulations on wireless networks. This discussion should provide a good insight to guide optimization engineers in the proper selection of the option sets for each application scenario based on availability.

16.5.3.1 TCP SACK Option for Wireless

TCP Selective Acknowledgement (SACK) is a TCP enhancement that allows receivers to specify precisely which segments have been received even in the presence of packet loss. TCP SACK is an Internet Engineering Task Force (IETF) proposed standard that is implemented for most major operating systems. SACK was added as an option to TCP by RFC 1072 and enables the receiver to give more information to the sender about received packets, allowing the sender to recover from multiple-packet losses faster and more efficiently. In contrast, TCP Reno and New Reno can retransmit only one lost packet per round-trip time because they use cumulative acknowledgements. The cumulative ACKs do not provide the sender with sufficient information to recover quickly from multiple packet losses within a single transmission window. Studies have shown that TCP enhanced with selective acknowledgment performs better than standard TCP in such situations [26]. The SACK option is implemented by the Santa Cruz TCP version, and the 'wireless TCP' (WTCP) [27]. The SACK option is particularly well suited to long fat networks such as on geo-satellite links with long round trip delays. In fact, TCP SACK is highly recommended for GPRS, and also for UMTS from an energy saving viewpoint. One of the proposals in the IETF draft on TCP for 2.5 and 3G systems was that 'TCP over 2.5G/3G SHOULD support SACK. In the absence of SACK feature, the TCP should use New Reno' [27]. The SACK option is not recommended when the window size is small (very small BDPs) and the loss rates are also high [19].

The SACK option is expected to reduce significantly the number of timer driven retransmissions in the wireless networks where high end-to-end delay variations and delay spikes are observed. In fact, when the SACK option was tried on a real busy GPRS network, the percentage of timer driven retransmissions dropped from 11.24% to only 1.73% [25]. This translates to a significant improvement in throughput. Other empirical studies involving large-scale passive analysis for TCP traces over different GPRS networks have shown that use of TCP SACK leads to much higher per-TCP connection utilization for all-size flow types [28]. The selective acknowledgement option lets the TCP sender cancel the timers of the segments it knows (via the SACK messages) that the receiving end has correctly received, and hence prevent spurious retransmissions. Therefore its use is also recommended in UMTS.

An Extension to SACK

An Extension to the Selective Acknowledgement (SACK) Option for TCP SACK, defined in RFC 2883, allows the receiver to indicate up to four noncontiguous blocks of received data. RFC 2883 defines an additional use of the fields in the SACK TCP option to acknowledge duplicate packets. This allows the sender of the TCP segment containing the SACK option to determine when it has retransmitted a segment unnecessarily and adjust its behavior to prevent future retransmissions. The reduced unnecessary retransmissions result in better overall throughput.

16.5.3.2 TCP Timestamp Option

Most common TCP implementations time only one segment per window to minimize the computations. Although this is expected to provide a good approximation to RTT for small windows (e.g., 4–8), it may not capture the fast round trip delay variations that can happen in the wireless networks. Moreover, it is not possible to accumulate reliable RTT estimates when time stamped segments get retransmitted (resulting in ambiguity over whether a receiver ACK is the response to the original or the retransmitted segment). This creates a retransmission ambiguity in the measurement of the RTT for the segment. The timestamp option helps to resolve this ambiguity problem.

The TCP timestamp option (also referred to as TCP echo option) allows the time stamping of every data segment, including the retransmitted segments, to provide more RTT samples for capturing the delay variations and provide more accurate estimation of the connection's retransmission time out. A more accurate Retransmission Time Out (RTO) should help avoid spurious retransmissions and the activation of congestion avoidance mechanisms. Nevertheless, the effect of the 12 byte overhead on TCP header should be considered in the decision to use the option.

The timestamp option will also allow the use of the Eifel algorithm [29], which allows a TCP sender to detect a posteriori whether it has entered loss recovery unnecessarily. The Eifel algorithm requires that the TCP Timestamp option be enabled for a connection. Eifel makes use of the fact that the TCP Timestamp option eliminates the retransmission ambiguity in TCP. Based on the timestamp of the first acceptable ACK that arrives during loss recovery, it decides whether loss recovery was entered unnecessarily. The Eifel detection algorithm provides a basis for future TCP enhancements. This includes response algorithms to back out of loss recovery by restoring a TCP sender's congestion control state.

16.5.3.3 Fast Retransmit and Recovery

When used together, the Fast Retransmit and Recovery options are designed to prevent the unnecessary triggering of congestion control via the slow start phase in the event of transient losses. The two mechanisms, when both are implemented such as in the TCP Reno version, interpret the receipt of a triple duplicate packet indicating the same sequence number as a sign of a packet loss without a break in the TCP flow due to excessive congestions. The TCP sender then retransmits the lost packet without waiting for the retransmission time out (fast retransmit), and then enters the congestion avoidance rather than the slow start congestion control phase (fast recovery). This will result in better utilization of the available bandwidth of the channel when occasional single packet losses occur.

The fast retransmit and recovery works well for single lost segments: It does not perform as well when there are multiple lost segments [30], which can happen for instance in hard handovers, deep fades, or channel switching phases in 3G. A new enhancement to the fast retransmit and recovery feature is implemented in the TCP New Reno (RFC 2582), which results in faster throughput recovery by changing the way that the sender increases the packet transmission rate during fast recovery when multiple segments in a window of data are lost and the sender receives a partial acknowledgement.

16.5.4 Conventional TCP Implementation Options

The conventional TCP implementations include TCP Tahoe, TCP SACK, TCP Reno, and TCP New Reno. The options provided by these implementations are shown in Table 16.1 for reference purposes.

16.5.5 Split TCP Solutions

The split TCP protocols isolate mobility and wireless related problems from the fixed network side. They do this by splitting the TCP connection in the base station (RNC) into two connections: a wired connection between the fixed host and the base station, and a wireless connection between the base station and the mobile terminal. In this way the wired wireless connection uses a TCP protocol modified to better suit the wireless link. The split connection mechanisms terminate the TCP connection at the base station, and thereby generally violate the semantics of the TCP in the strict sense. Examples of split TCP solutions developed for wireless networks are discussed in the following sections.

Table 16.1 The major TCP implementations and features.

	TCP implementation			
Feature	TCP Tahoe	TCP Reno	New Reno	TCP SACK
Slow start	Yes	Yes	Yes	Yes
Congestion avoidance	Yes	Yes	Yes	Yes
Fast retransmit	Yes	Yes	Yes	Yes
Fast recovery	No	Yes	Yes	Yes
Enhanced fast retransmit recovery	No	No	Yes	No
SACK	No	No	No	Yes

16.5.6 Indirect TCP (I-TCP)

Indirect TCP (I-TCP), developed in 1994–1995, is one of the earliest wireless TCP proposals [31]. I-TCP utilizes the resources of Mobility Support Routers (in the RNC) to provide transport layer communication between mobile hosts and hosts on the fixed network. Packets from the sender are buffered at the mobile support router until transmitted across the wireless connection. A handoff mechanism is proposed to handle the situation when the wireless host moves across different cells. A consequence of using I-TCP is that the TCP ACKs are not end-to-end, thereby violating the end-to-end semantics

16.5.7 Mobile TCP Protocol

Although mobile TCP (M-TCP) uses a split connection, it preserves the end-to-end semantics [32], and aims to improve throughput for connections that exhibit long periods of disconnection. M-TCP adopts a three level hierarchy. At the lowest level, mobile hosts communicate with mobile support stations in each cell, which are in turn controlled by a 'supervisor host'. The supervisor host is connected to the wired network and serves as the point where the connection is split. A TCP client exists at the supervisor host. The TCP client receives the segment from the TCP sender and passes it to a M-TCP client to send it to the wireless device. Thus, between the sender and the supervisor host, standard TCP is used, whereas M-TCP is used between the supervisor host and the wireless device. M-TCP is designed to recover quickly from wireless losses due to disconnections and to eliminate serial time outs. In the case of disconnections, the sender is forced into a 'persist' state by receiving *persist packets* from M-TCP. While in the persist state, the sender will not suffer from retransmit time out, it will not exponentially back off its retransmission timer, and it will preserve the size of its congestion window. Hence, when the connection resumes following the reception of a notification from M-TCP, the sender is able to transmit at the previous speed. TCP on the supervisor host does not ACK packets it receives until the wireless device has acknowledged them. This preserves the end-to-end semantics and preserves the sender time out estimate based on the whole round trip time.

16.5.8 Mobile-End Transport Protocol

Mobile-End Transport Protocol (METP) replaces the TCP/IP protocol over the wireless link by a simpler one with smaller headers. METP is designed specifically to directly run over the link layer [33]. This approach shifts the IP datagram reception and reassembly for the wireless terminal to the base station. The base station also removes the transport headers from the IP datagram. The base station acts as a proxy for TCP connections. It also buffers all the datagrams to be sent to the wireless terminal. These datagrams are sent using METP and a reduced header, which provides minimal information about source/destination addresses and ports, and the connection parameters. In addition, it uses retransmissions at the link layer to provide reliable delivery. When an inter-base station handover occurs, all the state information needs to be transferred to the new base station.

16.5.9 The Proxy Solutions

The proxy solutions insert a proxy between sender and receiver TCP hosts to help the TCP's performance. A well-known example of this approach is the Snoop protocol [34], which runs at the base station; Snoop cashes TCP packets at the base station and retransmits the ones that get lost over the wireless link. Snoop infers packet losses from duplicate ACKs or local time outs. It suppresses duplicate TCP ACKs from reaching the fixed host by sniffing all packets entering from the radio side before they are passed on to IP, and thus requires maintaining state information for all TCP connections passing through the base station. Snoop uses the conventional TCP from the fixed host side to the base station, whereas either TCP or some other reliable connection-oriented transport protocol is used between the base station and the receiver. The TCP sender is only affected by the congestion in the wired network and hence the fixed host sender is shielded from the wireless losses. This has the advantage that the transport protocol between the base station and mobile node can make use of its knowledge of the wireless link characteristics, or even of the application requirements.

Strategies based on explicit loss notifications (ELN) have also been proposed [35] to distinguish the wireless losses from packet losses due to congestion. ELN strategies are based on sniffing in the base station. The base station is used to monitor TCP packets in either direction. The base station checks the DUACKs from the TCP receiver to see if the indicated packet has been received. If so, the base station sets an ELN bit in the header of the ACK to inform the sender that the packet has been lost on the wireless link. The sender can then decouple retransmission from congestion control, i.e., it does reduce the congestion window. In contract to Snoop discussed in the above, the base station need not cache any TCP segments in this case, because it does not perform any retransmissions.

16.5.10 TCP End-to-End Solutions

The end-to-end wireless TCP solutions try to address and resolve the wireless losses at the transport layers of the sender and receiver. In some cases, these solutions simply make use of some of the features and options that are already provided within the available TCP implementations as we discussed in Section 16.5.1 on link layer solutions. For example, the TCP SACK option, which is implemented by most operating systems, allows the sender to recover from multiple-packet losses faster and more efficiently. SACK and its enhancement in RFC 2883 reduces the number of retransmission time outs significantly, which results in improved throughput and delay. The TCP with the timestamp option is another solution discussed earlier that helps to provide a more accurate estimate of round trip time, capture the delay variations more effectively, and hence prevent spurious time out caused retransmissions. In addition to SACK and the TCP timestamp option, there are a few other end-to-end TCP wireless solutions that have been proposed in the literature but require modifications to the TCP stack in the end-systems. For completeness of the discussion, we will briefly discuss the major ones in the following sections.

16.5.10.1 Probing TCP

The Probing TCP solution [36] is based on exchanging probing messages between the TCP sender and receiver to monitor and measure the network RTT when a data segment is delayed or lost. The probing is used as an alternative to retransmitting and reducing the congestion window size as is done in conventional TCPs. The probes are TCP segments with header options and no payload, which helps alleviate congestion because the probe segments are small compared to the data segments. The cycle terminates when the sender can make two successive RTT measurements with the aid of receiver probes. In cases of persistent errors, TCP decreases its congestion window and threshold. But for transient random errors, the sender resumes transmission at the same window size used before entering the probe cycle.

16.5.10.2 TCP Santa Cruz

TCP Santa Cruz [37] also makes use of the options field in the TCP header. This variation uses the relative delays of the data segments received in the forward direction to control the congestion window size. Therefore, the loss of ACKS and congestion in the reverse path do not affect the throughput of the connection. This merit is handy on highly asymmetric connections with little bandwidth on the reverse path. The scheme improves the RTT estimation compared to other algorithms and is able to include the retransmissions in the estimations and thus the RTTs in the congestion periods. TCP Santa Cruz also implements the selective acknowledgement feature.

16.5.10.3 Wireless TCP (WTCP99)

Wireless TCP (WTCP99) [27] uses rate-based transmission control as opposed to the TCPs self-clocking ACK based mechanisms. The transmission rate is determined by the receiver using a ratio of the average inter-packet delay at the receiver to the average inter-packet delay at the sender.

The sender transmits its current inter-packet delay with each data packet. The receiver updates the transmission rates at regular intervals based on the information in the packets, and conveys this information to the sender in the ACKs. WTCP99 computes an appropriate initial transmission rate for a connection based on a packet-pair approach rather than using slow start. This is useful for short-lived connections in the wide-area wireless networks where the round trips are large. WTCP99 achieves reliability by using selective ACKs. The protocol does not perform retransmissions triggered by time outs. Instead, the sender goes into 'blackout mode' when ACKs do not arrive at sender-specified intervals. In this mode, the sender uses probes to elicit ACKs from the receiver, similar to the Probing TCP, and is designed to cope with long fades. In WTCP99, the receiver uses the inter-packet delays as the congestion metric to control the transmission rate. The motivation for this approach is based on the considerations that the RTT calculation for a wireless WAN is highly susceptible to fluctuations due to the rapid delay variations discussed earlier.

16.6 Application Level Optimization

The web search applications make use of the hypertext protocol HTTP, which uses the services of TCP. HTTP version 1.1 can be used to expedite web downloading [38]. A web page usually contains the URL of a number of objects. Once the main web page is received, the client starts to request the objects embedded in the web page. And the request for a new object is not sent until the previous object has been received. Normally each object requires the setup of a different TCP connection, which incurs the overheads and the delays associated in the TCP connection set ups. When HTTP 1.1 is used, one 'persistent' TCP connection is kept alive during the whole download. Moreover, requests for multiple objects are pipelined so that the server will be able to start transmission of the next objects without further delays.

However the use of HTTP 1.1 can create certain complexities in the server load balancer when content is divided among multiple servers and URL switching is in place. In this case, the load balance needs to set up multiple TCP connections between it and the different serves and map them all to one TCP connection between itself and the user client. There are several ways that can be done, but all result in some loss of performance and incur new delays [39]. Therefore, when HTTP 1.1 is used, it is recommended not to use URL switching unless there is a clear need for it [39]. Some load balancers do not allow HTTP1.1 when URL switching is involved.

References

[1] Allman, M., Paxson, V. and Stevens, W., 'TCP Congestion Control', RFC 2581, April 1999.
[2] Postel, J., 'Transmission Control Protocol – DARPA Internet Program Protocol Specification', RFC 793, September 1981.
[3] Karn, P. and Partridge, C., 'Improving Round-Trip Time Estimation in Reliable Transport Protocols', *ACM Transactions on Computer Systems*, 1991, 364–373.
[4] Paxson, V. and Allman, M., 'Computing TCP's Retransmission Timer', RFC 2988, November 2000.
[5] Jacobson, V. and Karels, M.J., 'Congestion Avoidance and Control', *ACM SIGCOMM*, November 1988, 314–329.
[6] Karn, P. and Partridge, C., 'Improving Round-Trip Time Estimation in Reliable Transport Protocols, *ACM Transactions on Computer Systems*, 1991, 364–373.
[7] Stevens, W. R., 'TCP/IP Illustrated, Volume 1: The Protocols', Addison-Wesley Professional Computing Series, Reading, MA, 1994.
[8] Stevens, W., 'TCP Slow Start, Congestion Avoidance, Fast Retransmit, and Fast Recovery Algorithms', RFC 20002, Jan 1997.
[9] Kumar A., 'Comparative Analysis of Versions of TCP in a Local Network with Lossy Link', *IEEE/ACM Transactions on Networking*, August 1998, 485.
[10] Lefevre, F. and Vivier, G., 'Optimizing UMTS Link Layer Parameters for a TCP Connection', *Proceedings IEEE Conference on Vehicular Technology* (VTC) 2001, pp. 2318–2322.
[11] Chan, M.C. and Ramjee, R., 'TCP/IP Performance over 3G Wireless Links with Rate and Delay Variation', *Proceedings ACMMobiCom*, 2002, pp. 71–78.
[12] Inamura, H., Montenegro, G., Ludwig, R., Gurtov, A. and Khafizov F., 'TCP over Second (2.5G) and Third (3G) Generation Wireless Networks', RFC 3481, February 2003.
[13] Chakravorty, R., Cartwright, J. and Pratt, I., 'Practical Experience with TCP over GPRS', *Proceedings of IEEE Globecom*, Nov 2002, pp. 1–5.
[14] Teyeb, O., Wigard, J., Cellular Systems Group, AAU, Nokia Networks, 'Future Adaptive Communications Environment', Department of Communication Technology, Aalborg University, Aalborg, Denmark, 11 June 2003.
[15] Balakrishnan, H., Rahul, H.S., Seshan, S., et al., *The Effects of Asymmetry on TCP Performance*, ACM Mobile Networks and Applications, 1999.
[16] Fahmy, S., Prabhakar, V., Avasarala, S.R. and Younis, O.,' TCP over Wireless Links: Mechanisms and Implications', Technical report CSD-TR-03-004, Purdue University, 2003.
[17] 3GPP TS 25.322, 'RLC Protocol Specifications', 1999.
[18] Pavilanskas L., 'Analysis of TCP Algorithms in the Reliable IEEE 802.11b Link', *Proceedings 12th International Conference ASMTA* 2005, pp. 25–30.
[19] Balakrishnan, H., Padmanabhan, V.N., Seshan, S. and Katz, R., 'A Comparison of Mechanisms for Improving TCP Performance over Wireless Links', *IEEE Transactions on Networking*, December 1997, **5**(6), 756–769.
[20] Kojo, M., Raatikainen, K., Liljeberg, M., et. al., 'An Efficient Transport Service for Slow Wireless Telephone Links', *IEEE JSAC*, **15**(7), 1337–1348.
[21] Vacirca, F., Ricciato, F. and Pilz, R., 'Large-Scale RTT Measurements from an Operational UMTS/GPRS Network', *IEEE WICON 2005*, Budapest, Hungary, June 2005.
[22] Ludwig, R. and Rathonyi, B., 'Multi-layer Tracing of TCP over a Reliable Wireless Link', *Proceedings of ACM Sigmetrics*, 1999, pp. 144–154.

[23] Braden B., Clark, D., Crowcroft, J., et al., 'Recommendation on Queue Management and Congestion Avoidance in the Internet', RFC 2309, April 1998.
[24] Chan, M.C. and Ramjee, R., 'Improving TCP/IP Performance over Third Generation Wireless Networks', *Proceedings IEEE INFOCOM'04*, 2004.
[25] Sanchez, R., Martinez, J. and Jarvela, R., 'TCP/IP Performance over EGPRS Networks', europe.nokia.com/downloads/aboutnokia/*.../mobile_networks/MNW16.pdf.
[26] Fall, K. and Floyd, S., 'Simulation Based Comparison of Tahoe, Reno, and SACK TCP. *Computer Communication Review*, July 1996, **26**(2), 5–21.
[27] Sinha, P., Venkitaraman, N., Sivakumar, R. and Bharghavan, V., 'WTCP: A Reliable Transport Protocol for Wireless Wide-Area networks', *ACM Mobicom'99*, Seattle, WA, August 1999.
[28] Benko, P., Malicsko, G. and Veres, A., 'A Large-scale, Passive Analysis of End-to-end TCP Performance over GPRS', *Proceedings IEEE INFOCOM*, March 2004, pp. 1882–1892.
[29] Ludwig R. and Katz H., 'The Eifel Algorithm: Making TCP Robust Against Spurious Retransmissions', *ACM Computer Communications Review*, January 2000.
[30] Kumar, K., 'Comparative Performance Analysis of Versions of TCP in a Local Network with a Lossy Link', *IEEE ACM Transactions on Networking*, August 1998, **6**(4), 485–498.
[31] Bakre, A. and Badrinath, B.R., 'Handoff and System Support for Indirect TCP/IP', *Second USENIX Symposium on Mobile and Location-independent Computing Proceedings*, Ann Arbor, Michigan, April 1995.
[32] Brown, K. and Singh, S., 'M-TCP: TCP for Mobile Cellular Networks', *ACM Computer Communication Review*, **27**(5), Oct. 1997, 19–43.
[33] Wang, K.Y. and Tripathi, S.K., 'Mobile-end Transport Protocol: An Alternative to TCP/IP over Wireless Links', *INFOCOM*, San Francisco, California, March/April 1998, pp. 1046–1053.
[34] Balakrishnan, H., Seshan, S., Amir, E. and Katz, R.H., 'Improving TCP/IP Performance over Wireless Networks', *Proceedings of 1st ACM Conference on Mobile Computing and Networking*, Berkeley, California, November 1995.
[35] Balakrishnan, H. and Katz, R.H., 'Explicit Loss Notification and Wireless Web Performance', *Proceedings IEEE Globecom Internet Mini-Conference*, Sydney, Australia, Nov. 1998.
[36] Tsaoussidis, V. and Badr, H., 'TCP-Probing: Towards an Error Control Schema with Energy and Throughput Performance Gains', *Proceedings of ICNP*, 2000, pp. 12–21.
[37] Parsa, C. and Garcia-Luna-Aceves, J.J., 'Improving TCP Congestion Control over Internets with Heterogeneous Transmission Media', *Proceedings of the Seventh Annual International Conference on Network Protocols*, Toronto, Canada, Nov. 1999, pp. 213–221.
[38] Fielding, R., Gettys, J., Mogul, J., et. al., 'Hypertext Transfer Protocol-HTTP 1.1, RFC 2616.
[39] Kopparapu, C., *Load Balancing Servers, Firewalls, and Caches*, John Wiley & Sons, Inc., New York, 2002.

17

RAN Performance Root Cause Analysis and Trending Techniques for Effective Troubleshooting and Optimization

The different services in UMTS require a set of well-defined and measurable key performance indicators (KPIs) for each service category based on its characteristics and performance. These KPIs are then monitored to assess and trend the service quality perceived by the user. In turn, these KPIs are derived from or related to a set of basic measurements and indicators, which we will call network performance indicators (NPIs), which reflect an aspect of the network performance. The NPIs are required for root cause analysis and troubleshooting of performance problems perceived by the end-user (KPIs), as well as for the operator assessment of the network resource utilization and efficiency. The multitude of services in UMTS with different requirements and characteristics and the many inter-dependent service and network performance and design parameters pose a serious challenge for 3G network performance analysis and optimization. Efficient methodologies are required to pinpoint easily the inter-dependency among performance problems, identify cells that suffer from similar suboptimal configuration, and outline similar solutions for them. Moreover, the KPIs and NPIs should be measured and trended in such a way to reflect adequate statistics for input to the analysis. These issues are the subject of discussion in this chapter.

17.1 RAN KPIs

The Radio Access Network (RAN) plays a prominent role in overall system performance, because it typically represents the bottleneck in terms of available transmission resources and capacity on the air interface. The RAN performance and its impact on user perception of the service quality are evaluated through a number of KPIs. Depending on what aspect of the

RAN performance or user perception of the service quality is to be measured, different KPIs are defined. These KPIs may be direct measurements collected by certain counters in the Base Station Subsystem (BSS), or a combination of those counters based on well-defined mathematical formulas.

Certain KPIs are defined to assess and evaluate end-user perception of the service quality. For voice calls, for instance, the main KPIs include the call set up success rate or network accessibility, call drop rate or service retainability, call set up delays, and voice quality or integrity. Generally the KPIs for each service category must include indicators that reflect all aspects of the end-user perception of the service quality. In this regard, the KPIs for streaming services may include the service access time, streaming reproduction cut-off ratio, audio/video quality, and synchronization.

Those KPIs that measure the end-user's perception of the service quality are impacted by a number of factors that are indicative of certain aspects of the network performance. For example, the network accessibility or call set up success rate from the RAN perspective is mainly a function of radio resources (availability of traffic and signaling channels), signaling capacity over the access network, successful signaling as a result of good radio conditions, and BSS parameter configuration. Each of the stated factors reflects one aspect of the network performance. The availability of a traffic channel when needed is indicative of proper dimensioning of the network radio channels to handle the expected call traffic. The service retainability refers to the network ability to maintain the call or the session once it has been correctly established. Failures in coverage, problems with the signal quality due for instance to interference, and congestion have important impacts on this indicator and increase the call or session drop outs. Integrity refers to the service quality that is impacted by the radio channel conditions over the air interface. Poor radio conditions result in errors and losses in the user information, which can result in poor voice quality or delays in the transmission of data. Therefore, in order to analyze and discover the root cause of the user perceived KPIs, a number of NPIs should also be defined and monitored. These NPIs are used internally for troubleshooting and performance analysis within the operator organization. The NPIs may include such things as congestion indicators on traffic and signaling channels (TCH and SDCCH in GSM), load factors or noise rise in WCDMA (which are indicative of in-cell traffic level and interference from other cells), handover failure rates in GSM (which contribute to call drops), soft handover overhead in WCDMA (which impacts the efficiency of the network resource utilization as well as interference mitigation), frequency of cell reselections in GPRS (which impact data throughputs and delays), and so on.

17.2 Measurement Guidelines

17.2.1 Live Network Traffic

The radio network performance indicators are most readily measured through the statistical performance measurement counters imbedded in the base station equipment. The performance counters can be set to gather measurements on the live network traffic during specified time intervals and process the data through specific formulas to calculate various KPIs and other NPIs. This way of measuring the NPIs offers the advantage of providing sufficient statistics without generating any test traffic on the network, and helping to trend the network performance based on the actual user traffic pattern and distribution over time and area. The measurements should be collected over network peak traffic time in each area in order to best capture the effects of congestion on call blocking and interference. Care should be taken when comparing such measurements between different vendors' equipment as each may define the performance counters in different ways.

17.2.2 Drive Testing

Measurements collected by the live network do not always suffice for troubleshooting and performance trending. The network-collected measurements can provide overall statistics over a given area and an indication of the problematic areas of a network. The CDMA networks are heavily influenced by the site physical configurations and hence require heavy use of drive testing for optimization. At times, it is necessary to focus on specific hot spot locations and study the performance of the network with respect to pilot pollution, local call drops and service quality, and proper handover operations in different situations. Drive testing is also an essential operation in the service pre-launch phase to verify and test the network performance when there are still no commercial users on the network. Furthermore, drive testing is necessary at times to verify and find out the cause of specific performance problems and complaints indicated by customers in certain areas of the network coverage.

In the drive/walk test scenarios, the test calls are to be made at representative times of the day in proportion to the time distribution of real traffic. The survey should also be repeated at intervals over the reporting period to take account of changes with regards to time of the call demand and capacity of the network.

For the measurement of dropped calls, the test calls will need to be made for durations that are representative of the duration of real calls. When gathering measurements for trending statistical quantities such as call set up success rate in an area, a sufficient number of test calls should be made to reflect adequate statistics. This is normally specified through a confidence-interval or margin of error [1].

The margin of error expresses the amount of the random variation underlying the test results. This can be thought of as a measure of the variation one would see in reported percentages if the same drive test were taken multiple times. The larger the margin of error, the less confidence one has that the drive test reported percentages are close to the 'true' percentages, that is the percentages that would result if the number of test calls went to infinity. A margin of error is usually prepared for one of three different levels of confidence: 99%, 95%, and 90%. The 99% level is the most conservative, whereas the 90% level is the least conservative. The 95% level is the most commonly used. If the level of confidence is 95%, the 'true' drop call rate, for instance, would be within the margin of error around a particular test's reported percentage 95% of the time.

To determine the mathematical relationship between the margin of error and the number of test calls, each call is assumed to be independently treated by the network or undergo independent radio conditions. Let p be the expected call drop rate in the area over the time interval of the testing. Then the number of calls, X, that will drop is a random variable with a binomial distribution with parameters n and p. If n is large enough, then X is approximately normally distributed with expected value np and variance $np(1-p)$.

Therefore

$$Z = \frac{X - np}{\sqrt{np(1-p)}} \qquad (1)$$

is approximately normally distributed with expected value 0 and variance 1. Consulting tabulated percentage points of the normal distribution reveals that

$$P(-Z_c < Z < Z_c) = \text{Confidence level} \qquad (2)$$

where

$$Z_c = 1.96 \text{ for } 95\% \text{ confidence level}$$
$$= 2.576 \text{ for } 99\% \text{ confidence level}$$

Or, in other words, $P\left(-Z_c < \dfrac{X/n - p}{\sqrt{p(1-p)/n}} < Z_c\right) = $ confidence interval (that is, 0.95 or 0.99, etc.).
This is equivalent to

$$P\left(X/n - Z_c \sqrt{\frac{p(1-p)}{n}} < p < X/n + Z_c \sqrt{\frac{p(1-p)}{n}}\right) = \text{confidence interval} \qquad (3)$$

Replacing p in the first and third members of this inequality by the estimated value X/n seldom results in large errors if n is big enough. This operation yields

$$P\left(X/n - Z_c \sqrt{\frac{(X/n)(1-X/n)}{n}} < p < X/n + Z_c \sqrt{\frac{(X/n)(1-X/n)}{n}}\right) = \text{confidence level} \qquad (4)$$

The first and third members of this inequality depend on the observable X/n and not on the unobservable p, and are the endpoints of the confidence interval. In other words, the margin of error, E, with the specified confidence level is

$$E = Z_c \sqrt{\frac{(X/n)(1-X/n)}{n}} \qquad (5)$$

From Equation (5), the number of required test calls n is determined by repeating calls until the desired margin of error at the specified confidence level in the estimated call drop rate, X/n, is achieved. Likewise, for a certain number of test calls n the margin of error associated with the estimated drop call rate, X/n, is determined from the same equation.

For example, in 200 test calls, the number of drop calls observed was 2. In this case, Equation (5) results in a margin of error of 0.0138 with a confidence level of 95%. This means that the true call drop rate will differ from the estimated 2/200 (= 0.01) by at most 0.0138 with a 95% probability.

17.3 Correlation Based Root Cause Analysis

Network performance trending and troubleshooting requires the analysis of large volumes of data collected from network elements. The data complexity in 3G systems can considerably increase compared to 2G systems given the multitude of service classes with different parameters and requirements, not to mention the increased interaction between coverage, capacity, and quality. The operators' task is to filter the relevant information and present it in a format that can easily highlight the hidden dependencies and correlations between multiple network and service performance indictors. This is necessary in order to discover the root causes underlying performance problems and help develop the proper solution. Furthermore, viewing the correlations and the interactions among multiple network and service performance parameters helps to identify the tradeoffs that can be made to achieve service performance with the available network resources.

17.3.1 Correlative Analysis Based on a priori Knowledge

The traditional network performance analysis is handled by examining the KPIs separately on the screen. As a result, the number of graphs to be visualized increases with respect to the number of services, the number of network performance indicators, and the number of KPIs used in the analysis. Thus, the problem root cause analysis and optimization over a large set of performance elements requires the analysis of a large number of separate graphs, and the combination of their results to conclude the network state. The alternative is to form sets of KPI and NPI combinations, each of which can be used to study one aspect of the network performance such as interference, congestion, etc.

In the correlation analysis of network performance data based on a priori knowledge, the formation of sets of correlated KPI and NPIs is done based purely on the expertise of the person performing the task. However, with a level of expertise, it is a simple, intuitive, and quick approach that does not require the use of sophisticated software coding. In this approach, the KPIs and NPIs that have the most direct impact on each other are grouped into different *indicators-sets*. For example, we may identify interference, congestion, and handover as three different indicators with respect to which the relevant KPIs and network performance indicators can be identified and evaluated.

In GSM, high call drop rates may be caused by interference or low signals on the TCH or BCCH channels (if the call is on the BCCH carrier). If interference is the cause of the high call drops, we expect to see low voice quality indications and possibly high drops on SDCCH. In a high interference cell, frequent intra-cell handovers will be expected as well. Therefore, we can form an interference-coverage indicators-set in the following form:

$$\begin{aligned}\textit{interference-coverage indicators-set} = [&\text{Call\_Drop\_Rate},\\ &\text{Intra\_Cell\_Handover\_Rate}, \text{RxQualUL}, \text{RxQual\_DL}, \text{SDCCH\_Drop\_Rate}]\end{aligned} \quad (6)$$

Then to filter out the cells with possible interference or low signal strength problems, the cells within a cluster that are to be investigated for troubleshooting and optimization are organized into an Excel spreadsheet with the Cell Id and information in the first column, and the remaining columns indicating the five NPIs or KPIs that are shown in Equation (6). The indicator element values may be computed and averaged over peak hours of the week, or listed separately over the peak hour of each day in the week to study the variations over time. Then the Excel spreadsheet can be sorted in descending order with respect to the column indicating the call drop rates (the user perceived KPI) for each cell. In this way the top 10 worst performing cells within the cluster in terms of call drop rates are identified and resolved. Dealing with the worst ten cells in each optimization exercise, slowly resolves the problems and raises the overall quality of service available to subscribers. Then the remaining columns in the Excel spreadsheet will indicate which of the NPIs (such as quality, intra_cell handovers, and SDCCH drop rates) are also high when the call drop rates are high. By observing the correlations of the high call drop rate cells with the other NPIs in the spreadsheet, the possible causes of the high call drop rate can be narrowed down. For instance, if the cells with high call drop rates also indicate low voice quality, and high intra_cell handover rates, and possibly high SDCCH drop rates, interference or low signal strength is the most likely cause of the problem. If the high call drop rate results from calls on the interfered TCH channels, the drop rate on SDCCH may or may not be high. But if the high call drop rate results from calls on the BCCH carrier, and a combined signaling channel configuration is used in the cell, the SDCCH drops will most likely be high as well.

Likewise, a high call set up failure rate in GSM may be caused by signaling problems on the random access channel (RACH Failure), SDCCH establishment failures due to congestion on SDCCH or SCCP signaling link, interference-caused drops on SDCCH, TCH and transcoder blocking due to congestion, and/or reduced TCH or SDCCH availability. Therefore, to see if congestion on any of the stated channels could be a possible cause, a congestion *indicators-set* is formed in the following manner:

$$
\begin{aligned}
\textit{congestion indicators-set} = [&\text{Call\_Setup\_Failure\_Rate}, \text{RACH\_Failure\_Rate}, \\
&\text{SDCC\_Drops due to interference}, \text{TCH\_Blocking\_Rate}, \text{Transcoder\_Blocking\_Rate}, \\
&\text{TCH\_Congestion\_Time}, \text{SDCCH\_Establishment\_Failure\_Rate}, \\
&\text{SDCCH\_Congestion\_Time}, \text{TCH\_Availability}, \text{SDCCH\_Availability}]
\end{aligned} \quad (7)
$$

Then a separate Excel spreadsheet is used to organize the cells within the cluster along with the network or service performance indicators shown in Equation (7). As before, the average values of the indicator elements over the peak hour of each day of the week or their values over the busy hour of each day of the week for the cell may be used in the Excel spreadsheet. The spreadsheet is then sorted in descending order with respect to the first user perceived KPI, that is the Call_Set_Up_Failure Rate, and the top worst cells (say the worst 10 performing cells) are identified. By then checking the values for the other indicators in the spreadsheet, the possible cause of the high call set up failure cells may be attributed to RACH failures, drops on SDCCH due to interference that would affect RACH failure as well, or congestion on either TCH or signaling channels. If high call drops are correlated with high RACH failure rate, then interference or hardware problems are mostly likely the cause of the problem, and so on.

In similar fashion, the outgoing handover failure rate from a cell, together with congestion and some measure of interference in neighboring cells (such as call drop rates) may be grouped into a *handover problem indicators-set* and used to get an indication as to whether congestion or interference in a neighboring cell may be causing the high handover failure rates. If none of these two indicators are high, then the cell neighbor listing and other problems such as handover hysteresis or offset parameters, signal averaging filter length, and cell overlap adequacy are investigated as possible causes of the high handover failure rate.

$$\textit{handover problem indicators-set} = [\text{outgoing handover failure rate}, \\ \text{Call\_Drop\_Rate\_in Destination\_Cell, TCH\_Congestion\_Ind\_Destination\_Cell}, \\ \text{SDCCH\_Congestion\_Indication\_Destination\_Cell}] \quad (8)$$

The same basic procedure can be used in WCDMA to find out or narrow down the root cause of network performance problems. To determine the possible causes of a certain service performance indicator, those network performance or resource usage indicators that are expected to have an impact on the KPI in question are grouped together into an Excel spreadsheet along with the cell information for analysis. For instance, to study the root cause underlying low service quality in terms of FER (frame error rate) on the uplink in the cells, a service quality *indicators-set* is defined in the following form:

$$\textit{service quality indicators-set } UL = (\text{FER}, \text{Number\_of\_Users}, \text{ULNoise\_Rise}) \quad (9)$$

where FER represents the frame error rate and is used as a measure of the user perceived quality. Then an Excel spreadsheet is formed with the first column denoting the cell information and the remaining columns FER averaged over, say, the busy hours of the week, the number of users, and the average uplink noise rise over the period. The spreadsheet is then sorted in descending order of FER and the top worst performing cells with respect to quality are filtered out. The cause of the poor quality performance in each problem cell is then determined by examining high correlations with either the number of users and uplink noise rise, or just high uplink noise rise. Since certain performance indicators such as the number of users supported in the cell, can vary over a large range, it is important to cluster cells of similar site configuration and then normalize the indicator either with respect to the maximum value observed over the cell cluster or with respect to a limit threshold such as the uplink pole capacity for the cell based on its configuration. This will help to compare the cells within the cluster and see how the indicator varies from cell to cell and its correlation to the other performance indicators within the spreadsheet.

Then by looking at the observed correlation between indicators-sets for each cell in the spreadsheet, various scenarios can be sorted out. Specifically, if the normalized number of users is low, but the uplink noise rise is high, the problem is most likely interference from other cells. If both the number of users and uplink noise rise are high, the problem is intra-cell interference resulting from large number of users generating traffic in the cell. To better study the situation in problem cells, it is best to repeat this study using, say, busy hour data for each day of the week, and examine the behavioral pattern in time. It is possible that a problem cell may undergo different behavioral pattern over time. For instance, a cell may have high FER due to high number of users at certain times and high interference from other cells over other time periods. Moreover, the trend over time can reveal more about the correlation pattern between the various indicators under different scenarios of network traffic distribution and interference. This can be done more easily based on the automated correlative analysis, which uses the techniques of data clustering and cell behavioral histogram mapping as discussed in the remaining sections of this chapter.

However, the rather simple procedure discussed in the current section can be generalized and applied to other aspects of network performance and help to single out or narrow down the possible root causes of degraded service performance. The author applied this approach in his radio optimization work on a project in Indonesia, which resulted in significant simplification and expedition of the routine radio troubleshooting and performance tuning work for the engineers.

17.3.2 Correlative Analysis Based on Data Clustering

It was argued in the previous section that the many interacting variables impacting the performance of 3G networks and service quality require efficient means of handling multiple performance indicators and parameters simultaneously. Having efficient and effective means for service quality optimization is a critical issue for the operators, because the over provisioning of service quality can end up being costly and poor service quality will result in churn and the loss of business to the competition.

The simple techniques discussed previously will work when the interaction among the performance indicators are known to the engineer. In this section, we discuss more efficient and effective means based self-organized maps (SOM) and the K-means clustering algorithm to handle the simultaneous analysis of multiple performance indictors and identify their interaction and influence on each other. The techniques discussed in the present section will reduce the speed of troubleshooting efforts and increase the efficiency of the routine network optimization task [2–4].

The SOM is a very popular neural network based method for data mining and visualization of trends that are hidden in intense and complex data structures. Data mining is concerned with extracting relevant information from large amounts of data. Data mining typically includes clustering, classification, and trending out the interaction among various parameters and the anomalies hidden in the data. The SOM provides one efficient means for data mining applications and a 2-dimensional (2-D) visualization of the trends in multi-dimensional data. It provides a compression of the input data samples in a reduced data space representation with a topology-preserving mapping, which results in the visual display of cluster-like patterns in the data as discussed in the next section.

17.3.2.1 Data Reduction and Clusterization Based on Self-organized Maps (SOM)

The self-organizing map (SOM) invented by Kohonen, a professor of the Academy of Finland [5, 6], is a means of mapping multi-dimensional data to a much lower dimensional space, usually one or two. The basic process of reducing the dimensionality of data points, vectors, is known as vector quantization. However, the SOM arranges the information in such a way that any topological relationship or similarities in the input data are preserved and reflected in the lower dimensional classification of the data (say 2-D space). That is, it creates a meaningful map in which similar patterns in the n-dimensional data space are placed next to each other for easy visualization. Thus the mapping not only results in a clustering of data into rather distinct regions but also places the similar regions next to each other for easy visualization. SOMs make it easy to sort out and visualize the cluster patterns and trends hidden within vast amounts of data.

Implementation

A 2-D SOM that is commonly used is chosen to illustrate how the concept is implemented. The map is created from a 2-D lattice of 'nodes', each of which is fully connected to the input layer. The input layer is assumed to be an n-dimensional vector space. Each node in the lattice has a specific x, y coordinate and contains a vector of weights of the same dimension as the input vectors. That is to say, if the training data consist of vectors, V, of n dimensions made of V_1, V_2, V_3...V_n, the weight vector associated with each node will take the form W, where W has the components W_1, W_2, W_3...W_n. The input vector components could consist of, for instance, the key service performance indicators perceived by the subscribers (such as call drops rates, call blocking, call quality) and the NPIs that could affect them (TCH drop rates, TCH congestion, SDCCH drop rates, RACH Failure rate, etc.). Drawing from a genetic application of SOM, the input vector components could consist of measurements obtained on different features of a gene. Then the dimension of the input vector and hence the weight vector associated with each node in the lattice map would be determined by the number of distinct features that would be required to represent a gene in a genetic behavioral classification study.

Therefore the learning algorithm for the lattice works by using the input vector data to stimulate nodes with similar weight vectors and then adjust its weight vector and those of its neighboring nodes to match more closely the input vector presented to the lattice. This process continues until all the input vector data are presented to the lattice and the lattice nodes adjust their weights iteratively. Then, any new, previously unseen input vectors presented to the network will stimulate nodes in the zone with similar weight vectors.

The training occurs in several steps and over many iterations as follows:

1. Each node's weights are initialized to small standardized random values.
2. A vector is chosen at random from the set of training data and presented to the lattice. It is important to scale each component in the input vector to represent small standardized values.
3. Every node is examined to calculate which one's weights are most like the input vector, using for instance the Euclidean distance measure. The winning node is commonly known as the Best Matching Unit (BMU).
4. The radius of the neighborhood of the BMU is now calculated. This is a value that starts large, typically set to the 'radius' of the lattice, but diminishes exponentially at each time-step. Any nodes found within this radius are deemed to be inside the BMU's neighborhood.
5. Each neighboring node's (the nodes found in step 4) weights are adjusted to make them more like the input vector. The farther a node is from the BMU, the less its weights get altered using an exponential decay function with distance.
6. Step 2 is repeated for N iterations.

The algorithm results in the grouping of similar vector data into neighboring nodes, and thus produces cluster-structured datasets. In other words, after the training is completed, most of the SOM nodes neighboring each other in the lattice are placed next to each other also in the input space. This would make it simple to estimate the clustering structure in the input data. Then a K-means clustering algorithm (discussed in the next section) can be used over the SOM data code words (represented by the weights of the lattice nodes) and obtain well-defined clusters of data sets. The computational complexity for the implementation of SOM is N.M.d as discussed in Reference 7, where N is the number of input data vectors, d is the dimension of the data vectors, and M is the number of nodes (map units) chosen for the SOM. A mathematical formulation for the implementation of the SOM algorithmic steps given in above is provided in the Appendix at the end of this chapter.

17.3.2.2 Clustering of SOM Data Based on the K-means Algorithm

The K-means algorithm was introduced by J.B. MacQueen in 1967, and is one of the most common clustering algorithms that groups data with similar characteristics or features together [7–11]. These groups of data are called *clusters*. The data in a cluster has similar features or characteristics that are *dissimilar* from the data in other clusters.

The K-means algorithm accepts the number of clusters to group data into and the *dataset* to cluster as input values. It then finds a set of k points (k chosen) in d-dimensional space that minimizes the mean squared distance from each of n given data points to its nearest centre. The algorithm for finding the k cluster centers is implemented in the following steps:

1. Choose k initial cluster center points randomly.
2. Cluster data by assigning each data point to the nearest cluster (the cluster to which it is most similar) using a measure of distance or similarity such as the *Euclidean distance* measure or some other *distance metric*.
3. Calculate the arithmetic mean of each cluster (using only points within each cluster) and use that as the new center point for each cluster.
4. Re-cluster all data using the new center points. This could result in some data points being placed in a different cluster.
5. Repeat steps 3 and 4 until the center points have moved so that in step 4 no data points are moved from one cluster to another, or step 3 does not create new clusters as the cluster center, the or Arithmetic Mean of each cluster formed is the same as the old cluster center, or alternatively until some other convergence criteria is met.

Choosing k
Since the correct or best number of clusters is not known, the optimal value for k is defined using a number of different techniques as follows:

- Basing it on some prior knowledge about the data structure, such as normal and abnormal genes, etc. This is not always practical if the input vector dimension is large.
- Using insight obtained from the SOM of the original input data. The SOM can be used to organize and map the n-dimensional data into a 2-D clustered-like display and then use the resulting cluster pattern to estimate the best value for k.
- Running the algorithm with several different values for K and picking the one that appears to result in the best cluster set based on visual inspection, or the least clustering error, or based on some cluster validity index such as the Davies- Bouldin one [12].

It is to be noted that the above algorithm can cluster any set of n data points. But the resulting clusters may not be meaningful, or may be indicative of a significant clustered structure in the data. A cluster can be formed even when there is no similarity between clustered patterns. This occurs because the algorithm forces k clusters to be created. Therefore an expert human visual check is required to see if the patterns formed are relevant.

The following criteria may be used to examine if the resulting clusters reveal a meaningful structure in the data and thus evaluate the quality of the clustering:

- The diameter (size) of the cluster versus the inter-cluster distance.
- The distance between the members of a cluster compared to distance from the cluster center.

Algorithm Complexity

The computational complexity of the K-means algorithm is said to be proportional to $\sum_{k=1}^{C_{max}} N.k$, where N is the number of data samples and C_{max} is the maximum number of clusters in which the data is to be clustered [2, 3].

17.4 Applications to Network Troubleshooting and Performance Optimization

The data clustering techniques discussed in previous sections provide a highly visual and non-manual means of examining the correlation and interaction among multiple performance indicators. This is in contrast to the traditional way of analyzing network performance where only a few KPIs are analyzed separately at a time. The data clustering techniques and correlation analysis based on SOM and the K-means algorithm have been applied to other areas of engineering such as process control and data analysis [13].

To apply the data clustering techniques for network performance correlation analysis, four major steps are required:

1. Formation of vector performance indicators
2. Data scaling
3. Clustering of performance data
4. Clustering of cells into behavioral classes

These are discussed in the following four sub-sections based on the analysis provided in Reference 3.

17.4.1 Formation of Vector PIs

A vector performance indicator (PI) is formed for each KPI that is under study. The PI vector components will consist of the KPI and those network performance indicators that are known to directly or indirectly impact it. Since this correlation analysis is based on data clustering and trending, the vector PI may include PIs whose impact to the KPI under consideration may not be certain to the user. Vector PIs may also be formed from a number of KPIs, PIs, and cost functions derived from them whose interaction and correlation is to be investigated in the process of root cause analysis of performance problems. In essence, this is done along the lines discussed in Section 17.3.1. Basically the selection of performance indicators is dependent on what aspect of the network performance is to be investigated. As an example, if the goal is to analyze and increase the network capacity on the uplink in UMTS, the performance indicators that are selected should be related to counters and measurements that affect the uplink capacity such as the noise rise, quality measures, etc. Because not all the PIs and KPIs are equally relevant, this consideration should be taken into account when forming a vector PI to unravel possible performance patterns with respect to interacting determinant factors.

The data for the vector PIs should be selected from measurements obtained from the network cells over sub-intervals of time over the period under consideration. This will generate sufficient data that can reflect the behavioral pattern of the cells over the period. By capturing the form of the value distribution of the different KPIs and PIs inside the period, their interaction and correlations is revealed more effectively. For example, instead of using weekly or daily averages of a PI, the hourly or even fractional hourly values over an interesting part of the days in the week is used. This allows the detection of various anomalies that occur as a result of different value combinations of the network parameters over time, and helps to prevent suboptimal solutions based on long time averaging of different interacting data variables.

17.4.2 Data Scaling

Since the different performance indicators represented by the different components of a vector PI may span data ranges that may be quite wide apart, it is important to first balance the different variables through scaling. Scaling helps to make all variables equally and independently important on the measurement unit. One method of balancing the different performance indicators is to normalize the variance, using the a priori knowledge on the variables ranges of each, to 1 [3]. Prior to the scaling process, any anomalies or missing data which can result in the network configuration processes should be removed. If there are outliers (extreme values) in the data, they should also be removed if the average normal behavior is of interest. In the case when the outliers may carry interesting information such as signs of serious network performance problems, their large values can be reduced through some sort of conversion function in order to prevent them from dominating the analysis. This can be done by a logarithmic or trigonometric function such as a tanh(a.x) where the parameter a can be selected to control the mapping. The mapping will then reduce the effect of the outliers and help to emphasize the range that is of more practical interest. For example, if the frame error rate is under consideration, it can in principle range anywhere from 0 to quite large values. But in practice anything above a 2% frame error rate may be taken to be just as bad as a 2% FER for a particular service. In that case a tanh(a.x) conversion function can be used with proper selection of the parameter a to provide the mapping, for instance, from the interval {0, ∞) to {0,2}. This will help to emphasize the range from 0 to 2, which is the more interesting part of the frame error rate distribution.

17.4.3 Clustering of Performance Data (Building Performance Spectrum)

The objective here is to cluster the trends seen in the aggregate performance data obtained over many time sub-intervals and cells for the components of a vector PI and build a performance spectrum [3]. In other words, the goal is to identify various rather distinct (clusters) combinations of value ranges observed for the PIs and observe the trends in their interaction. Thus, the performance spectrum, built in this way, partitions the data into naturally occurring clusters, each of which describes different performance or problem scenarios.

This is done by clustering the time-series measurement data collected for the vector PI from a large number of cells in the network using a combination of SOM and the K-means algorithm. At first, a SOM with M lattice nodes is trained using the data vectors. These data vectors may also include a set of own-defined performance target data vectors to help generate wanted behavioral clusters. Next, the resulting set of M weight vectors of the SOM are clustered [14] into several different numbers of clusters using the K-means algorithm as discussed in Section 17.3.2.2. The guidance for arriving at the best value for k was also discussed in the same section. Since the parameter M can generally be selected to be much smaller than the number of time-series data for the vector PI, the SOM provides a much reduced data set to the K-means clustering algorithm compared to using the original data. Thus, when the K-means clustering is applied to the resulting weight vectors of the SOM lattice nodes, instead of the original dataset, the computational cost of clustering is reported to reduce to $NM + \sum_{k=1}^{C_{max}} M.k$ in which C_{max} is the maximum number of clusters in the K-means clustering algorithm, thus making the analysis computationally a tractable solution.

In this way, a performance spectrum for the selected vector PI is obtained. The performance spectrum will indicate the different combination range values observed for the components of the selected vector PI and hence indicate the possible interactions among the different performance indicators. This information can help find the root cause behind performance problems indicated by a KPI and guide the optimization tradeoffs.

17.4.4 Clustering Cells into Behavioral Classes

Once a performance spectrum is obtained for a particular vector PI based on aggregate data over many time subintervals and cells in the network, as discussed in the previous section, it can be used to classify the behavioral pattern of individual cells in the network. This can be done statically or dynamically. In the static case, the performance data for the cell is averaged over a long time and the resulting single data point is mapped into one of the clusters in the performance spectrum. This does not reveal much information about the cell's behavior over time, which results from the different conditions that it undergoes over the course of time (such as the traffic level, the interference resulting from surrounding cells, the propagation conditions, etc.). In the dynamic case, the performance data for the cell is collected over many sub-intervals of time over an extended period (several days for instance) and used to build a hit histogram [3]. The hit histogram of a cell is formed from the distribution of the number of times that the observed performance for the cell falls into each of the clusters within the performance spectrum. Thus, the dynamic classification of the cell's performance will reveal its behavioral pattern over the course of time. The hit histogram of a cell that is formed from time series data over a period of time provides a characterization of the different scenarios of traffic, interference, etc. that the cell undergoes over time. It also provides temporal information about the cell's behavior.

Moreover, the hit histograms of the cells can be used as input into another similar combination of the SOM and K-means clustering processes to generate clusters of similarly behaving groups of cells. This is done by first training a SOM using the hit histogram vectors of each cell. Then, the codebook vectors of the SOM are clustered into an optimum number of clusters using the K-means algorithm as discussed in Section 17.3.2.2. As a result of this process, each cell is automatically assigned to a behavioral cluster, where the number of clusters will be well under the number of cells in the network. Each behavior cluster resulting from this process can then be described by the rules from certain clusters in the performance spectrum.

The optimization phase can then concentrate on the optimization/automation of the resulting, similarly behaving, cell groups rather than optimizing each individual cell with its own selection of configuration parameter settings. The clustering of cells into similarly behaving groups helps to visualize [15] different ranges of performance for performance indicators-sets (or combinations) and identify the problematic cases. This greatly simplifies the troubleshooting process because many problematic cells can be identified at once from the clusters to which they belong. Furthermore, optimal settings of cell configuration parameters are identified from well performing clusters and applied in the optimization of the poor performing cell groups. Thus optimization process is made much simpler and efficient by working with groups of cells with similar problems than with individual cells, as each cell in a cell-group will also use the same configuration parameter set. Case studies based on the techniques discussed in this section are given in Reference 3.

Appendix

This appendix provides a mathematical formulation for the implementation of the SOM algorithm in Chapter 17.

The best matching unit (BMU) in the self-organized map (SOM) learning is found by calculating the Euclidean distance between each node's weight vector and the current input vector. The node with a weight vector closest to the input vector is marked as the BMU.

The Euclidean distance is given as

$$d = \sqrt{\sum_{i=0}^{n} (W_i - V_i)^2} \qquad (A.1)$$

where V is the current input vector and W is the node's weight vector. Once the BMU is determined, the next step is to adjust the weight vector of the BMU and all nodes in its neighborhood. Thus the node with the weight with the smallest distance is chosen and the neighbors are determined. Then the weight of the neighboring nodes are adjusted (the *learning process*) to become more like the sample. The farther away a neighbor is, the less it learns. Thus in the learning process, the weight of the nodes neighboring the winning node (the node with the closest distance to the input data sample) are pulled towards that of the current input sample, resulting in clustering of similar data patterns onto the lattice.

The neighborhood may be defined with an exponential decay coefficient in the following form:

$$\Omega(t) = e^{-\frac{d^2}{2\sigma(t)^2}} \qquad (A.2)$$

where t is the current time step in the iterations (t = 0, 1, 2, 3, ...), d is a measure of the distance of the node from the BMU (say the Euclidean distance, and $\Omega(t)$ is in turn an exponential time decay function given as

$$\delta(t) = \delta_0 e^{-\frac{t}{\lambda}} \qquad (A.3)$$

which has the effect of making the neighborhood shrink with time. Therefore, over time the neighborhood will shrink to the size of just one node, that is, the BMU itself. σ_0 denotes the width of the lattice at time step $_0$ (initially).

The decay of the learning rate after each iteration is realized through another coefficient:

$$L(t) = L_0 e^{-\frac{t}{\lambda}} \qquad (A.4)$$

in which L_0 is a constant coefficient and λ is a time constant that is dependent on σ_0 and the number of iterations chosen for the algorithm to run.

Then every node in the lattice adjusts its weight at each iteration step t (t = 0, 1, 2, 3, ...) through the equation

$$W(t+1) = W(t) + \Omega(t).L(t).(V(t) - W(t)) \qquad (A.5)$$

This equation simply states that the new adjusted weight for the node is equal to the old weight W(t), plus a fraction of the difference (L) between the old weight and the input vector V(t). By examining the expressions given for $\Omega(t)$ and L(t) in Equations (A2) through (A.4), it is seen that not only does the learning rate decay over time, but also the effect of learning is proportional to the distance that a node is from the BMU.

References

[1] Hayter, A.J., *Probability and Statistics for Engineers and Scientists*, PWS Publishing, Boston, 1996.
[2] Raivio, K., Simula, O. and Laiho, J., 'Neural Analysis of Mobile Radio Access Network', *IEEE International Conference on Data Mining*, San Jose, CA, 29 November–2 December, 2001, pp. 457–464.
[3] Laiho, J., Raivio, K., Lehtimäki, P., Hätönen, K. and Simula, O., 'Advanced Analysis Methods for 3G Cellular Networks', Tech. Rep. A65, Helsinki University of Technology, Laboratory of Computer and Information Science, 2002.
[4] Lehtimäki, P., Raivio, K. and Simula, O., 'Mobile Radio Access Network Monitoring using the Self-organizing Map', *European Symposium on Artificial Neural Networks*, Bruges, Belgium, 24–26 April, 2002, pp. 231–236.
[5] Kohonen, T., *Self-Organizing Maps*, Springer-Verlag, Berlin, 1995.
[6] Kohonen, T., Oja, E., Simula, O., Visa, A. and Kangas, J., 'Engineering Applications of the Self-organizing Map', *Proceedings of the IEEE*, October 1996, **84**(10), 1358–1384.
[7] Kaufman, L. and Rousseeuw, P., *Finding Groups in Data*, John Wiley and Sons, Inc., New York, 1989.
[8] Fukunaga, K., *Introduction to Statistical Patterns Recognition*, 2nd edition, Academic Press Inc., San Diego, 1990.
[9] Duda, R.O., Hart, P.E. and Stork, D.G., *Pattern Classification*, Wiley- Interscience, New York, 2001.
[10] Balakaishnan, P.V., Cooper, M.C., Jacob, V.S., et al., 'A Study of the Classification Capabilities of Neural Networks using Unsupervised Learning: A Comparison with K-means Clustering', *Psychometrika*, 1994, 59(4), 509–525.
[11] Everitt, B.S. *Cluster Analysis*, Arnold, London, 1993.
[12] Davies, D.L. and Bouldin, D.W., 'A Cluster Separation Measure', *IEEE Transactions on Pattern Analysis and Machine Intelligence*, April 1979, **1**(2), 224–227.
[13] Alhoniemi, E., Hollmen, J., Simula, O. and Vesanto, J., 'Process Monitoring and Modeling using the Self-organizing Map', *Integrated Computer Aided Engineering*, 1999, **6**(1), 3–14.
[14] Vesanto. J. and Alhoniemi, E., 'Clustering of the Self-organizing Map', *IEEE Transactions on Neural Networks*, May 2000, **11**(3), 586–600.
[15] Flexer, A., 'On the Use of Self-organizing Maps for Clustering and Visualization', in *PKDD99*, volume **1704** of *LNAI*, Prague, Czech Republic, 1999, pp 80–88.

Abbreviations

2G	Second Generation
3G	Third Generation
3GPP	Third Generation Partnership Project
4G	Fourth Generation

A

AAL	ATM Adaptation Layer
Abis	GSM interface between BTS and BSC
ABR	Available Bit Rate
AC	Admission Control
ACIR	Adjacent Channel Interference Ratio
ACK	Acknowledgement
ACLR	Adjacent Channel Leakage ratio
ACP	Adjacent Channel Protection
ACS	Adjacent Channel Selectivity
AICH	Acquisition Indication Channel
AM	Acknowledgement Mode
AMR	Adaptive Multi-Rate (voice coding)
API	Application Program Interface
APN	Access Point Name
ARQ	Automatic Repeat Request
AS	Access Stratum, Active Set
ATM	Asynchronous Transmission Mode
AWGN	Additive White Guassian Noise

B

BCCH	Broadcast Control Channel
BDP	Bandwidth Delay Product
BER	Bit Error Rate
BH	Busy Hour (of network traffic)
BICC	Bearer Independent Call Control
BLER	Block Error Rate
BMC	Broadcast/Multicast Channel

BMU	Best Matching Unit
BPSK	Binary Phase Shift Keying modulation
BSC	Base Station Controller
BSS	Base Station Subsystem
BTS	Base Transceiver Station
BW	Bandwidth
BYE	Session termination in TCP

C

C/I	Carrier-to-Interference Ratio
CBR	Constant Bit Rate
CC	Call Control
CCCH	Common Control Channel
CCH	Common Channel
CCPCH	Common Control Physical Channel
CCTrCH	Coded Composite Transport Channel
CDMA	Code Division Multiple Access
CIR	Committed Information Rate
CN	Core Network
CNWD	Congestion Window Size (in TCP protocol)
CPICH	Common Pilot Channel
CRC	Cyclic Redundancy Check
CS	Circuit Switched
CSCF	Call Service Control Function
CTCH	Common Traffic Channel

D

DCH	Dedicated Channel
DHCP	Dynamic Host Client protocol
DL	Downlink
DNS	Domain Name Server
DPCCH	Dedicated Physical Control Channel
DPCH	Dedicated Physical Channel
DPDCH	Dedicated Physical Data Channel
DRNC	Drifting RNC
DRX	Discontinuous Reception
DSCH	Downlink Shared Channel
DSCP	Differentiated Service Code Points
DTX	Discontinuous Transmission
DUACK	Duplicate ACKs
DXU	Digital Cross-Connect Unit

E

E1	European Standard 2 Mbps Transmission Line
E_b/N_0	Signal to noise ratio per bit
EDGE	Enhanced Data Rate for GSM Evolution

EFR	Enhanced Full Rate (speech coding)
EIRP	Equivalent Isotropic Radiated Power
ELN	Explicit Loss Notification
ETSI	European Telecommunication Standards Institute

F

f factor	ratio of the own cell interference power to the total multi-user interference power
FA	Foreign Agent (in Mobile IP protocol)
FACH	Fast Access Channel
FBI	Feedback Information
FCS	Frame Check Sequence
FDD	Frequency Division Multiplexing
FER	Frame Erasure Rate
FR	Full Rate (speech coding)
FTP	File Transfer Protocol

G

Gb	interface between BSC and SGSN in GPRS
Ge	Geometric Factor
GERAN	GSM Edge RAN
GGSN	Gateway GPRS Support Node
GMM	GPRS MM
GOS	Grade of Service
GPRS	General Packet Radio System
GPS	Global Positioning System
Grx	Receiver Antenna Gain
GSM	Global System for Mobile Communication
Gtx	Transmitter Antenna gain

H

HA	Home Agent
HARQ	Hybrid ARQ
HLR	Home Location Register
HO	Handover
HOL	Head of the Line (queuing)
HR	Half Rate (speech coding)
HSDPA	High Speed Downlink Packet Access
HSS	Home Subscriber Server
HSUPA	High Speed Uplink Packet Access
HTTP	Hyper Text Transfer Protocol

I

ICDM	Inter-Cell Dependency Matrix
IE	Information Element
IEE	Institution of Electrical Engineers

IEEE	Institute of Electrical and Electronic Engineers
IETF	Internet Engineering task Force
IM	Inter-Modulation
IMD	Inter-Modulation Distortion
IMS	IP Multimedia Subsystem
IMSI	International Mobile Subscriber Identity
IMT	International Mobile Telecommunication
IP	Internet Protocol
IP4	IP version 4
IP6	IP version 6
ISCP	Interference Signal Code Power
ISDN	Integrated Services Digital Network
ISP	Internet Service Provider
ITU	International Telecommunication Union
Iu	Interface between RNC and core network
Iub	Interface between Node B and RNC
Iur	Interface between two RNCs

K

Kbps	Kilo bit per second
KPI	Key Performance Indicator

L

LA	Location Area
LAC	Location Area Code
LC	Load Control
LLC	Logical Link Control
LOS	Line-of-Sight

M

MAC	Media Access Control
MAP	Mobility Specific Application Part
MCC	Mobile Country Code
MCGF	Media Control Gateway Function
MCL	Minimum Coupling Loss
MCLP	Maximal Coverage Location Problem
MDC	Macro-Diversity Combining
MEHO	Mobile Evaluated Handover
METP	Mobile End Transport Protocol
MGC	Media Gateway Control
MGW	Media Gateway
MHA	Mast Head Amplifier
MIMO	Multiple Input Multiple Output
MM	Mobility Management
MNC	Mobile Network Code
MPLS	Multi-Protocol Label Switching

MS	Mobile Station
MSC	Mobile Switching Center
MSS	Maximum Segment Size
MTU	Maximum Transmission Unit size

N

NAS	Network Access Stratum
NEHO	Network Evaluated Handover
NLOS	Non-Line-of-Sight
NM	Network Management
NMS	Network Management System
NMT	Nordic Mobile Telephone
Node B	radio base station in WCDMA
NP	Network Performance
NPI	Network Performance Indicator
NRT	Non-Real Time

O

O&M	Operating and Maintenance
OFDM	Orthogonal Frequency Division Multiplexing
OMA	Open Mobile Alliance
OPEX	Operating Expenditures
OSI	Open System Interconnection
OVSF	Orthogonal Variable Spreading Factor

P

PACCH	Packet Associated Control Channel
PAGCH	Packet Access Grant Channel
PC	Power Control
PCS	Personal Communication System
PCCCH	Packet Common Control Channel
PCCH	Paging Control Channel
P-CPICH	Primary CPICH
PCU	Packet Control Unit
PDCP	Packet Data Convergence Protocol
PDP	Packet Data Protocol
PDSCH	Physical DSCH
PDU	Protocol Data Unit
PF	Proportional Fair (a form of queue scheduling)
PI	Performance Indicator
PICH	Paging Indication Channel
PLMN	Public Land Mobile Network
PN	Pseudo-Noise
PPCH	Packet Paging Channel
PPP	Point to Point Protocol
PRACH	Physical Random Access Channel

PS	Packet Switched
P-SCH	Primary Synchronization Channel
PSTN	Public Switched Telephony Network
PVC	Permanent Virtual Circuit

Q

QAM	Quadrature Amplitude Modulation
QoS	Quality of Service
QPSK	Quadrature Phase Shift Keying

R

RA	Routing Area
RAB	Radio Access Bearer
RACH	Random Access Channel
RAI	Routing Area Indicator
RAKE	multipath resolving and combining receiver for CDMA systems
RAN	Radio Access Network
RANAP	Radio Access Network Application Part
RAT	Radio Access Technique
RF	Radio Frequency
RFC	Request For Comment
RLC	Radio Link Control
RLCP	Radio Link Control Protocol
RNC	Radio Network Controller
RNP	Radio Network Protocol
RNS	Radio Network Subsystem
RNTI	Radio Network Temporary Identity
RRC	Radio Resource Control
RRH	Remote Radio Head
RRM	Radio Resource Management
RSCP	Received Signal Code Power
RSSI	Received Signal Strength Indicator
RSVP	Resource Reservation Protocol
RT	Real Time
RTO	TCP Retransmission Time Out parameter (in TCP)
RTT	Round Trip Time
Rx	Receive
RxD	Receiver Diversity
RWND	Receiver Window size (in TCP)
Rxlev	Received signal Level

S

SACK	Selective Acknowledgement option of TCP
SAP	Service Access Point
SCCH	Shared Control Channel

SCH	Synchronization Channel
S-CCPCH	Secondary CCPCH
S-CPICH	Secondary CPICH
SCTP	Streamed Control Transmission Protocol
SDU	Service Data Unit
SF	Spreading Factor
SGSN	Serving GPRS Support Node
SHO	Soft Handover
SIGTRAN	Signaling Transport
SIM	Subscriber Identity Module
SIP	Session Initiation Protocol
SIR	Signal to Interference Ratio
SLA	Service Level Agreement
SMG	Special Mobile Group
SMS	Short Message Services
SNDCP	Sub-Network Data Convergence Protocol
SNR	Signal to Noise Ratio
SOM	Self-Organized Networks
SRNC	Serving RNC
SS7	CCITT Signaling System no 7
S-SCH	Secondary Synchronization Channel
STTD	Space-Time Transmit Diversity
SVC	Switched Virtual Circuit
T	
T1	American Standard 1.543 Mbps Transmission line
TCAP	Transaction Capability Application part
TCP	Transmission Control Protocol
TDD	Time Division Duplex
TDM	Time Division Multiplexing
TDMA	Time Division Multiple Access
TE	Terminal Equipment
TF	Transport Format
TFC	Transport Format Combination
TFCI	Transport Format Control Identifier
TFCS	Transport Format Control Specification
TFI	Transport Format Identifier
TIA	Telecommunication Industry Association
TOS	Type Of Service field of IP4
TPC	Transmitter Power Control
TR	Technical Recommendation
TrCH	Transport Channel
TS	Technical Specification
TSG	Technical Specifications Group
TSTD	Time Switched Transmit Diversity
TTI	Transmission Time Interval

TU3	Typical Urban 3 km/h (standard channel type specified in 3GPP)
Tx	Transmitter

U

UBR	Unspecified Bit Rate
UDP	User Datagram Protocol
UE	User Equipment (in WCDMA)
UL	Uplink
UM	Unacknowledged Mode
UMA	3GPP Unlicensed Mobile Access standards
UMTS	Universal Mobile Telecommunication System
UPH	UE transmission Power Headroom
URA	UMTS Routing Area
URL	Uniform Resource Locator
USIM	UMTS SIM card
UTRAN	Universal Terrestrial Radio Access Network
Uu	radio interface between UTRAN and UE

V

VBR	Variable Bit Rate
VLR	Visitor Location Register
VoIP	Voice over IP

W

W	UMTS spreading bandwidth (i.e., the chip rate of 3.84 Mcps)
WAP	Wireless Application Protocol
WB	Wideband
WI	Walfisch-Ikegami
WCDMA	Wideband Code Division Multiple Access
WLAN	Wireless Local Area Network

X

X-pole	Cross-pole (antennas)

Index

Access stratum, 8, 9
ACIR, 117
ACLR, 108, 116, 117
ACS, 116, 117
Active set, 149
   optimal range, 156
   size, 156, 157
   update, 151
   weighting coefficient, 156
Add window, 156
Admission control, 33, 139, 150, 160–162, 164
Admission threshold, 160, 161, 164, 165
AICH, 23, 26
AMR, 205–212
AMR-wideband, 212
Antenna
   antenna tilt, 88, 89, 188, 183
   configuration, 83, 182, 195–200
   diversity, 21, 76, 97, 176, 177
   isolation requirements, 187
   sharing, 4, 192
Area coverage probability, 52, 94, 95, 110–112, 132
Asymmetry
   in TCP performance, 283
ATM
   parameters, 252, 253
   QoS classes, 268, 269
   sizing, 252, 253
   virtual circuits, 219, 252
   virtual path, 218

Band-width delay product, 278, 279
BCCH, 30, 111, 133, 138, 303
BCH (broadcast control channel), 25, 29–31, 111

Beam forming
   gains, 185
   implementation considerations, 184, 185
Bearer reconfiguration, 166, 167
Bearer service, 8, 262–264, 267
BER
   residual bit error ratio, 263, 264
Best server, 108, 110, 112, 156
BLER, 38, 48, 142, 144, 281, 282
BMC, 31, 37
Broadcast control channel, 30. *see* BCH

Call control (CC), 9, 31, 242, 243
Capacity improvement, 76, 77, 127, 176, 177
Carrier sharing, 126, 127
CCTrCH, 19, 28
Cell grouping, 311, 312
Cell isolation, 4, 5, 17
Cell overlap optimization, 155–157
Cell reselection, 36, 111, 277, 281, 282
Cell selection, 34, 202
Channelization
   codes, 13, 18, 167, 168, 171
   code tree, 167, 185, 188
Channel models, 48, 49
Channel state transition, 34–36, 167
Channel switching, 159, 166, 167, 168, 282, 292
Chip rate, 13, 14, 43, 93, 96
Code resource allocation, 168–170
Coherence bandwidth, 43, 44
Coherence time, 46, 48
Congestion control, 159, 164–166
Congestion detection, 165
Congestion resolving actions, 165, 166
Convolutional decoder, 143

Core network
  architecture, 232–237
  design and dimensioning, 245–250
  elements, 233, 235, 247, 248
  function, 231, 232
  protocol overheads, 250
  transport technology, 250–255
Co-siting, 4, 194, 196
COST 231 Hata model, 55
COST 231 Walfisch-Ikegami, 52, 56, 57, 130
Coverage
  area, 82, 99, 186, 187, 244
  coverage–capacity tradeoff, 73
  probability, 52, 90, 94, 95, 110–112, 132
  threshold, 93
CPCH, 29, 30, 36, 143, 170
CTCH, 31

DCH, 29, 31, 112, 166, 170
Delay analysis (end-to-end), 267, 268
Delay spikes, 282, 290
Delay spread, 42–44, 71, 103, 183
DiffServ, 265
Diversity
  gain, 15, 16, 42, 76, 97
  macro-diversity combining gain (MDC), 96, 137, 142, 153, 165
  spatial, 46, 73, 76
  transmit, 177–179
Doppler effect, 45
Doppler frequency, 45, 47, 139
DPCCH, 19–21, 28, 138, 139, 141, 142, 170
DPCH, 23, 25, 141, 147
DPDCH, 19, 20, 28, 142, 209
Drop window, 152, 156
DTX, 27, 208
Ducting effect, 62, 63
Dynamic threshold, 149

E-DCH, 29
Eifel algorithm, 291
End-to-end
  delay, 113, 212, 261, 267, 268, 288
  significance, 9

FACH, 25, 29, 31, 85, 111
Fast fading, 47, 48, 96, 144, 148, 153, 154
FDD, 11, 29, 138, 194
FER, 86, 113, 143, 212, 305
f factor, 68, 162

Frame alignment, 10. *see* Synchronization
Frame timing, 10, 28
Framing, 19
Frequency accuracy, 11
Frequency allocation, 81
Frequency correction, 131
Frequency hoping, 14
Frequency re-use, 4, 66, 68, 87
Frequency stability, 10, 11

Geometric factor, 68, 74
GGSN, 3, 224, 233, 234
  dimensioning, 247
GPRS
  core network protocol architecture, 235, 236
  MS modes of operation, 239
GPS, 11

Hadamard matrix, 168
Handover
  gain estimation, 94
  inter-frequency handover, 144, 146, 149
  inter-system handover, 113, 146, 191, 202, 244, 258
  intra-system handover, 147
Handover procedure, 148, 149
  hard, 127, 145, 149, 277, 281, 292
  soft/softer, 18, 19, 129, 153, 161
Hata models, 55, 56
Header compression, 37, 287
Headroom, 96, 106. *see* Power control
Hierarchical cell structure(s), 147
HSDPA, 3, 6, 127, 164, 172, 173
HS-DPCCH, 20, 21, 170
HS-DSCH, 20, 27, 28, 30, 164, 166, 170, 171, 175, 282
HS-PDSCH, 23, 27, 28, 101, 102, 177
HS-SCCH, 23, 27, 28, 30, 177
HS-USCH, 170
HTTP, 271, 295, 296, 317
Hybrid-ARQ acknowledgement, 21

IMD, 194, 318
IMS (IP multimedia subsystem), 5, 240
Instant messaging, 237, 241, 262
Interference
  geometry, 4, 73, 74, 101, 129, 157
  inter-cell, 10, 73, 74, 87, 89, 118, 133–135
  inter-operator, 116–119
  intra-cell, 17, 83

margin, 94, 96
multi-user, 4, 13, 65–69, 71, 76, 77, 81–83, 90, 91, 94, 102–104, 137, 175, 184, 185
narrowband, 14, 15
other-cell interference, 108
parameters, 41, 42
Inter-modulation distortion (IMD), 191, 192, 194, 318
Inter-symbol interference, 6, 42–44
IP address allocation, 239, 240
IP multimedia subsystem, 3, 240. *see* IMS
ISCP, 318
Isolation requirements, 187, 191–193

Keenan-Motley, 58, 59

Layered approach, 126
Layered cell architecture, 126–128
Layered networks, 149
Layered radio access, 175
Link dimensioning, 223–230, 248
Load control, 150, 164, 173
Load factor, 67, 69, 73, 103, 122, 123, 162
Location area, 238
Logical channel, 19, 30, 31
Lognormal fading, 51, 52, 94, 95

MAC, 19, 21, 31, 34
Macro-diversity combining (MDC), 96. *see* Diversity
Maximal coverage location problem, 82
Maximal ratio combining (MRC), 18, 96, 106, 153
Maximum segment size (MSS), 274, 289
Maximum transmission unit (MTU), 274, 289
Measurement intervals, 161
Measurement reporting
  event triggered, 166
  periodic reporting, 146, 151, 166
Measurement types
  periodic reporting of, 146
MHA (mast head amplifier), 175, 179–181, 195
MIMO, 5, 6, 77
Mobility management
  user states, 238, 239
Model resolution, 62
Model tuning, 59–61, 132
Morphographic data, 57
Morphological class, 60
MPLS, 255

Multi-carrier, 71, 77, 119, 127, 177
Multi-code transmission, 27, 172, 177, 284
Multipath channel
  3GPP model, 48, 49
  effect, 45–46
  fading, 15, 45, 48, 148, 154
  ITU model, 49, 50

Near-far problem, 17, 18, 137, 139
Neighbor, 10, 36, 68, 78, 100, 111, 116, 146
  cell list, 304
  set, 149
Noise figure, 67, 73, 99, 101, 105, 201
Noise floor, 13, 73, 90, 91, 93, 99, 101, 105, 201
  ambient noise measurement, 201
Noise rise, 67, 68, 94, 97, 100, 125, 127, 310
Non-access stratum, 9, 31, 34

OCNS, 113
Okumura-Hata model, 52, 54–58
Orthogonality factor, 17, 66, 74–75, 101–104, 108, 112, 124

Packet scheduling, 5, 102, 159, 170, 171, 282
Path loss models, 51, 52, 89, 131
PCCH, 319
P-CCPCH, 23, 25, 30, 84, 85
P-CPICH, 138, 139, 145–148, 150–152
PDCP, 31, 32, 36, 37, 40
PDP address, 233, 235, 236
PDP context, 9, 202, 236, 247, 248, 266, 267
Physical layer measurements, 37
PICH, 23, 27, 30, 36, 84, 85
Pilot coverage, 83, 85, 202
Pilot pollution, 84, 87–89, 111, 134, 301
Pole capacity, 68, 70, 73, 74, 126, 305
Power classes, 27
Power control, 84, 87–89, 111, 134, 201
  algorithms, 145
  dynamic range, 138, 141
  fast closed-loop PC, 138, 139
  headroom, 96, 106, 140, 154
  in compressed mode, 144
  inner-loop PC, 138, 139
  open-loop PC, 138, 139
  optimization, 145
  outer-loop PC, 138, 144
  power drifting, 142, 150
  step size, 139–140, 145, 152
Power delay profile, 43

Power rise, 140, 142, 183
PRACH, 22, 26, 30, 39, 138
Preamble, 22, 138, 141
Pre-launch optimization, 113
Primary common control physical channel, 25, 30, 111. *see* P-CCPCH
Primary common pilot channel, 24. *see* P-CPICH
Priority queuing, 229, 259
Processing gain, 13–15, 69, 94, 96, 112, 145
Propagation model, 48, 59, 82, 99, 130–132
Protocol architecture, 31, 32, 40, 235, 236
P-SCH, 23, 84

QAM, 166, 171, 172, 177
  16-QAM, 27
QoS, 2, 5, 8, 34, 113, 122, 127, 145, 172
  fundamental concepts, 258
  inter-working, 267
  mechanisms, 267, 269
  monitoring, 259
  parameter mapping, 264, 266, 270, 277
  signaling, 265
  significance, 257
QoS categories, 258–260
  background class, 260, 261, 263, 264
  conversational class, 260, 263
  interactive class, 261, 263
  streaming class, 260, 261, 263
QOS requirements, 5, 122, 212, 215, 227, 245, 257–259
Quality of service, 257. *see* QOS
Queue scheduling algorithm, 229

RACH, 22, 29, 31, 34, 36, 86, 87, 166, 170, 304, 307
Radio access bearer, 8, 9, 33, 263–265
Radio link budgeting (RLB), 89, 215
Radio link control (RLC), 31
Radio network signaling, 10
Radio resource control (RRC), 31, 266, 282
Rake, 15, 18, 19, 42, 43, 45, 76, 85, 102, 153, 155
  fingers, 15, 42, 43, 155
  receiver, 15, 18, 42, 43, 76, 78, 85, 182, 183
  reception, 15
Rayleigh fading, 46, 47, 136, 137
Ray tracing models, 52, 57, 58, 62, 130
Receiver sensitivity, 61, 76, 93, 94, 96, 98
Refractive index, 62
Remote radio heads (RRH), 175, 181
Repeater, 176, 185–188
  impact, 187
  isolation requirements, 187
  operating characteristics, 186
Replacement window, 152, 156
Reporting event, 150–153, 156
Residual bit error ratio, 263
Resource manager, 150, 169, 264
Rician channel, 46
RLC, 31–34, 167
  retransmission timer, 285
RNC, 8–10, 18–21, 33, 36, 37, 71, 142, 149, 150, 226, 292
  planning and dimensioning, 215
  RNC-Node B dimensioning, 224
RNP tool, 105, 114
RNTI, 35
Roaming, 3, 5, 18, 203, 237, 240–242
  mobility handling mechanism, 242
Routing area, 234, 238, 239, 247
RSCP, 37–39, 85, 138, 202
RSSI, 38, 133, 134, 147

S-CCPCH, 25, 31, 111
Scrambling codes, 17, 115, 116, 169, 188
SCTP, 237
Sectorisation/sectorization, 87, 175, 177
  gain, 175
Self-organizing networks, 188, 189
Sensitivity, 93, 94, 96, 100, 106, 108, 114, 179, 180, 205, 246
Service access point (SAP), 31, 262
Service interaction, 104, 121, 122
Service strategy, 113
Session initiation protocol, 240, 241, 267. *see* SIP
SGSN, 229, 233–236, 238–240, 242
  function, 234
  SGSN dimensioning, 247
  SGSN-RNC link dimensioning, 227
SIP, 241, 243
SIR, 39, 83, 139, 141, 143
Site, 4, 5, 41, 52, 54–56, 61, 63, 66, 82, 83, 87, 119, 129, 191, 196, 216
  co-location, 4
  location/locating, 81–83
  overlap, 87–89
  planning, 81
  site engineering, 83
Soft handover, 18, 19, 84, 132, 133, 140, 145–149
  overhead, 215

diversity gain, 94–96, 155, 165
soft handover optimization, 154
Soft switching, 242–245
  benefits of, 243
  elements, 243
  transition to soft switching, 244
Space-time coding, 178
Spectrum allocation, 12
Speech codecs, 205–207
  channel mode, 205
  codec mode, 205–210
  codec mode parameter, 211
  mode adaptation, 205, 208, 211, 212
  tradeoffs, 209, 285
Spreading factor, 16, 17, 21, 26, 168, 170
Spread spectrum, 12
SS7oIP, 237
S-SCH, 23
Static analysis, 90
State transitions, 34–36, 282
Static simulation, 105, 106
Synchronization, 26, 47, 108, 145, 225
  frame synchronization, 10
  frequency synchronization, 11
  inter-Node B synchronization, 10
  RNC-Node B synchronization, 10
  slot synchronization, 12

TCP, 273–277, 279–296
  congestion avoidance, 276, 277
  congestion window (CNWD), 276
  fast recovery, 279, 281
  fast retransmit and recovery, 291, 292
  goodput, 285
  parameter tuning, 288
  performance optimization, 284
  receiver advertised window, 275, 276, 287, 288
  RLC influence, 287, 288
  round trip time (RTT), 275, 277

  selecting proper options, 290
  selective acknowledgement, 280, 281, 290, 291, 295
  short lived connections, 295
  slow start, 280, 281, 283, 290
  WTCP99, 295
TDD, 11, 12
TFCI, 21–23, 25, 26, 85, 86
Throughput, 5, 113, 164, 172, 226, 248, 255, 273, 283–285, 293
  GGSN, 248
  RNC, 215, 216
  SGSN, 247
Traffic center of mass, 216
Traffic handling priority, 262, 263
Traffic modeling, 113, 114
Transport format (TF), 21, 29, 34
Transport format set, 21, 29
Turbo-decoder, 143

UMTS, 1–3, 7, 8, 10–13, 66, 70, 75, 89, 96, 113, 119, 129–135, 191–193, 201–203, 221, 231–233, 235, 257, 260, 264–267, 284, 299
  bearer service attributes, 262, 263
  high level architecture, 8
  quality of service, 2, 9, 257, 270. *see* QOS
Uplink load factor, 91–93, 162
URA, 322
UTRAN, 8, 11, 25, 35–38, 143, 150–152, 166, 167, 264
  architecture, 9, 10
  elements, 8

Viterbi decoder, 143

Wideband antennas, 193, 197
  pros and disadvantages, 197
Wideband power, 138, 162–166
  uplink wideband power, 161